Sustainable Development Goals Series

The **Sustainable Development Goals Series** is Springer Nature's inaugural cross-imprint book series that addresses and supports the United Nations' seventeen Sustainable Development Goals. The series fosters comprehensive research focused on these global targets and endeavours to address some of society's greatest grand challenges. The SDGs are inherently multidisciplinary, and they bring people working across different fields together and working towards a common goal. In this spirit, the Sustainable Development Goals series is the first at Springer Nature to publish books under both the Springer and Palgrave Macmillan imprints, bringing the strengths of our imprints together.

The Sustainable Development Goals Series is organized into eighteen subseries: one subseries based around each of the seventeen respective Sustainable Development Goals, and an eighteenth subseries, "Connecting the Goals", which serves as a home for volumes addressing multiple goals or studying the SDGs as a whole. Each subseries is guided by an expert Subseries Advisor with years or decades of experience studying and addressing core components of their respective Goal.

The SDG Series has a remit as broad as the SDGs themselves, and contributions are welcome from scientists, academics, policymakers, and researchers working in fields related to any of the seventeen goals. If you are interested in contributing a monograph or curated volume to the series, please contact the Publishers: Zachary Romano [Springer; zachary.romano@springer.com] and Rachael Ballard [Palgrave Macmillan; rachael.ballard@palgrave.com].

Huadong Guo

Big Earth Data in Support of the Sustainable Development Goals (2023)—China

Science Press
Beijing

Springer

Huadong Guo
International Research Center of Big
Data for Sustainable Development
Goals (CBAS)
Chinese Academy of Sciences
Aerospace Information Research
Institute
Beijing, China

ISSN 2523-3084 ISSN 2523-3092 (electronic)
Sustainable Development Goals Series
ISBN 978-981-95-2635-2 ISBN 978-981-95-2636-9 (eBook)
https://doi.org/10.1007/978-981-95-2636-9

Jointly published with Science Press.
The print edition is not for sale in China mainland. Customers from China mainland please order the print book from: Science Press.
ISBN of the Co-Publisher's edition: 978-7-03-080830-1

This project is co-published with Science Press, Beijing, China.

This Springer imprint is published by the registered company Springer Nature Singapore Pte Ltd.
The registered company address is: 152 Beach Road, #21-01/04 Gateway East, Singapore 189721, Singapore

Editorial Board

Principal Authors

Preface

From 2015 to the present, the implementation of the United Nations' "Transforming Our World: The 2030 Agenda for Sustainable Development" (hereinafter referred to as the "2030 Agenda") has reached its halfway point. During this period, progress on the Sustainable Development Goals (SDGs) has encountered serious challenges and difficulties despite the overall advancements. The international community has gained deeper insights into long-term issues such as climate change and biodiversity loss, but further efforts are needed to enhance implementation. Unexpected events like the COVID-19 pandemic and regional conflicts have had significant and far-reaching impacts on society, economy, and environment.

It is important for the international community to learn both the experience and lessons of the first half of the 2030 Agenda so that we can enhance the implementation of the SDGs in the second half and explore directions for future sustainable development. The slow progress during the initial phase underscores the global inadequacy in addressing long-term risks as well as short-term crises in the process of promoting sustainable development. The lack of timely and accurate data remains a weakness in our response to both long-term and short-term issues and is a bottleneck hindering the implementation and monitoring of the SDGs and the formulation of science-based decisions.

In September 2021, Chinese President Xi Jinping proposed the Global Development Initiative (GDI) during the 76th United Nations General Assembly. The GDI aims to deepen international cooperation, accelerate the implementation of the 2030 Agenda, and promote stronger, greener, and healthier development. It places a special emphasis on advancing collaboration in areas such as digital connectivity in the digital age to expedite the realization of the 2030 Agenda. The GDI is committed to building a global community of development, and a crucial foundation in this endeavor is the promotion of digital technology. Big Earth Data, as a representative form of such technology, can play a vital role in filling the gaps in current SDG statistical data and spatio-temporal information.

2024 marks the third anniversary of President Xi Jinping's announcement of the establishment of the International Research Center of Big Data for Sustainable Development Goals (CBAS) in China, and also the fifth consecutive year of publishing the report *Big Earth Data in Support of the Sustainable Development Goals (2023)—China* (hereinafter referred to as "the report"). Over the past few years, the research team has made use of the increasingly refined Big Earth Data platform and leveraged the advantages of objective scientific data and rich spatio-temporal information to assess the mid-term progress on sustainable development in China and globally. The team's focus has been on exploring the bottlenecks hindering sustainable development and identifying future development directions.

The report of this year focuses on the mid-term evaluation of sustainable development in China and globally through Big Earth Data. It aims to further expand data products of SDG indicators, broaden the depth of indicator contents, complete progress evaluations of all environment-related SDG indicators in China, and actively provide public data products to serve all countries, particularly developing countries in terms of evaluating the implementation of the 2030 Agenda. It also provides recommendations on science-based decision-making in the future, accelerating sustainable development processes, building the big data acquisition capability, and adjusting and optimizing indicators. The report stands as scientific support for the implementation of the 2030 Agenda.

The report is a collaborative effort, written by over 160 researchers from more than 50 organizations, including research institutes and universities in the field of sustainable development and big data. The project has received strong support from the Chinese Academy of Sciences and various government departments, and the team members have dedicated significant efforts to this endeavor. We express our heartfelt gratitude to all those involved in making the report a reality.

Huadong Guo
Beijing, China

Director-General of the International
Research Center of Big Data
for Sustainable Development Goals

Former Member, the 10-Member group
of the United Nations Sustainable
Development Goals Technology Facilitation Mechanism

Member, International Science Council Global Commission
on Science Missions for Sustainability

Competing Interests The author has no competing interests to declare that are relevant to the content of this manuscript.

Executive Summary

The report gives full play to the advantages and characteristics of Big Earth Data. It focuses on 16 targets related to 7 SDGs, including SDG 2 (Zero Hunger), SDG 6 (Clean Water and Sanitation), SDG 7 (Affordable and Clean Energy), SDG 11 (Sustainable Cities and Communities), SDG 13 (Climate Action), SDG 14 (Life Below Water), SDG 15 (Life on Land), as well as the interactions among SDG indicators in 28 research cases. From three aspects including data products, methods, and models, and decision support, the report shows the results of monitoring and mid-term evaluation of relevant SDG indicators at two scales: national and representative areas of China. The report represents an innovative practice of big data in support of the implementation of SDGs and can provide a scientific reference for decision-makers.

Regarding Zero Hunger (SDG 2), the report focuses on two main themes: sustainable food production and the assessment of related policies and their impacts. At the national scale, we developed large-scale terraced field mapping methods based on remote sensing, a model for assessing the soil conservation benefits of large-scale terraced fields, a zoned and layered mapping method for extracting irrigated dryland, and an evaluation method for the effectiveness of well-facilitated farmland construction. These efforts resulted in the generation of 30 m spatial resolution datasets of terraced field distribution in China for 2015 and 2022, datasets of irrigated dryland in China for 2015 and 2022, and a dataset of changes in irrigated dryland in China from 2015 to 2022. At the typical regional scale, we developed a corn residue coverage estimation model using spatio-temporal migration random forest (RF) techniques, which generated a 30 m resolution dataset of corn residue coverage in the black soil region of Northeast China for 2010 and 2013–2022. The report evaluates the contribution of terraced fields to reducing hydraulic erosion in Chinese farmland, the changes in corn residue coverage, the distribution and changes of irrigated dryland, and the effectiveness of well-facilitated farmland construction projects in key agricultural areas across provinces. It summarizes China's experience in farmland protection and provides recommendations for addressing the risks and challenges associated with future well-facilitated farmland construction.

Regarding Clean Water and Sanitation (SDG 6), the report conducts a case study on the population coverage rate of public toilets in China from 2015 to 2020. By integrating point of interest (POI) data, remote sensing, and statistical data, a public toilet population coverage evaluation model was developed to conduct a comprehensive assessment of the progress of SDG 6.2.1a. The research provides a demonstration of how Big Earth Data technology can be used to address the issue of missing sanitation facility service indicator data, offering significant value for wider application.

Regarding Affordable and Clean Energy (SDG7), the report focuses on two indicators and conducts a mid-term progress evaluation of China. The study found that significant progress has been made in two Chinese SDG 7 indicators evaluated. By 2022, the proportion of people in China who have access to clean cooking energy and technology (SDG 7.1.2) had reached 83.55%; China's transition toward renewable energy (SDG 7.2.1) has made significant progress, with the amount of renewable energy transmitted through direct current ultra-high voltage (UHV) transmission in 2022 equivalent to 1.84 times that of 2016.

Regarding Sustainable Cities and Communities (SDG 11), the report focuses on four indicators and conducts monitoring and a comprehensive assessment of the progress of SDG 11 indicators on the national scale. The study found that it is necessary to strengthen the construction of inclusive and sustainable cities and improve the sustainable capacity of human settlement planning and management. From 2010 to 2022, the overall level of $PM_{2.5}$ across the country decreased. China's comprehensive air pollution prevention and control actions have achieved remarkable results, and the risk of Chinese residents being exposed to $PM_{2.5}$ is gradually decreasing. From 2017 to 2022, in China's five major urban agglomerations, both the proportion of urban public open spaces and the proportion of per capita urban available public open spaces generally increased. China's urban planning attaches greater importance to the construction of public spaces, and the urban environment has become more livable. The differences in urban and rural economic development in China have shown a narrowing trend to varying degrees under different dimensions, and the level of integrated urban-rural development has been significantly improved. The localized urban comprehensive index of SDG 11 shows characteristics of differentiation in different geographical regions and urban scales. These studies will provide important support for the refined analysis of the progress of indicators in different regions and for scientific decision-making regarding sustainable cities and communities.

Regarding Climate Action (SDG 13), the report focuses on three indicators and conducts a mid-term progress assessment of China. Research has found that China has achieved certain results in disaster prevention and reduction. From 2010 to 2021, the number of people affected by floods and waterlogging disasters in China (SDG 13.1.1) showed a significant downward trend. After the implementation of the Sendai Framework for Disaster Risk Reduction 2015–2030 (2016–2021), the annual average number of deaths, missing persons, affected persons, crop affected areas, crop failure areas, and direct economic losses per 100,000 people in China due to floods and waterlogging disasters have significantly decreased compared to before the implementation (2010–2015); after 2020, China has set clear goals for "Carbon Peak" and "Carbon Neutrality" (namely the "Dual Carbon" target), released a national strategy for adapting to climate change, and coordinated the national adaptation work and strategic system for addressing climate change (SDG 13.2.1/13.b.1); China's climate change education (SDG 13.3.1) is still in its early stages, and the proposal of the "Dual Carbon" target has a significant impact on public awareness of climate change issues. The public on Chinese social media has confidence in achieving the country's "Carbon Peak" and "Carbon Neutrality" goals and has a positive attitude toward mitigating climate change as a whole.

Regarding Life Below Water (SDG 14), focusing on specific targets such as SDG 14.1 to reduce marine pollution, SDG 14.2 to protect marine ecosystems, and SDG 14.7 to sustainably manage aquaculture, we analyzed the spatio-temporal variations of relevant indicators including microplastics, seagrass beds, bay ecosystem health, and offshore raft aquaculture in China's coastal waters. The study found that from 2017 to 2021, the average abundance of microplastics in China's coastal waters was at a low to medium level, with small interannual variations. From 1985 to 2020, the distribution of seagrass beds in China showed a trend of first declining and then slightly rebounding. From 2010 to 2021, the health of the Jiaozhou Bay ecosystem has continued to improve overall. From 2017 to 2021, the area of offshore raft aquaculture in China has entered a phase of orderly growth and tends to be located farther from the coast. The monitoring methods, data products, and suggestions related to these indicators can provide important references for better achieving SDG 14.

Regarding Life on Land (SDG 15), the report focuses on two specific targets and conducts assessments in the Qinghai-Xizang Plateau, the arid region of Northwest China, and the Saihanba region in Chengde. The study reveals that, in biodiversity conservation, the centroids of the suitable habitats for the black-necked crane were in the southern part of Qinghai Province from 1980 to 2020, showing a general trend of migration toward lower latitudes in the south and the interior of the Qinghai-Xizang Plateau. Conservation priority for the black-necked crane in the Qinghai-Xizang Plateau is generally higher in the eastern region compared to the central and western regions. In combating desertification, the oasis area expanded slowly between 2000 and 2005, grew rapidly from 2005 to 2015, and experienced the highest growth rate between 2015 and 2020, demonstrating significant achievements in oasis expansion in the arid region of Northwest China. Regarding the overall SDG 15 indicators, precipitation in the Saihanba region increased at a rate of 1.69 mm/a from 1991 to 2020 compared to 1961 to 1990, with a cumulative increase of approximately 50 mm over 30 years. The increase in water resources has had a significant positive impact on the local ecological environment. The optimal groundwater depth for deciduous broad-leaved forests is 2–6 m, while for herbaceous plants, it is 1–3 m.

Regarding SDGs' integrated assessment and interactions, the report examines two themes: integrated evaluations and interactions among SDGs in Chinese cities at the prefectural level and synergies for "One Health" in China. The report finds significant spatial differences in the comprehensive scores and balance scores of SDGs across China's cities at or above the prefecture level, with urban clusters outperforming non-cluster areas in the development of SDGs. The analysis also reveals that there are more synergies than trade-offs among SDGs in different cities, with more intensive interactions observed in urban clusters such as the Yangtze River Delta, Chengdu-Chongqing, and the Pearl River Delta. SDGs in Chinese cities are showing an upward trend. In the future, it is necessary to focus on key goals like SDG 4 (Quality Education), SDG 11 (Sustainable Cities and Communities), SDG 12 (Responsible Consumption and Production), and SDG 15 (Life on Land) to promote the overall achievement of SDGs. The restoration and improvement of ecological conditions can have a positive impact on human health.

Contents

Acronyms

3DED	3D edge density
3DLPI	3D largest patch index
3DPLand	3D percentage of landscape area
ANTL	Average nighttime light
CEADs	China Carbon Emission Accounting Databases
CLUD-A	Annual China land use dataset
DEM	Digital elevation model
DL	Deep Learning
DMSP	Defense Meteorological Satellite Program
DNTL	Deviation of nighttime light
GLC_FCS	Global land cover with a fine classification system
GMIA	Global map of irrigation areas
GPS	Global positioning system
HISTARFM	Highly scalable time-adaptive reflectance fusion model
IEA	International Energy Agency
IUCN	International Union for the Conservation of Nature
Larea	Total lit area
MODIS	Moderate-resolution imaging spectroradio-meter
MR	Mixed rural area
MU	Mixed urban area
NDVI	Normalized differential vegetation index
NPP	National Polar-orbiting Partnership
NPP-VIIRS	National polar-orbiting partnership-visible infrared imaging radiometer suite
NTL	Nighttime light
OSM	Open Street Map
PCA	Principal component analysis
PE	Polyethylene
PERSIANN-CDR	Precipitation estimation from remotely sensed information using artificial neural networks—climate data record
PET	Polyethylene terephthalate
PMS	Panchromatic and multispectral system
POI	Point of Interest
PP	Polypropylene
PS	Polystyrene
PSR	Pressure-State-Response

RMSE	Root mean square error
RSEI	Remote sensing based ecological index
RU	Rural area
SDGs	Sustainable Development Goals
SRTM	Shuttle radar topography mission
TNTL	Total nighttime light
UNSDSN	United Nations Sustainable Development Solutions Network
UPOS	Urban public open space
UR	Urban area
WHO	World Health Organization

1 Introduction

The year 2023 marks the midterm of the 2030 Agenda. It is imperative to assess the progress of current SDGs with more precise data to gain a comprehensive understanding of the challenges faced globally and by individual nations. This assessment will facilitate the exploration of scientifically feasible solutions and provide directions for the implementation phase of the 2030 Agenda.

However, based on the analysis of the 17 goals in the United Nations' "The Sustainable Development Goals Report" and research results of over 230 indicators, it is evident that approximately half of the countries worldwide still severely lack progress data for these indicators. Additionally, the data updates are not timely, there is a lack of geospatial information, and these deficiencies significantly impede the monitoring of indicators and informed decision-making (United Nations 2022, 2023). The midterm evaluation of SDGs presents an opportunity to assess the progress in implementing SDGs, identify shortcomings, and to enhance data acquisition capability. According to the project team, out of the currently over 230 SDG indicators, 35% remain at Tier II, where methodologies exist but data are unavailable. Moreover, for those indicators with available data, their distribution across different countries and sectors is highly uneven. This is particularly challenging for many countries, which find it difficult to achieve large-scale data computation in the short term and urgently require high-quality global data products.

The United Nations has promoted the "Roadmap for Digital Cooperation" aimed at accelerating sustainable development through digital technologies (United Nations 2020). Big data is a crucial method and tool within digital technologies. In the report, we integrate Big Earth Data from multiple sources, including satellite observations, station records, survey statistics, online media, and foundational geographic data, to compensate for the absence of statistical data. This integration also provides rich spatio-temporal information, illustrating the spatial disparities and progress of indicators. The "Sustainable Development Science Satellite 1" (SDGSAT-1), launched in November 2021, is dedicated to serving the 2030 Agenda. Currently, satellite data are being shared globally to further enhance the acquisition and service capabilities of SDG data.

From 2019 to 2023, we have continuously published the report *Big Earth Data in Support of the Sustainable Development Goals* (http://www.cbas.ac.cn/en/publications/reports/), aiding the realization of SDGs in China and along the Belt and Road Initiative. The report continues to focus on the progress of indicators at the scales of China and the Belt and Road from the perspectives of data products, methods and models, and decision support. Additionally, the report emphasizes the midterm progress assessment of indicators within China and the Belt and Road. Among the

H. Guo, *Big Earth Data in Support of the Sustainable Development Goals (2023)—China*, Sustainable Development Goals Series, https://doi.org/10.1007/978-981-95-2636-9_1

more than 230 SDG indicators, over one-third are related to the environment (United Nations Environment Programme 2021). These environment-related indicators are also the areas where Big Earth Data's spatio-temporal advantages are most evident. As Big Earth Data research deepens, the data available for indicator computation are continuously increasing and being updated.

At the national level in China, based on the datasets from the report released consecutively from 2019 to 2023, as well as certain national and United Nations' statistical data, the report evaluates and analyzes the progress of 98 environmentally focused SDG indicators in China from 2010 to 2022. The partial quantitative conclusions regarding the implementation progress of China's SDGs presented in the report are exploratory research outcomes based on the application of key Big Earth Data processing and analysis technologies and innovative methods. The results indicate that the evaluated indicators have generally shown sustained and stable improvement from 2010 to 2022. Between 2010 and 2015, 76.5% of the evaluated indicators continuously improved, while 14.7% deteriorated. From 2015 to 2022, the improvement trend further intensified, with 81.0% of the indicators continuously improving and no indicators showing deterioration. Among the 98 indicators assessed using Big Earth Data, only 13 indicators had achieved their 2030 targets by 2015. By 2022, more than half (51 indicators, 52.04%) had already met their 2030 targets ahead of schedule. Since the implementation of the sustainable development agenda in 2015, China's environmentally related SDG indicators have significantly improved. Specifically, in three goals—SDG 2 (Zero Hunger), SDG 13 (Climate Action), and SDG 15 (Life on Land)—over 60% of the participating indicators have already achieved their targets in 2030.

Big data offers advantages in the computation of globally consistent indicators, including rapid updates, repeatability, and extensive coverage, which play a crucial role in addressing data deficiencies in some countries (Guo 2020, 2021, 2022). However, throughout the long-term research and tracking of indicators, from the perspective of big data analysis and computation, there are three areas where the definition and evaluation standards of certain SDG indicators can be further refined and improved.

1. Limited data comparability for some indicators: Certain indicators do not account for differences in population, development levels, and geographical environments between countries. For example, SDG 13.2.2, which measures annual greenhouse gas emissions, does not consider disparities in population size and economic scale among nations.
2. Unclear evaluation standards for some indicators: The standards for some indicators are not quantifiable, making it difficult to measure progress or determine whether the goals have been achieved. For instance, SDG 14.a.1, which pertains to the proportion of the budget allocated to marine research activities, lacks a clear threshold for what constitutes the achievement of the goal. Additionally, increasing the budget proportion for marine research may lead to reduced budgets for other research areas such as agriculture and climate change.
3. Difficulty in obtaining spatial data for some indicators: Certain indicators require differentiation by gender, disability status, etc., such as SDG 11.2.1 (proportion of the population with access to public transportation) and SDG 11.7.1 (proportion of urban public open spaces). While the intention may be to emphasize the protection of vulnerable populations, obtaining spatial distribution data for different genders and disabled populations within cities is challenging. Relying solely on statistical data fails to capture spatial disparities.

In response to the current disparities in data acquisition capabilities among countries and the issues in indicator definitions, we recommend that, on the one hand, countries strengthen support to continue enhancing their ability to compute indicators using big data; on the other hand, the United Nations should adjust the indicator settings by clarifying definitions and standards and increasing considerations for data accessibility. These measures will better leverage the role of globally consistent big data, advancing toward a framework that is "more equitable and accessible".

References

Guo HD (2020) Big earth data in support of the sustainable development goals (2019). Science Press and EDP Sciences, Beijing

Guo HD (2021) Big earth data in support of the sustainable development goals (2020). Science Press and EDP Sciences, Beijing

Guo HD (2022) Big earth data in support of the sustainable development goals (2021). Science Press and EDP Sciences, Beijing

https://www.unep.org/resources/publication/measuring-progress-environment-and-sdgs [2024-08-01]

United Nations (2020) Road map for digital cooperation: implementation of the recommendations of the high-level panel on digital cooperation. https://digitallibrary.un.org/record/3864685 [2024-08-01]

United Nations (2022) The sustainable development goals report 2022. https://unstats.un.org/sdgs/report/2022/ [2024-08-01]

United Nations (2023) The sustainable development goals report 2023: special edition. https://unstats.un.org/sdgs/report/2023/ [2024-07-24]

United Nations Environment Programme (2021) Measuring progress: environment and the SDGs

SDG 2 Zero Hunger

2

2.1 Background

Sustainable management of land, soil, and water resources is the foundation for ensuring nutrition, diverse diets, and building a resource-saving value chain during the transition toward sustainable consumption patterns. About 90% of the calories and 80% of the protein consumed by humans originate from cropland (Kastner et al. 2012). However, due to population growth, the per capita arable land area has decreased by approximately 18% between 2000 and 2020 (FAO 2022). At the same time, evidence shows that the growth rate of agricultural productivity is slowing down, and high pollution and high emissions have pushed production capacity to the limit, causing land and environmental degradation (FAO 2021).

China is a major agricultural country, and its improvement in land management and agricultural production methods plays an important role in achieving SDG 2. To address the above challenges and promote sustainable agricultural development, China has taken a series of effective measures in promoting terraced farming, strengthening the management of irrigated land, carrying out black soil conservation, and advancing the construction of well-facilitated farmland. Approximately two-thirds of China's territory consists of mountainous regions, where terraces not only assist in controlling water erosion but also play a crucial role in soil conservation. While ensuring food production, they also have significant ecological benefits. Clarifying the spatial distribution and soil conservation benefits of terraces is crucial for planning and managing agricultural land use. Irrigated land, as an important part of irrigated cropland, has a high yield and good benefits, but irrigation also brings ecological and environmental problems such as water and soil pollution and soil erosion. Establishing a zoning and layering mapping method for irrigated dryland and monitoring its distribution and changes can support precise and scientifically cropland use and management. Due to long-term high-intensity development and utilization, the black soil resources in the cold regions of the Northeast China are facing a crisis of degradation. Residue cover, as an effective protective farming measure, can effectively reduce wind erosion and hydraulic erosion, restore and increase soil organic matter, and stably increase the thickness of the black soil layer. Utilizing multi-source remote sensing data to monitor the implementation of protective farming practices on black soil aids in formulating strategies for protecting black soil resources. China's construction of high-standard farmland improves agricultural production efficiency, promotes the conservation of land, water, fertilizers, and pesticides, and increases farmers' income, making an important contribution to "sustainable agriculture". Systematically investigating and summarizing China's methods

H. Guo, *Big Earth Data in Support of the Sustainable Development Goals (2023)—China*, Sustainable Development Goals Series, https://doi.org/10.1007/978-981-95-2636-9_2

of cropland protection and well-facilitated farmland construction can better understand the effectiveness of well-facilitated farmland construction and provide a scientific basis for sustainable agricultural-related strategic decision-making.

Against the backdrop of an increasingly severe global food security situation, promoting the comprehensive benefit enhancement of multi-objective collaborative cultivation land use through scientific planning and comprehensive goal guidance will be the key to achieving the goal of zero hunger. This chapter focuses on two aspects, namely monitoring sustainable food production systems and evaluating sustainable food production policies and benefits, summarizing China's recent progress in promoting the realization of the zero hunger goal.

2.2 Main Contributions

This chapter conducts monitoring and evaluation of SDG 2 indicators based on two specific objectives through five case studies, highlighting China's achievements in three areas: data products, methods and models, and decision support for SDG 2 indicator monitoring (Table 2.1).

2.3 Case Study

2.3.1 Distribution of Terraced Fields in China and Their Soil Conservation Benefits

Target: SDG 2.4 By 2030, ensure sustainable food production systems and implement resilient agricultural practices that increase productivity and production, help maintain ecosystems, strengthen capacity for adaptation to climate change, extreme weather, drought, flooding, and other disasters, and progressively improve land and soil quality.

- **Background**

Terraced farming has been an important agricultural practice in China, given that approximately two-thirds of China's territory is covered by mountains. Terraces are fields constructed along contour lines in mountainous regions, which not only ensure food production but also provide significant ecological benefits. In the sustainable development of agriculture, cropland water erosion is a major threat. China is one of the countries most severely affected by water erosion, and constructing terraced fields is the most effective measure for controlling water erosion on sloped cropland. In recent years, as soil and water loss control on sloped cropland has become a key focus of land consolidation in China, terraces have become an increasingly important type of cropland, playing a critical role in the soil conservation in China. Therefore, studying the spatial distribution and soil conservation benefits of terraced fields at the national scale is of great significance for the understanding of terrace effects, the evaluation of water erosion, and the management of agricultural land use. This case study aims to explore the large-scale remote-sensing-based terraced field mapping method, produce the China's terraced field distribution map, develop the large-scale terraced field soil conservation benefit evaluation model, quantify the water erosion reduction benefits of terraced fields, and provide support and guidance for achieving SDGs.

- **Data**

- Landsat images at 30 m resolution in 2015 and 2022.
- Shuttle Radar Topography Mission (SRTM) Digital Elevation Model (DEM) data.
- GlobeLand30 data.
- Annual China Land Use Dataset (CLUD-A) for 2015 and 2022.
- Chinese meteorological element station observation data.

Table 2.1 Case studies and their major contributions

Targets	Tiers	Case studies	Contributions
SDG 2.4 By 2030, ensure sustainable food production systems and implement resilient agricultural practices that increase productivity and production, that help maintain ecosystems, that strengthen capacity for adaptation to climate change, extreme weather, drought, flooding, and other disasters, and that progressively improve land and soil quality	Tier II	Distribution of terraced fields in china and their soil conservation benefits	Data product: 30 m spatial resolution terraced field distribution data for China in 2015 and 2022 Method and model: large-scale terraced field mapping methods based on remote sensing, and models for large-scale terraced field soil conservation benefit assessment Decision support: the findings highlight the importance and spatial heterogeneity of terraced fields and their soil conservation benefits, thereby providing a basis for decision-making in soil conservation and land planning
	Tier II	Spatio-temporal characteristics of irrigated dryland distribution in China (2015–2022)	Data product: irrigated dryland datasets for China in 2015 and 2022, and a dataset of changes in irrigated dryland in China from 2015 to 2022 Method and model: a method for extracting irrigated dryland by integrating environmental factors and vegetation index Decision support: this approach facilitates the adjustment of farmland structures based on local conditions, harmonizing the conflicts among food production, water use, and environmental sustainability, thereby ensuring food security and the sustainable development of resources and the environment
	Tier II	Estimation of corn residue coverage and its spatio-temporal analysis of the black soil in Northeast China	Data product: remote sensing-estimated corn residue coverage dataset for the black soil region in Northeast China in 2010 and from 2013 to 2022, with 30 m spatial resolution Method and model: a corn residue coverage estimation model based on the RF regression algorithm Decision support: the data can be used for monitoring and evaluating the implementation extent and scope of conservation tillage measures in the black soil region, thereby supporting decision-making for the sustainable development of black soil
	Tier III	Policy leadership and experience sharing on cropland protection in China	Data product: a compilation of major cropland protection policy documents in China from 1949 to 2022 Method and model: the "Leaf Model" based on literature analysis, which examines the evolution of cropland protection policies over time Decision support: based on national conditions, this approach facilitates the implementation of regulatory, constructive, and incentive-based protection measures for cropland, thereby achieving a comprehensive protection framework that integrates quantity, quality, and ecological sustainability. This effort contributes valuable insights to the global fight against hunger and the promotion of world peace

(continued)

Table 2.1 (continued)

Targets	Tiers	Case studies	Contributions
	Tier III	Comprehensive evaluation of the benefits of well-facilitated farmland construction in China	Data product: data on the implementation effectiveness of well-facilitated farmland construction projects in key agricultural areas across China's provinces Method and model: evaluation methods for assessing the effectiveness of well-facilitated farmland construction Decision support: while prioritizing the enhancement of grain production capacity, this approach also promotes farmers' income, resource efficiency, and environmental sustainability, providing a model for sustainable and high-quality agricultural development

Note The indicators are divided into three levels. Tier I: Indicators are conceptually clear there are an internationally established methodology and standards, and data are regularly produced by countries for at least 50% of countries and of the population in every region where the indicators are relevant. Tier II: Indicators are conceptually clear, there are an internationally established methodology and standards, but data are not regularly produced by countries. Tier III: No internationally established methodology or standards are yet available for the indicators, but methodology/standards are being (or will be) developed or tested. (As of the 51st session of the United Nations Statistical Commission, the global indicator framework does not contain any Tier III indicators.)

- WorldClim data.
- Moderate-resolution Imaging Spectroradiometer (MODIS) Normalized Difference Vegetation Index (NDVI) data.
- Basic soil property dataset of high-resolution China Soil Information Grids.

- **Methods**

This case study conducted the evaluation of China's terraced field distribution mapping and soil conservation benefits. For China's terraced field distribution mapping, we developed the large-scale remote-sensing-based terraced field mapping method. Based on the Landsat remote-sensing imagery and DEM data, we extracted multi-temporal spectral features and topographic features, constructed the optimized feature sets, obtained samples through visual interpretation and sample transfer techniques, designed appropriate classification algorithms, and developed China's terraced field distribution dataset at 30 m spatial resolution. For the soil conservation benefit evaluation of China's terraced fields, we constructed the large-scale terraced field soil conservation benefit evaluation model based on the Revised Universal Soil Loss Equation (RUSLE) to quantify the soil erosion reduction benefits of terraced fields in China.

- **Results and Analyses**

1. **Distribution mapping of terraced fields in China**

Terraced fields are widely distributed across China, accounting for about a quarter of the national cropland area. The spatial distribution of terraced fields in China is closely related to the topography, mainly distributed in plateau mountainous regions (e.g., the Loess Plateau, the Yunnan-Guizhou Plateau), hilly regions (e.g., the central and southeastern hills), and mountain-basin transition regions (e.g., the eastern Sichuan Basin) (Fig. 2.1). From 2015 to 2022, the provinces with a relatively high proportion of newly added terraced fields were mainly located in areas with numerous and dense pre-existing terraced fields, such as Gansu and Shaanxi on the Loess Plateau in northwestern China, and Sichuan and Yunnan in southwestern China. About 46.7% of the newly added terraced fields were in areas with slope gradients of 6°–15°. The conversion of sloping land to terraces in

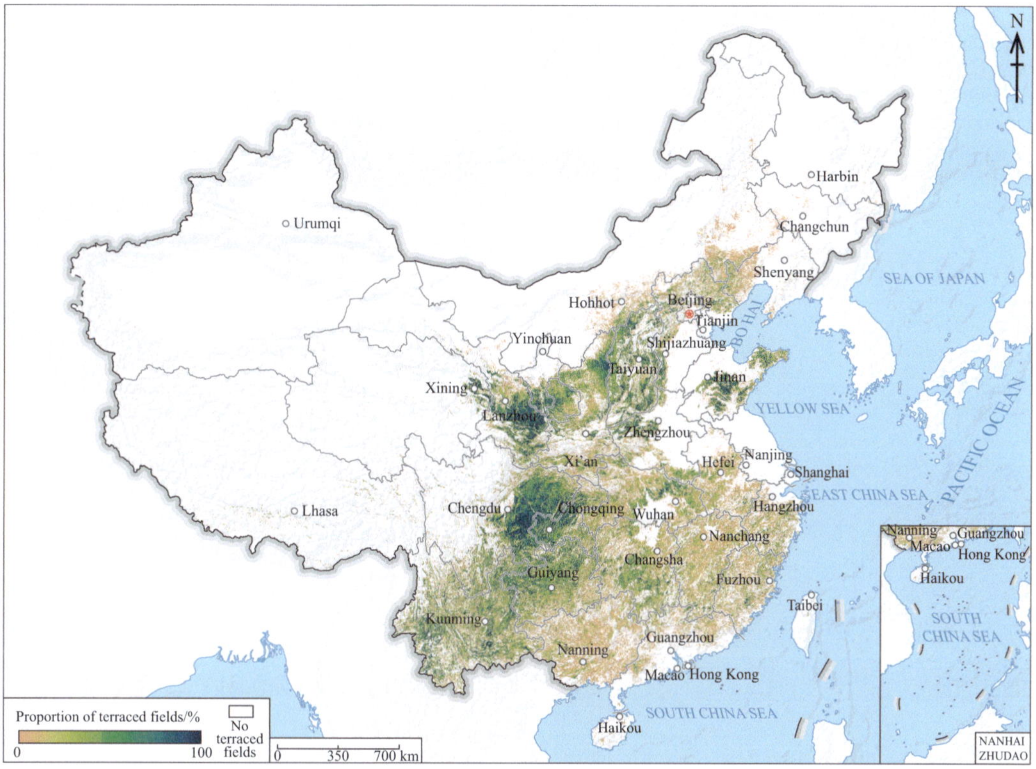

Fig. 2.1 Distribution map of terraced fields in China in 2022 *Note* The pixel value in this figure represents the percentage of terraced fields within each 1 km × 1 km grid cell, calculated based on the results of the 30 m resolution terraced field mapping

areas with relatively high slope gradients plays a positive role in mitigating soil erosion resulting from improper land utilization.

2. **Evaluation of soil conservation benefits of terraced fields in China**

The soil conservation benefits of terraced fields in China are significant, and the existing terraced fields can reduce the total water erosion of cropland in China by about 50%. The benefits of terraced fields exhibit spatial heterogeneity. Terraced fields demonstrate significant soil conservation benefits in agricultural regions dominated by mountainous and hilly landscapes. In southwestern and northwestern China, where the terraced field area is large and the potential erosion rate is high, terraced fields contribute most to erosion reduction. In the eastern and southern hilly regions of China, terraced fields also show considerable erosion reduction benefits per unit terraced field area. In regions with fewer terraced fields and lower potential erosion amounts per unit terraced field area, such as the northeastern region, the soil erosion reduction benefits of terraced fields are relatively low. In the areas with slope gradients ranging from 6° to 25°, terraced fields lead to the greatest reduction in cropland erosion in terms of both the total amount reduction and percentage reduction. In the areas with slope gradients greater than 25°, terraced fields demonstrate the largest erosion reduction per unit terraced field area. The results indicate that terraced fields play a crucial role in soil conservation on steep-slope cropland (Fig. 2.2).

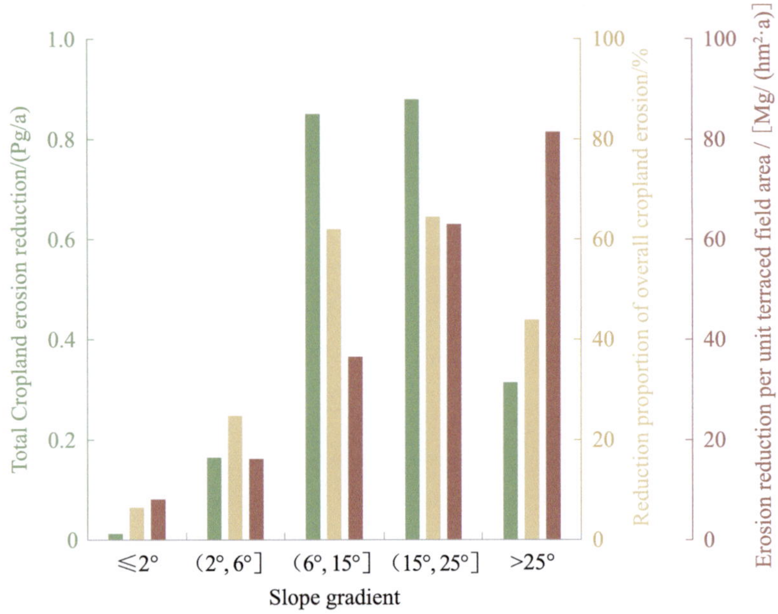

Fig. 2.2 Water erosion reduction benefits of terraced fields in areas with different slope gradients

Highlights

- The large-scale terraced field mapping method based on remote sensing was developed, and the China's terraced field distribution datasets in 2015 and 2022 at 30 m spatial resolution were produced. The mapping results indicate that terraced fields are widely distributed in China, accounting for approximately a quarter of the national cropland area, and about 46.7% of newly added terraced fields from 2015 to 2022 were in areas with slope gradients of 6°–15°.
- The large-scale terraced field soil conservation benefit evaluation model was constructed, and the soil conservation benefits of terraced fields in China were quantified. The modeling results show that the soil conservation benefits of terraced fields in China are significant, and the existing terraced fields can reduce the total water erosion of cropland in China by about 50%, and the terraced field benefits exhibit spatial heterogeneity.

- **Discussion and Outlook**

This case study focused on mapping the distribution of terraced fields in China and evaluating their soil conservation benefits. A large-scale terraced field mapping method based on remote sensing was developed, and the China's terraced field distribution dataset at 30 m spatial resolution in 2015 and 2022 was produced. A large-scale terraced field soil conservation benefit evaluation model was constructed, and the spatial heterogeneity of soil conservation benefits was studied. Based on the research results, further decision-making suggestions can be proposed for SDG 2.4, starting from the multiple actors of "central government-local government-ordinary farmers", with a parallel approach of "monitoring-incentive-regulation". First, remote sensing technologies such as satellite imagery and unmanned aerial vehicle (UAV) should be used to establish a regularly updated terraced field monitoring system. This system can be integrated into agricultural statistics and land-use planning to monitor the status and changes of terraced fields, as well as the implementation and maintenance of "slope-to-terrace" project engineering. Second, appropriate reward and incentive mechanisms, including subsidies for terraced farming and mechanization, should be explored, and the technical support and guidance for cultivation and management should be provided. The mechanisms and support can enhance farmers' willingness to protect existing terraced fields and ensure their continued presence and soil conservation functions. Finally, the current fiscal system should be innovated by creating a special fund for terraced field compensation, consisting of subsidies from central and local governments. The evaluation of terraced fields' soil conservation benefits should also be incorporated into cropland planning and the local government performance assessment system.

2.3.2 Spatio-Temporal Characteristics of Irrigated Dryland Distribution in China (2015–2022)

Target: SDG 2.4 By 2030, ensure sustainable food production systems and implement resilient agricultural practices that increase productivity and production, help maintain ecosystems, strengthen capacity for adaptation to climate change, extreme weather, drought, flooding, and other disasters, and that progressively improve land and soil quality.

- **Background**

In the State of the World's Land and Water Resources for Food and Agriculture: Systems at Breaking Point, the Food and Agriculture Organization of the United Nations (FAO) states that the majority of pressure on global land, soil, and water resources stems from agriculture itself. It advocates for climate-smart land management practices that align with the variability in soil and water resources (FAO 2021). This underscores the severe conflicts among food production, arable land, and water resources, necessitating the development of precise and scientifically informed cropland management strategies (Gabr et al. 2020; Gleeson et al. 2012). Irrigated dryland, as an essential part of irrigated croplands, refers to cultivated land with upland crops with assured irrigation in general, and often concentrated in the severely water-scarce regions, like Northwest and North China, accounting for over 90% of the nation's irrigated dryland (National Bureau of Statistics 2023). These regions largely depend on groundwater extraction for irrigation. However, due to factors such as excessive fertilization (Jia et al. 2018) and severe soil erosion (Shan and Wang 2015; Jia 2010), irrigation is often accompanied by ecological challenges, including water and soil pollution and soil erosion. Compared to paddy fields, which are predominantly located in water-rich regions, irrigated dryland is more critical for food security, agricultural production, water resource utilization, and ecological sustainability.

Given these challenges, it is imperative to clarify the current distribution patterns of irrigated dryland, along with their evolution and underlying causes, to implement precise and scientifically informed cropland management measures. In China, existing cropland data are broadly categorized into paddy fields and non-paddy fields, with irrigated dryland falling under the latter category (Zhang et al. 2014; Kuang et al. 2022), or into irrigated croplands and rainfed croplands (Liu et al. 2020), with irrigated dryland classified as the former. These classifications, however, do not explicitly differentiate irrigated dryland from other types, resulting in a lack of precise spatial distribution data. To address this gap, this case study developed a zoned and layered mapping methodology for irrigated dryland based on time-series remote sensing data. The distributions and changes in irrigated dryland in China from 2015 to 2022 were monitored using this methodology. This approach can bridge the gap in accurate data on irrigated dryland helping to estimate crop yields precisely and provide a better understanding of food security. Additionally, it facilitates the rational estimation of agricultural irrigation water demands and the coordination of regional water resource allocation. Furthermore, it enables the precise monitoring of agricultural non-point source pollution and soil erosion, supporting the implementation of localized ecological protection measures. Ultimately, this work can contribute to the sustainable development of food, water, and environmental systems.

- **Data**

- Landsat imagery at 30 m spatial resolution: Data from 2014 to 2016 and from 2021 to 2023.
- National DEM data and statistical data (including economic statistical yearbooks, agricultural statistical yearbooks, and rural statistical yearbooks at the provincial or municipal level).
- Daily land surface temperature data: MYD11A1 V6 product data from 2013 to 2016 and from 2020 to 2023.
- Daily precipitation data: PERSIANN-CDR (Precipitation Estimation from Remotely Sensed Information Using Artificial Neural Networks-Climate Data Record) from 2013 to 2016 and from 2020 to 2023.
- Land use data: Datasets for 2015 and 2020 provided by the Aerospace Information Research Institute, Chinese Academy of Sciences.
- Sample data obtained from field surveys in 2018 and 2019 in areas such as Zhangjiakou in Hebei Province and Xinzhou in Shanxi Province. The collected sample data include geographic coordinates, irrigation status, major crops, and cropping systems.

- **Methods**

Given China's vast territory and significant environmental variability, this case study employs a zoned approach to extract irrigated dryland from non-paddy fields. First, based on precipitation, topography, cropping systems, and major crops, the study area is divided into five zones: the Northwest Irrigation Zone, the Northern Plateau Single-Cropping Zone, the Northeastern Plain Single-Cropping Zone, the Central-Eastern Plain Double-Cropping Zone, and other zones. Second, all drylands in the Northwest Irrigation Zone are classified as irrigated dryland since precipitation in this zone is barely sufficient to sustain crop growth, while irrigated dryland identification is conducted for the remaining four zones. Third, leveraging the principle that crops in irrigated dryland experience no significant drought stress during dry seasons while vegetation indices in rainfed dryland are notably affected by drought stress (Sun and Kafatos 2007; Zhu et al. 2022), optimal identification features and their combinations are proposed for each zone. These features include intra-annual vegetation indices, environmental factors, and their composite curves, as well as additional features such as slope and texture. Based on these feature combinations, the Random Forest model is employed to generate classification results for 2015 and 2022, along with the optimal feature combinations for each zone. Subsequently, changes in the optimal identification features for irrigated dryland from 2015 to 2022 are used to identify the types of irrigated dryland changes during this period directly. Due to the limited sample size for 2022, the classification results are only preliminary results, and the final irrigated dryland distribution in 2022 is derived by combining the irrigated dryland distribution data in 2015 with the identified change types in this period.

This methodology mitigates the impact of environmental variability through zoning, achieving higher accuracy in irrigated dryland identification. By directly extracting change types, the approach avoids potential error

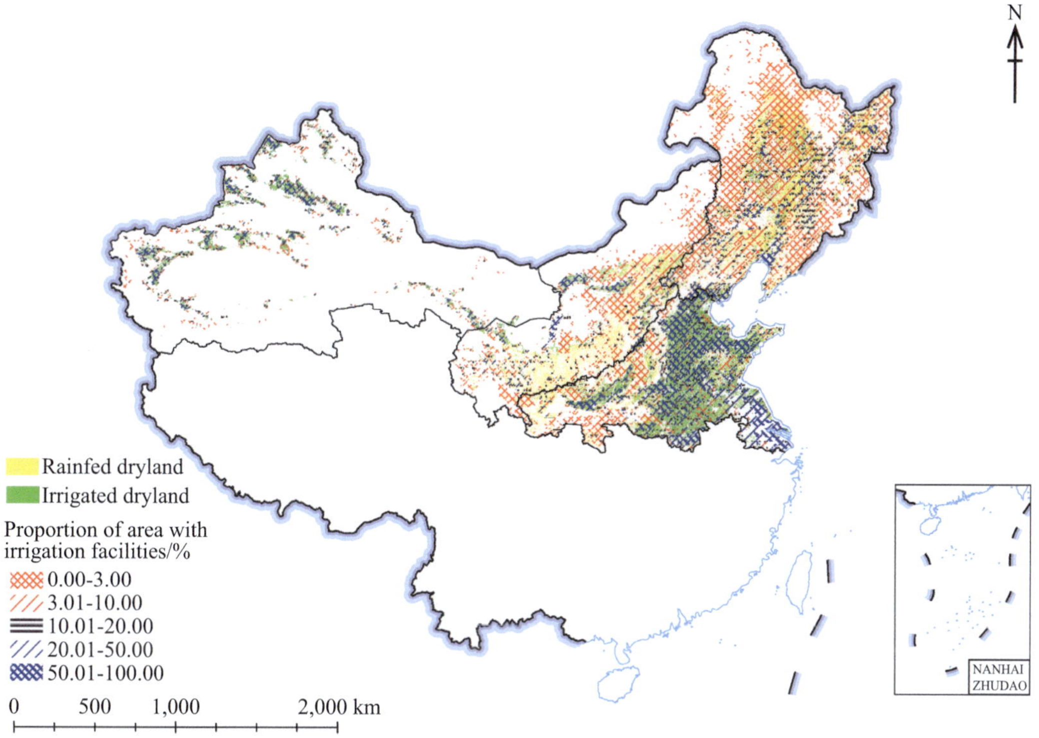

Fig. 2.3 Comparison of extraction results with GMIA for the northern area in 2015. *Note* GMIA refers to global map of irrigation areas

accumulation from overlay analysis of the results from two time periods. Meanwhile, accounting for the spatio-temporal dynamics of pixels results in more accurate change information. As a result, the overall accuracy and Kappa coefficient for each zone exceed 80% and 0.65, respectively, and the results align well with the statistical data from typical counties and cities in each zone. Compared with existing spatial distribution data on irrigated croplands (Figs. 2.3 and 2.4), the results of this case study exhibit broad agreement in the northern zones with concentrated irrigated dryland, while demonstrating improved extraction of small-scale irrigated dryland in the central-northern region. In the southern region, where irrigated croplands are predominantly paddy fields and irrigated dryland constitutes a negligible proportion, no comparison is made.

- **Results and Analysis**

1. **Spatial distribution patterns of irrigated dryland**

The classification results of China's irrigated dryland in 2015 (Fig. 2.5) show that irrigated dryland is distributed sporadically in the southeast, concentrated in the north, mixed in central and northeastern areas (Sanjiang Plain in Northeast China, Loess Plateau, Sichuan Basin, and Yunnan-Guizhou Plateau), and discontinuous clusters in the northwest. Excluding the North China Plain, the proportion of irrigated dryland

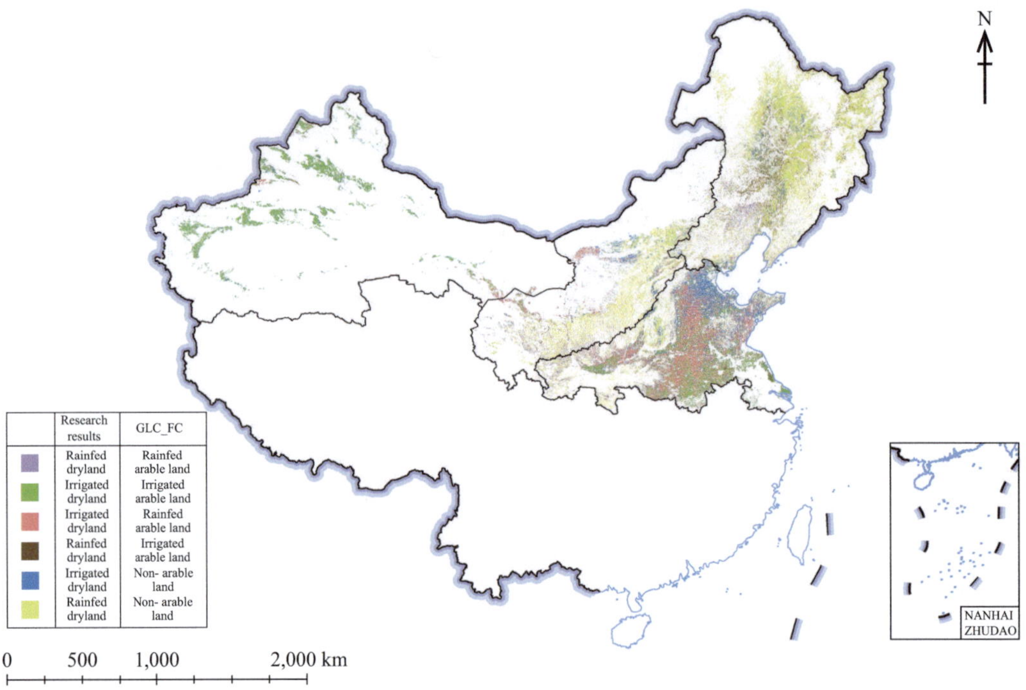

Fig. 2.4 Comparison of extraction results with GLC_FCS for the northern area in 2015. *Note* GLC_FCS refers to global land cover with a fine classification system

negatively correlates with precipitation levels, showing a gradual transition from the southeast to the northwest. Despite receiving a relatively high amount of precipitation, the North China Plain experiences severe structural water shortages, especially during the winter and spring seasons. Overall, the distribution of irrigated dryland closely aligns with the spatial patterns of precipitation during the growing season and water resource availability.

Due to the predominance of monsoon climates across most areas of China, crop water requirements during the dry seasons (winter and spring) are often unmet. In the eastern region, favorable topography and abundant water resources support large-scale irrigation. The southeastern areas, in particular, are predominantly characterized by irrigated paddy fields, with only a small proportion of irrigated dryland. In contrast, the North China Plain in the east-central region is dominated by irrigated upland crops, where irrigated drylands are concentrated. This results in the pattern of sporadical irrigated drylands in the southeast transitioning into concentrated distributions as one moves northward.

The zone extending from the Sanjiang Plain in Northeast China, through Loess Plateau, Sichuan Basin, to Yunnan-Guizhou Plateau, represents a mixed distribution of irrigated dryland and rainfed dryland. In this region, crops experience long-term or seasonal drought stress, and water resources are insufficient to support large-scale irrigation, unlike in the eastern regions. Consequently, irrigated dryland is developed only in low-lying areas with favorable topography. Concentrated irrigation areas have only formed in river valleys with natural water sources, such as the Yangwu River irrigation area in Yuanping, Xinzhou (Fig. 2.6a) and Wenbi Town in Hequ County, Xinzhou (Fig. 2.6b) in Shanxi Province.

Additionally, advances in technology and supportive agricultural policies have enabled some firms or farmers to drill deep wells and construct hydraulic engineering projects, leading to the development of irrigated dryland in arid

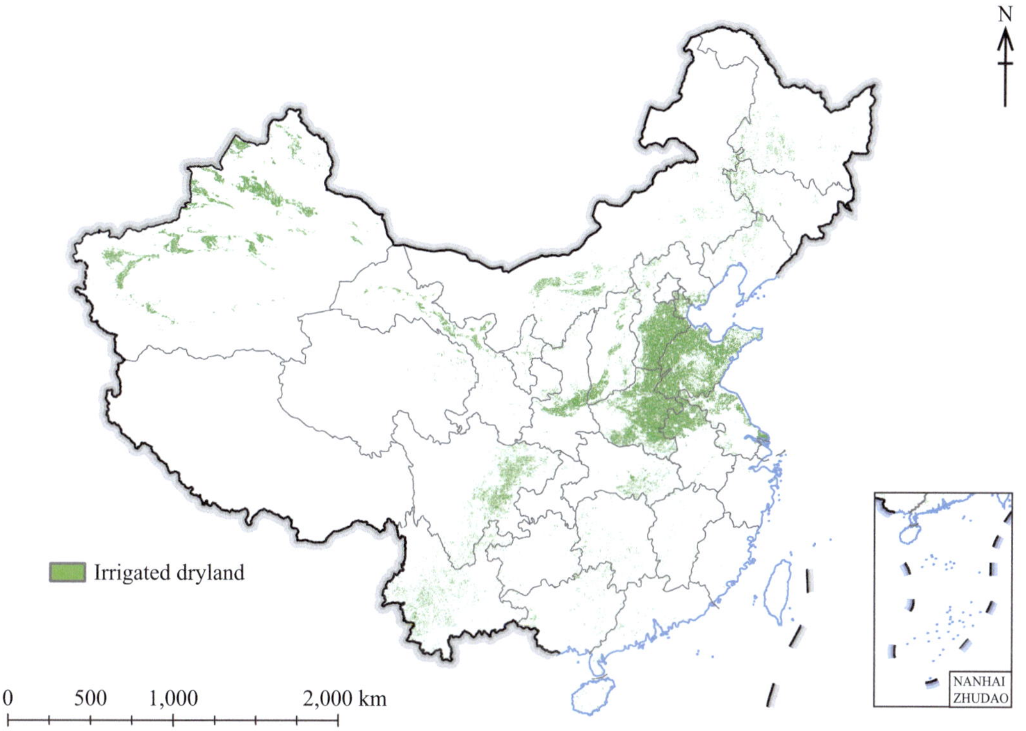

Fig. 2.5 Distribution of irrigated dryland in China in 2015

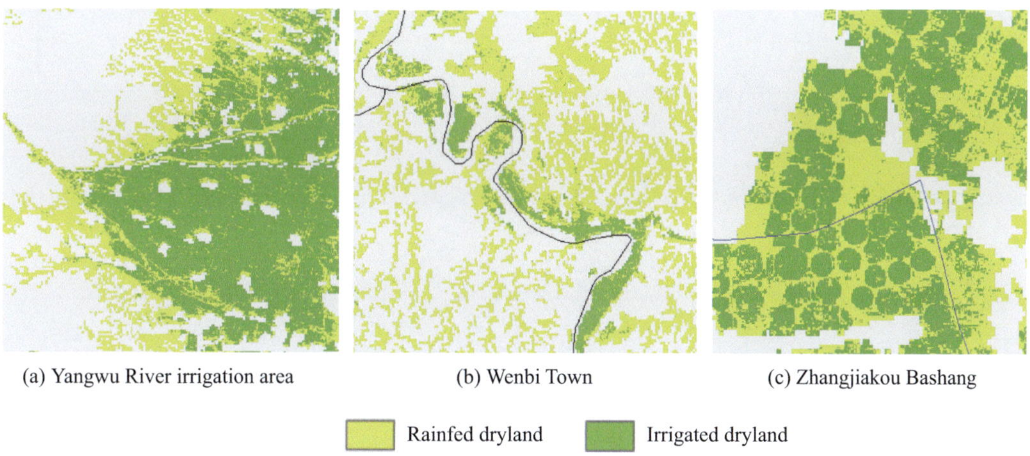

Fig. 2.6 Typical irrigated dryland

regions. For example, in the Bashang region of Zhangjiakou, located on the Nei Mongol Plateau, irrigated dryland mixes with rainfed dryland, forming a mixed distribution pattern (Fig. 2.6c).

In the northwest, annual precipitation is typically below 300 mm, which is insufficient to meet water requirements for crop growth. Irrigated drylands constitute a significant proportion of croplands in these regions, and they are primarily distributed around oases with available water sources. Furthermore, advancements in technology and funding, coupled with increasing food demand, have driven the continuous expansion of irrigated drylands in this region. For instance, in Xinjiang, nearly 100% of the arable land is irrigated, while in the Bayannur City of Nei Mongol, large irrigation areas fed by the Yellow River ensure that irrigated dryland accounts for 95.63% of the total arable land.

2. **Analysis of changes in irrigated dryland**

The overall spatial pattern of irrigated dryland in 2022 remains consistent with that of 2015 (Fig. 2.7), with a slight increase in the total area and observable changes in quantity and distribution in certain local regions (Fig. 2.8). First, in the central-eastern plain regions, irrigated dryland has become more densely distributed. Second, due to advancements in technology and economy, the western regions have experienced rapid expansion of water infrastructure projects in recent years, making them the fastest-growing areas for irrigated dryland in China. Lastly, some irrigated drylands have been converted into rainfed drylands due to the reduction or disappearance of irrigation water sources. In some cases, local communities have proactively implemented "reverting irrigated dryland to rainfed dryland" projects, recognizing the incompatibility of irrigated farming with local natural and ecological conditions.

Overall, the changes in irrigated dryland during this period have contributed positively to balancing economic development with resource and environmental conservation. The stable growth of irrigated dryland supports future food security, and the rapid development of irrigated dryland in western regions has stimulated local agricultural productivity and the overall economic growth. In water-scarce areas, both the natural disappearance of irrigated dryland and the deliberate implementation of conversion projects have fostered water conservation and ecological protection. Moving forward, policies for cropland management should be tailored to

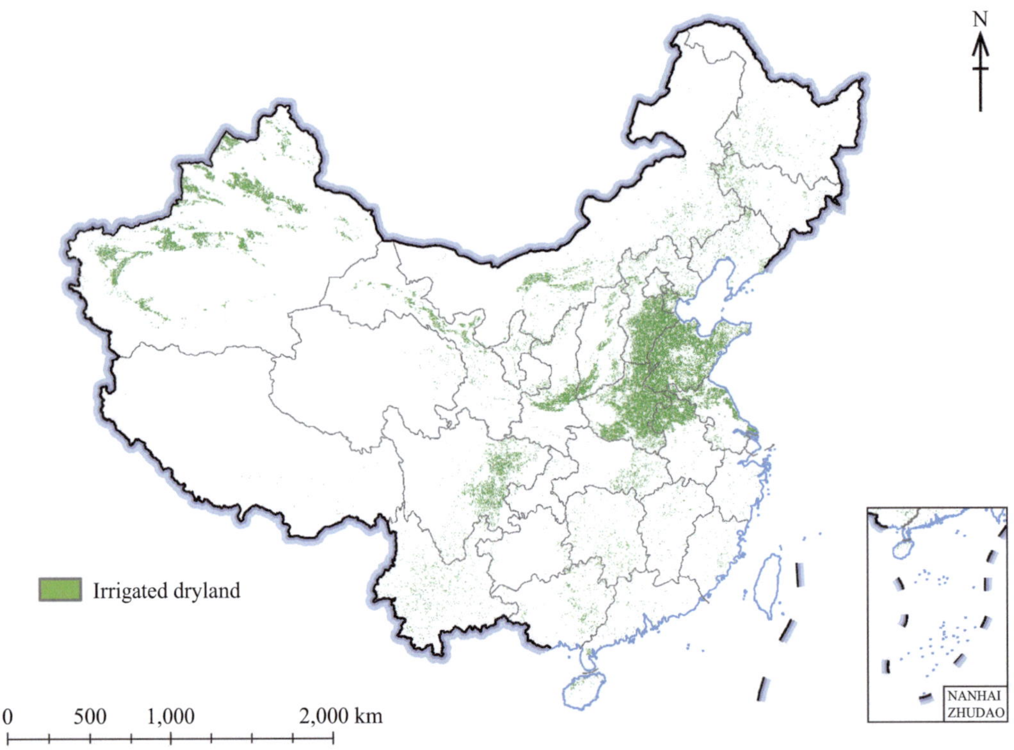

Fig. 2.7 Distribution of irrigated dryland in China in 2022

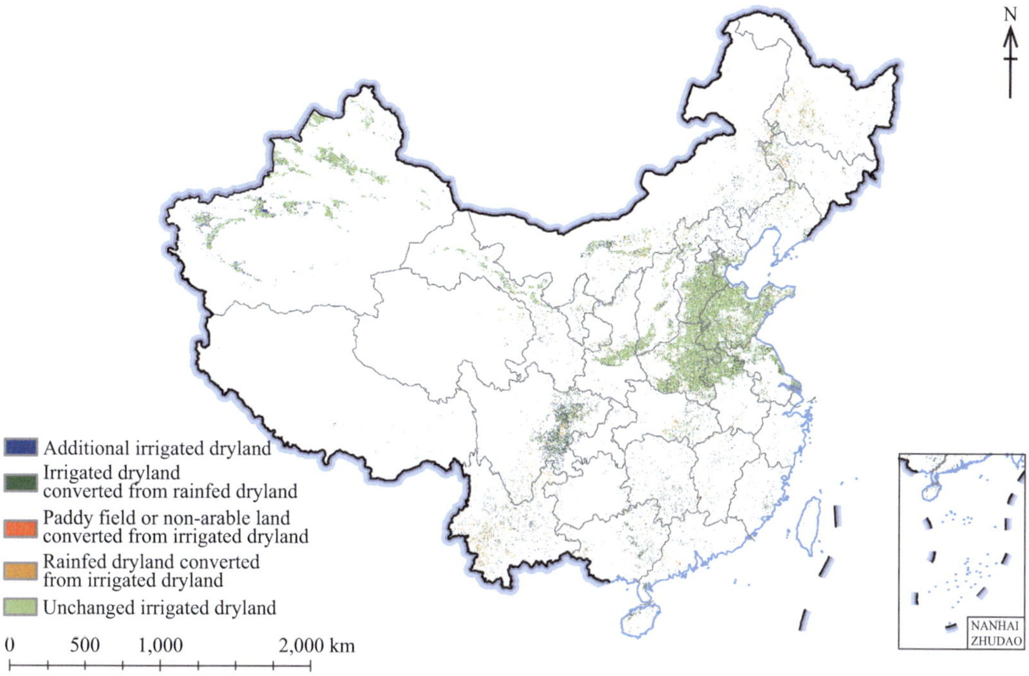

Fig. 2.8 Changes in irrigated dryland distribution in China in 2015–2022

the precise spatial and temporal dynamics of irrigated dryland, ensuring sustainable development in food production, water resource management, and environmental protection.

Highlights

- The spatial distribution of irrigated dryland in China is sporadical in the southeast, concentrated in North China, mixed in central areas (from Sanjiang Plain in Northeast China, Loess Plateau, Sichuan Basin, and Yunnan-Guizhou Plateau), and clumped and scattered in the northwest.
- From 2015 to 2022, irrigated dryland experienced certain development, with distributions becoming denser and shifting toward the northwest. The stable development of irrigated dryland has ensured China's food security, contributed to global food security, and provided valuable insights into high-quality farmland construction for other countries.

- **Discussion and Outlook**

This case study utilizes long-time-series Landsat data and other auxiliary data to realize the monitoring of the cropland sub-type-irrigated dryland in SDG 2.4, analyzes the distribution and dynamic patterns of irrigated dryland in China, and produces a national-scale 30 m resolution dataset for irrigated dryland, providing basic data support for achieving sustainable development of food, water, and environment.

This case study yielded the following findings.

(1) The primary factors influencing the distribution of irrigated dryland are growing-season precipitation, topography, and accessibility to water resources. These factors shape the distribution patterns at different scales. On a large scale, a negative correlation exists between growing-season precipitation and the proportion of irrigated dryland. On a medium scale, relatively flat and low-lying areas are typically the primary regions for concentrated

irrigated dryland. On a small scale, water sources and hydraulic infrastructure are the main influencing factors. Additionally, with economic and technological development, human factors are increasingly affecting the distribution of irrigated dryland.

(2) The overall spatial pattern of irrigated dryland in 2022 remained consistent with that of 2015, with a slight increase in total area, denser distribution patterns, and a shift toward the northwest. Furthermore, in areas where irrigation water sources are no longer reliable or ecological concerns have been recognized, irrigated dryland has either disappeared naturally or been deliberately converted into rainfed dryland. These changes contribute positively to balancing economic development with resource and environmental conservation.

2.3.3 Estimation of Corn Residue Coverage and Its Spatio-Temporal Analysis of the Black Soil in Northeast China

Target: SDG 2.4 By 2030, ensure sustainable food production systems and implement resilient agricultural practices that increase productivity and production, help maintain ecosystems, strengthen capacity for adaptation to climate change, extreme weather, drought, flooding, and other disasters, and that progressively improve land and soil quality.

- **Background**

SDG 2.4 relates to the monitoring and evaluation of sustainable agriculture, and the protection of cropland is one of the overall strategic issues related to the sustainable development of China. The black soil in Northeast China is the most fertile and precious agricultural resource for planting with a high organic matter content. Unfortunately, the black soil in Northeast China is facing challenges including fertility declining, organic matter decreasing, coupling with wind erosion, and hydraulic erosion after high-intensity reclamation and planting. And the black soil is becoming thinner, leaner, harder, and dirtier. Professor Baoguo Li from China Agricultural University reported to the General Secretary Xi Jinping on how to protect black soil through the "Lishu Model" on July 22, 2020. The Black Soil Protection Law of the People's Republic of China was passed by the 35th Meeting of the Standing Committee of the 13th National People's Congress on June 24, 2022. All these national measurements and actions deliver the objectives of China to protect the black soil.

Crop residue covering is an effective measure to reduce wind and water erosion, increase the organic matter content of soil, and improve the physical properties of black soil. Crop residue coverage is an indicator that characterizes the implementation of conservation tillage on black soil, and analyze the potential for sustainable development of black soil. To characterize the protection of black soil in Northeast China, we have retrieved the corn residue coverage (CRC) in the black soil zone of Northeast China since 2010 using multi-source remote sensing data.

- **Data**

- Landsat 5/7/8/9 imagery ranging in 2010 and 2013–2022, and reflectance images of MODIS ranging in 2010 and 2013–2022.
- Conservation tillage data from the China Agricultural Mechanization Yearbook in 2013–2020.
- Panchromatic and multispectral system (PMS) imagery from Gaofen-2 from 2017 to 2022, and high-resolution images from Google Earth Pro for labeling CRC.
- UAV data with different corn cover types in Jilin Province and Liaoning Province in 2015, 2017, and 2019 for labeling CRC.

- **Methods**

The CRC is estimated by the RF regression modeling with the labeling from the high spatial-resolution images including Google Earth Pro, UAV, and Gaofen-2 PMS in the black soil zone in Northeast China. The workflow is as follows. (1) The CRC labels are collected using the OTSU algorithm and binarization algorithm of optimal global thresholding from the high spatial-resolution images including Google Earth Pro, UAV, and Gaofen-2 PMS. (2) Monthly image fusion is done based on a highly scalable time-adaptive reflectance fusion model (HISTARFM) using Landsat series images and MODIS surface reflectance images. (3) RF regression modeling is done for estimating CRC in the black soil Zone of Northeast China (Martínez-Ferrer et al. 2022; Dong et al. 2023) using the collected samples and the harmonized reflectance images and their derived residue spectral vegetation index (Yue et al. 2020) and textural features. (4) The CRC estimation is validated using the coefficient of determination R^2 and root mean square error (RMSE).

RF regression algorithm is an important application branch of RF. The RF regression model establishes multiple unrelated decision trees by extracting samples and features randomly, obtaining the prediction results through parallelism. Each decision tree can produce a prediction result through the extracted samples and features, and the regression prediction result of the whole RF is obtained by averaging the results of all the decision trees. In the training phase, RF uses Bootstrap sampling to collect multiple different sub-training datasets from the input training dataset to train multiple different decision trees sequentially; in the prediction phase, RF averages the prediction results of multiple internal decision trees to obtain the final result. The feature importance (n_k) during model training and prediction is calculated as follows.

$$n_k = w_k \times G_k - w_{\text{left}} \times G_{\text{left}} - w_{\text{right}} \times G_{\text{right}} \tag{2.1}$$

where $w_k, w_{\text{left}}, w_{\text{right}}$ are the ratios of the number of training samples to the total number of training samples in node k and its left and right children nodes, respectively, and $G_k, G_{\text{left}}, G_{\text{right}}$ are the impurity of node k and its left and right children nodes, respectively. Once the importance of each node is known, the importance of a feature can be derived from the following equation:

$$f_i = \frac{\Sigma n_j}{\Sigma n_k} \tag{2.2}$$

Here, the importance f_i of feature i is obtained by adding the importance of all the nodes that split on the feature and dividing it by the sum of the importance of all the nodes. f_i is the importance of feature i, j is the node that splits on feature i, Σn_j is the sum of the importance of all the nodes that split on feature i, Σn_k is the sum of the importance of all the nodes.

To make the values of the importance of all features add up to equal 1, the importance of each feature needs to be normalized and the formula used is

$$f_{ni} = \frac{f_i}{\Sigma f_j} \tag{2.3}$$

where f_{ni} is the normalized importance of feature i, f_i is the importance of feature i, Σf_j is the sum of the importance of all features.

- **Results and Analysis**

The CRC was estimated using the RF regression model in the black soil zone of Northeast China ranging from 2010 to 2022 (You et al. 2023; Li et al. 2024; Dong et al. 2024) as shown in Fig. 2.9. The estimation results showed that corn residue was mainly distributed in the Songnen Plain in northern Liaoning Province, central Jilin Province, southern Heilongjiang Province, as well as western Liaoning Province and northeastern Heilongjiang Province of the Sanjiang Plain. The CRC was categorized as low, medium, and high coverage by (0, 0.3], (0.3, 0.7], and (0.7, 1], respectively. In 2010, the percentages of

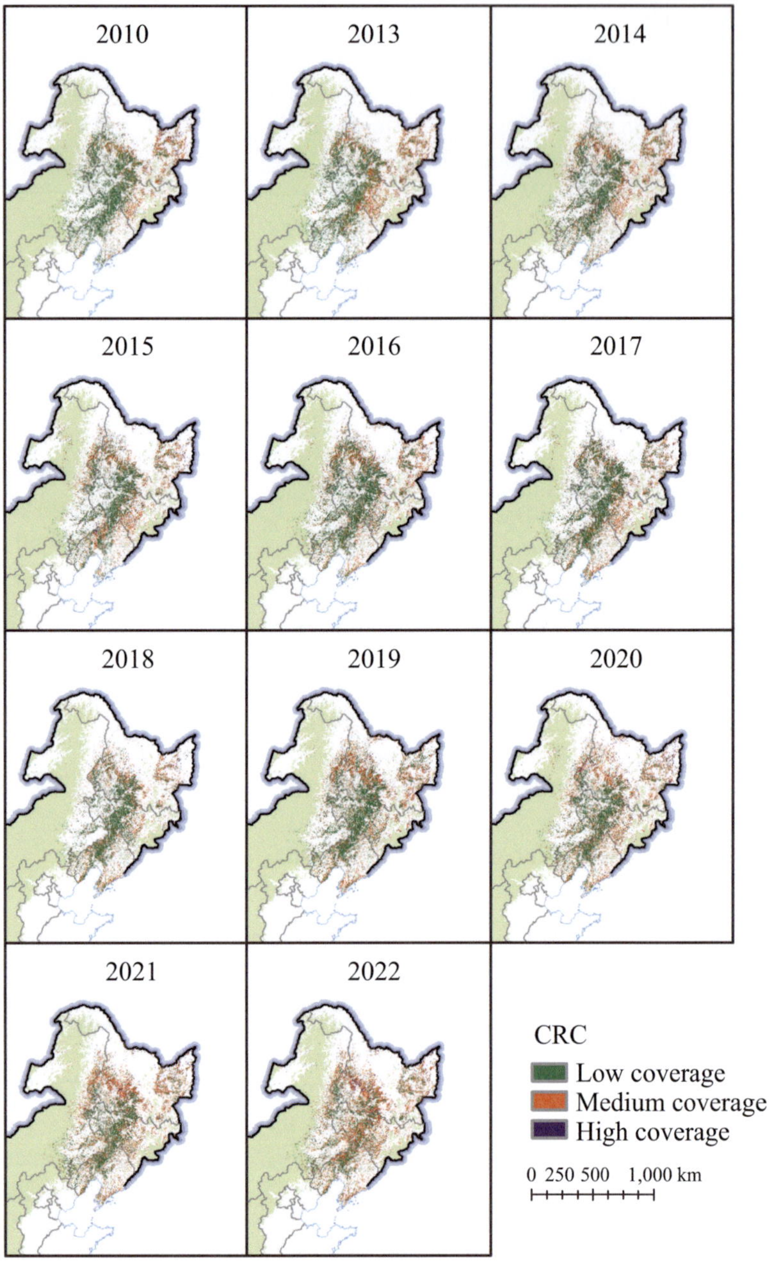

Fig. 2.9 Spatio-temporal distribution of CRC in Northeast China from 2010 to 2022. *Note* Due to the unavailability of Landsat-8 satellite imagery and the absence of a validation dataset for the years 2011–2012, CRC during this period was not mapped in this study

the three types of coverage were 64.9% for low coverage, 32.0% for medium coverage, and 3.1% for high coverage, while in 2022, the percentages of the three types of coverage were 35.3% for low coverage, 60.3% for medium coverage, and 4.4% for high coverage. In Fig. 2.9, it can be seen that in 2010, the low-coverage areas were mainly distributed in the Songnen Plain and the western part of Liaoning Province, the medium-coverage areas were mainly distributed in the Changbai Mountain area in the eastern part of Jilin Province, the central and southwestern part of Heilongjiang Province, and some parts of the Sanjiang Plain, and the high-coverage areas were mainly in the northeastern part of the Sanjiang Plain. In 2022, the CRC areas were

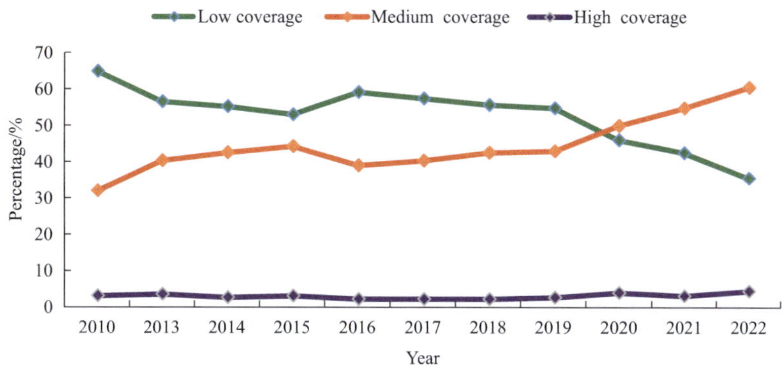

Fig. 2.10 Spatio-temporal changes of CRC in Northeast China from 2010 to 2022

mainly distributed in the central part of Liaoning Province, the four eastern league areas of Nei Mongol, the northern part of Songnen Plain, and part of Sanjiang Plain; the medium coverage areas were widely distributed in the study area; the high coverage areas were mainly distributed in the northern part of Heilongjiang Province. With the increase of latitude, the CRC showed an increasing trend.

The spatio-temporal evolution of CRC in Northeast China is shown in Fig. 2.10. From 2010 to 2015, the area with CRC above 0.3 (medium coverage and high coverage) increased year by year, but showed a decreasing trend in 2016, which was because the state carried out crop cultivation structure adjustment in Northeast China before 2016, and the area of corn cultivation was reduced by 23 million mu[1] compared with that in 2015, leading to the decrease in the area covered by corn residue. From 2016 to 2022, the area with CRC more than 0.3 showed a rising trend, which was due to the national implementation of conservation tillage methods based on returning corn residue to the soil.

Highlights

- The CRC indicating the protection measures for the black soil is estimated by developing the RF regression model using the 3,185 CRC samples collected from multi-source remote sensing images. The further spatio-temporal transfer of the RF regression model is done by the agroecological zoning for producing the CRC data products in Northeastern China in 2010 and 2013–2022.
- The CRC estimation results reveal that the total area of corn residue covering in Northeast China has increased year by year since 2013. The definition of conservation tillage is the implementation of crop residue covering with CRC more than 0.3 by FAO. Our CRC estimation results show that the area of CRC more than 0.3 increases by 68,000 hm^2 with a growing percentage of 21.3%. This indicates that the area of cropland with crop residue covering has increased significantly, and the implementation of conservation tillage measures in Northeast China has been effective.

- **Discussion and Outlook**

In this case, a CRC model based on the RF regression algorithm was developed to meet the demands of black soil protection and sustainable development in China. The RF regression model was migrated spatially and temporally according to the implementation process of agroecological zoning and protective measures in the three northeastern provinces, and the products of CRC in 2010 and 2013–2022 were produced. Moreover, the spatio-temporal evolution patterns of CRC guided by the national black soil protection policy and natural geographic factors were comprehensively analyzed, and the spatial scope

[1] 1 mu ≈ 666.67 m^2.

and intensity of black soil protection were explored.

The monitoring system by integrating the satellite, drone, and field measurement for protecting the black soil should be established in the future, ensuring the sustainable development of the black soil. Firstly, establish a monitoring system using the satellite-UAV-ground observing data, and build a regularly updated monitoring system for the implementation of black soil protection. Secondly, explore the mechanism of crop residue covering for improving the organic matter content of soil, decrease wind erosion and hydraulic erosion, develop reasonable crop residue covering strategies, and incorporate the evaluation of the benefits of crop residue covering on black soil protection into the cropland planning system.

2.3.4 Policy Leadership and Experience Sharing on Cropland Protection in China

Target: SDG 2.4 By 2030, ensure sustainable food production systems and implement resilient agricultural practices that increase productivity and production, help maintain ecosystems, strengthen capacity for adaptation to climate change, extreme weather, drought, flooding, and other disasters, and progressively improve land and soil quality.

- **Background**

Food security is an important guarantee for world peace and development, and it is related to the sustainable development and future destiny of human beings. Food security is also an important basis for national security, and is related to the development of national economy and social stability. The key to ensure food security is to ensure food production capacity. Cropland is the foundation of food production, and is the fundamental carrier to provide food sources and ensure food supply Given the basic national conditions of low per capita cropland, generally low quality of cropland, and shortage of cropland reserve resources, China has implemented the food crop production strategy based on farmland management and technological application, resolutely staying above the red line of China's cropland of 1.8 billion mu, and set out a path of cropland protection with Chinese characteristics. This case study systematizes the policies and systems related to the protection of cropland and the construction of well-facilitated farmland since the founding of the People's Republic of China, demonstrating to the world China's wisdom of ensuring food security and promoting sustainable development.

- **Data**

From 1949 to 2022, the documents and reports on the cropland protection and utilization policies issued by the CPC Central Committee, the State Council, provinces (autonomous regions and municipalities directly under the Central Government), ministries and commissions and agencies directly under the State Council.

- **Methods**

The methods of literature analysis and historical analysis were employed to search, organize, and summarize the policies related to the field of cropland utilization and protection in China since the founding of the People's Republic of China. The main content and development process of these policies were introduced, and a complete timeline of the history of cropland protection in different development stages was created, showcasing the experience of China's cropland protection and ensuring food security.

- **Results and Analysis**

Food is the paramount necessity of people, and land is the foundation of food. In 2020, China's per capita cropland area was 0.09 hm^2, which was less than half of the world's per capita cropland area (0.26 hm^2). With 7% of the world's cropland, China is feeding 22% of the world's population, leaving valuable treasure in the cause of combating hunger and promoting world peace. Since the founding of the People's Republic of China, the state has attached great importance to the protection and utilization of cropland, and has implemented "the strictest system for the protection of cropland in the world". In response to the issues in the protection, utilization, and management of cropland, specific policies and measures have been continuously adjusted and improved. The concept has evolved from focusing on "quantity" to "quantity + quality", and further to the trinity governance of "quantity + quality + ecology". In terms of the system, the level of "national policy-basic national policy-lifeline-red line" has been progressive. The means have changed from simple administrative means to a combination of administrative, economic, and engineering means, and then to a large number of technical means. The main body of governance has transformed from a single subject (the government) to a complex and diversified subject (including the government, the society, and farmers). Overall, all of the above reflect that the protection of cropland is constantly moving toward integration, refinement, and systematization (Niu and Fang 2019; Wang et al. 2020a).

In 1950, China promulgated the Land Reform Law of the People's Republic of China, which realized the transformation of land ownership, greatly stimulated the enthusiasm of farmers in production, and carried out various forms of farmland construction activities such as land reclamation, farmland infrastructure construction, farmland consolidation, and farmland water conservancy construction. In 1986, China promulgated the Circular on Strengthening Land Administration and Stopping Unauthorized Occupation of Cropland, clearly stating that the protection of cropland was a basic state policy. The State Land Administration was established, and the "Land Administration Law of the People's Republic of China" was promulgated, thus bringing the protection of cropland into the stage of legalization. In 1994, China promulgated the Regulations on the Protection of Basic Farmland, and implemented the basic farmland protection system. In 1997, China issued the Circular on Further Strengthening Land Administration and Effectively Protecting Cropland, which proposed to implement the dynamic balance of the total amount of cropland and the balance between the occupation and replenishment of cropland. In 1998, the Ministry of Land and Resources was established, and the Land Administration Law of the People's Republic of China was revised in the same year, with a chapter entitled "Protection of Cropland" containing 12 articles. Up to this point, China's current land management system, centered on the dynamic balance of the total amount of cropland and the land-use regulation system, has taken shape. The Outline of the National General Land Use Planning in 1997–2010 delineates the basic cropland protection areas throughout the country as "life-saving lands" and "lifelines". At this stage, the state has issued policies on the protection and utilization of the quantity of cropland at various levels. A relevant system has been established at the legal level, and the state has continued to deepen policies on the protection and utilization of the quantity of cropland from multiple aspects by using economic and technical means, forming a relatively systematic policy system for the protection and utilization of cropland.

In 2004, the "Decision of the State Council on Deepening Reform and Strengthening Land Administration" was promulgated, which made special arrangements for the protection of cropland. For the first time, it was proposed that the quantity and quality of supplemental cropland be converted according to grades, preventing the possession of more land to make up for less land, and occupying the superior land to make up for the inferior land. Meanwhile, it established a system for assessing the responsibility of cropland protection and a system of land supervision. In 2006, the CPC Central Committee issued the Outline of

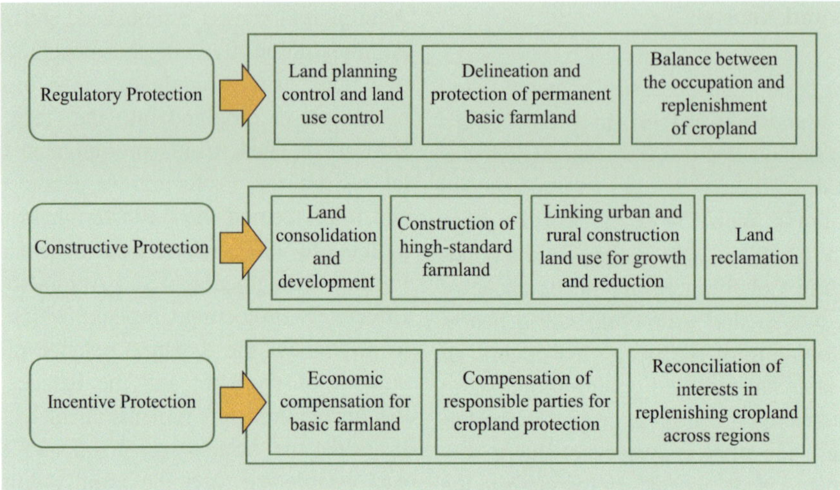

Fig. 2.11 Analysis of the main measures of regulatory, constructive, and incentive protection

the 11th Five-Year Plan for National Economic and Social Development of the People's Republic of China, which proposed a binding target of 1.8 billion mu of cropland. In 2008, the Outline of the Master Plan for National Land Use (2006–2020) reaffirmed that the red line of 1.8 billion mu of cropland must be guarded. The Outline of the National Territorial Space Plan (2021–2035) specifies the tasks of protecting 1.865 billion mu of cropland and 1.546 billion mu of permanent basic farmland, and the tasks have been broken down and sent down to the localities step by step. In 2023, the state established a reward-punishment mechanism for the assessment of cropland protection responsibility targets, and formed a dual protection system of "quantity + quality" of cropland. It is coordinated by the Ministry of Natural Resources, based on the Land Administration Law of the People's Republic of China, the Agricultural Law of the People's Republic of China and the Criminal Law of the People's Republic of China as the legal basis, with policies such as the balance between the occupation and replenishment of cropland, land use control, and basic farmland protection as the core, guaranteed by the performance appraisal of cropland protection and the technology of cropland classification (Han 2022).

In 2012, the Ministry of Land and Resources issued the Circular on Enhancing the Level of Cropland Protection and Comprehensively Strengthening the Construction and Management of Cropland Quality, marking that China's policy of cropland protection has formally entered the "trinity" stage of quantity control, quality management, and ecological management and protection. In 2017, China promulgated the Circular on Improving the Management to Implement the Balance between the Occupation and Replenishment of Cropland Effectively, establishing a new pattern of cropland protection based on the "trinity" of quantity, quality and ecology, and constructing a new mechanism for cropland protection based on the combination of control, construction, and incentive measures (Fig. 2.11) (Lu and Wang 2022). In 2018, China established the Ministry of Natural Resources, laying down an institutional guarantee for the "trinity" protection of cropland resources. In 2019, the amended Land Administration Law of the People's Republic of China realized the transformation of the focus of cropland protection from quantitative balance to the "trinity" of quantity, quality, and ecology. And it has elevated "basic farmland" to "permanent basic farmland". In 2020, the Circular of the General Office of the State Council on Resolutely Preventing the "Non-agricultural" Behavior of Cropland and the Opinions of the General Office of the State Council on Preventing the

"Non-grain Production" of Cropland to Stabilize Grain Production were issued, to further implement the strictest system of cropland protection, requiring that the red line of 1.8 billion mu of cropland be firmly guarded.

At the technical level, since 1984, China has successively issued technical regulations such as Technical Guidelines for Land Use Survey, National Land Classification (Trial Implementation), Land Use Classification, Regulations for Farmland Grading, and Regulations for Farmland Valuation, and conducted three large-scale surveys on current land use at the national level, which provided information on the quantity and quality distribution, and dynamic changes of cropland. The work of uploading maps into the database of well-facilitated farmland construction projects has been carried out throughout the whole country, which allows for digitized management. An information management platform for land consolidation has been established to realize the unification of "land, data, and records". With the demarcation of "three types of spatial zones and alert lines" and the implementation of territorial space planning, all localities will fully decompose and issue the tasks of cropland protection and permanent basic farmland protection step by step with the location to ensure that the "quantity, line and map" are consistently implemented to specific parcels and incorporated into the "one map" management system of territorial space. The tasks are used as rigid indicators for the implementation of the target assessment of the responsibility for cropland protection. Relying on the comprehensive monitoring and supervision platform for natural resources that cover the five layers of "sky, space, ground, people, and network", a grid-based supervision system covering five levels, namely national, provincial, municipal, county, and township, was established to implement cropland protection responsibilities in terms of area and location.

One billion mu of well-facilitated farmland had been built by the end of 2022, and more than half of the 1.918 billion mu of cropland was well-facilitated farmland. The average cropland quality of the completed well-facilitated farmland can be improved by about one grade, and the average grain production capacity per mu can be increased by 10%-20%. The project area can save 13.8% of fertilizers and 19.1% of pesticides. The effect of saving cost and increasing income is obvious, providing significant support for the national grain output to stay above 6.5×10^{12} kg for years running (Fig. 2.12). The National Well-facilitated Farmland Construction Plan (2021–2030) divides the national well-facilitated farmland construction into seven regions, including the Northeast Region, the Huang-Huai-Hai Region, the Middle and Lower Reaches of the Yangtze River Region, the Southeast Region, the Southwest Region, the Northwest Region, and the Qinghai-Xizang Region, and puts forward the priorities of construction in each region and the tasks of construction in each province (Sun et al. 2022).

Highlights

- The concept of cropland protection deepens. Based on the laws, regulations, policies and documents on cropland protection issued since the founding of the People's Republic of China, especially since the Reform and Opening-up, the history of cropland protection has been divided into the stage of exploration (1949–1978), the stage of simultaneous emphasis on quantity and quality (1979–2011), and the stage of trinity protection pattern, integrating quantity, quality, and ecology (2012–2022).
- The continuous enrichment of measures for cropland protection is being implemented. Based on a review of cropland protection, it is believed that China's cropland protection, adhering to both problem and goal-orientation approaches, has carried out constructive, regulatory and incentive protection of cropland as well as capacity building for cropland protection, which has resolutely curbed the "non-agricultural" behavior of cropland, and prevented the "non-grain" behavior of cropland, and strictly adhere to the red line of cropland.

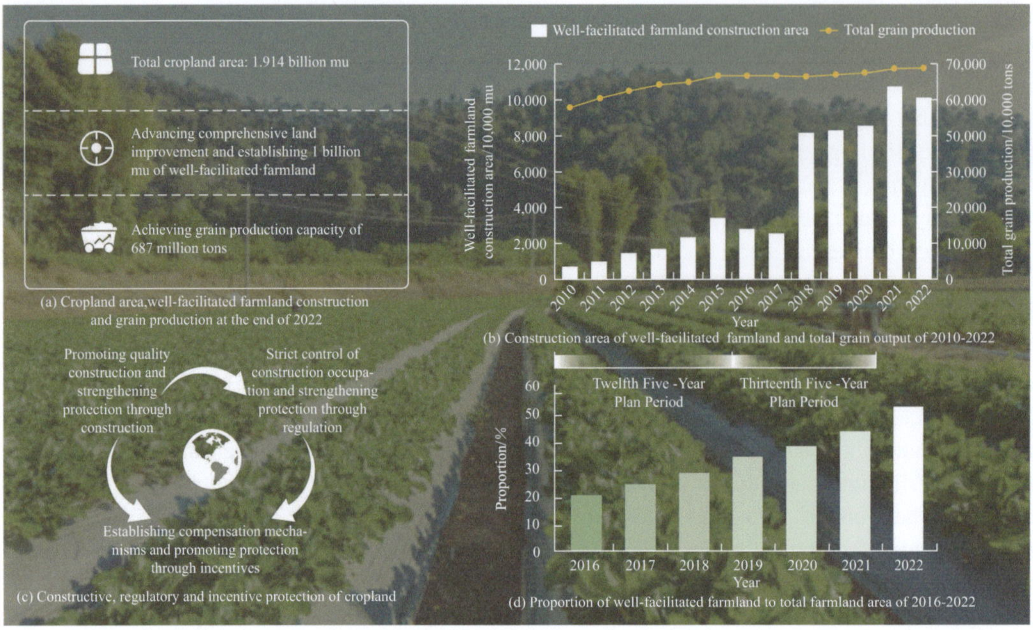

Fig. 2.12 Statistical analysis of the effectiveness of cropland utilization and protection

- **Discussion and Outlook**

The case study analyzes the cropland protection policies over the past 70 years since the founding of the People's Republic of China, delineates the stages of cropland protection, the problems it faced, and the measures it took in different periods, and describes the development process of cropland protection by timeline, showing China's efforts to guarantee food security and promote sustainable agricultural development. Currently, more than half of China's cropland is well-facilitated farmland. By 2025, 1.075 billion mu of well-facilitated farmland will be built and 105 million mu of well-facilitated farmland will be renovated and upgraded, in order to guarantee that well-facilitated farmland will provide more than 5.5×10^{11} kg of grain production capacity. In the next stage, more attention will be paid to building a comprehensive protection system for the quantity, quality, ecology, and capacity of cropland, as well as establishing and implementing a multi-functional synergistic management mechanism for cropland protection.

2.3.5 Comprehensive Evaluation of the Benefits of Well-facilitated Farmland Construction in China

Target: SDG 2.4 By 2030, ensure sustainable food production systems and implement resilient agricultural practices that increase productivity and production, help maintain ecosystems, strengthen capacity for adaptation to climate change, extreme weather, drought, flooding, and other disasters, and that progressively improve land and soil quality.

- **Background**

"Sustainable agriculture" (SDG 2.4) is one of the important goals of SDGs (United Nations 2015), and China's farmland construction has played a pivotal role in promoting sustainable agriculture (Wang et al. 2019). Well-facilitated farmland refers to farmland with flat and consolidated land, perfect facilities, supportive agricultural infrastructure, fertile soil, healthy ecosystems, strong ability to resist disasters, high and stable yields less affected by droughts and floods, which is compatible with modern agricultural production and management methods, and designated as permanent basic farmland. By improving or eliminating major constraints, the farmland in the regions such as major areas for rural land improvement and basic farmland preparation is organized and constructed according to the land remediation planning and overall planning for land utilization. Sustainable agriculture achieves the harmonious symbiotic development between human beings and nature by optimizing the resources used, improving the efficiency of agricultural production, conserving resources, and reducing environmental pollution. Sustainable agriculture involves a variety of indicators, including inputs and outputs, and there is no shared large-scale data on the benefits of well-facilitated farmland in terms of these indicators. This poses a great challenge to scientifically assessing the benefits of well-facilitated farmland construction in sustainable agriculture across China. Consequently, there is an urgent need to utilize a comprehensive set of data on inputs and outputs of well-facilitated farmland in sustainable agriculture, scientifically established evaluation indicators, and systematically compare and analyze the benefits of inputs and outputs of well-facilitated farmland in China's major agricultural areas on national scale (Wang et al. 2020b).

Therefore, this case study focuses on the construction of well-facilitated farmland in China's major agricultural areas, which is an important measure for sustainable agriculture, and uses questionnaire data from a wide range of typical project areas to map out the benefits of well-facilitated farmland in terms of saving land, water, fertilizer, pesticide, and improving farmers' income. This will provide a scientific basis for future strategic decisions related to sustainable agriculture, and contribute to the realization of SDG 2.4 with China's experience in the construction of well-facilitated farmland for sustainable agriculture.

- **Data**

A questionnaire on the implementation benefits of well-facilitated farmland construction projects in China's major agricultural areas was employed (survey time: 2017). The questionnaire can serve as a reliable indicator of the benefits of the recent construction of well-facilitated farmland for the following reasons: (1) The construction of well-facilitated farmland is a pivotal initiative aimed at fortifying and enhancing grain production capacity, thereby ensuring national food security. It operates under established standards that remain consistent over a specified period. (2) We selected the project areas for investigation based on a comprehensive consideration of multiple factors, including the representativeness of the area, the spatial heterogeneity, the benefit of construction, the construction area, the construction time, etc. (3) The complement of well-facilitated farmland construction results in long-term benefits for decades.

- **Methods**

The specific steps of the questionnaire on the construction benefits of well-facilitated farmland are as follows.

(1) Inquiry the overall construction of high-standard farmland in the sampling counties from the heads of relevant agricultural departments.
(2) Collect the information in the four directions of the sampling project area through the global positioning system (GPS).
(3) Collect the basic information of the project area, record the implementation year of the project area, the basic situation of construction, and the benefits of capacity improvement.
(4) Take photos and videos using UAV.
(5) Inquiry the characteristics, experience, modes, and problems of the well-facilitated farmland construction areas from the staff and the production and management subjects of the agricultural departments of the relevant provinces, cities, and counties through visits and talks.

The evaluation indicators for the well-facilitated farmland construction project are constructed by combining the input and output assessment objectives of the project with the relevant statistical content of the questionnaire. The evaluation indicators specifically include the following: the land-saving rate, the water-saving rate, the fertilizer-saving rate, the pesticide-saving rate, and the improvement rate of farmers' net income per mu.

1. **The land-saving rate**

The land-saving rate is the ratio of crop cultivation area to the total area of the project area. The improvement of the land-saving rate after the completion of well-facilitated farmland construction project can reflect the use efficiency of land resources. The statistic content is the land-saving rate before and after the implementation of the well-facilitated farmland construction project (%).

2. **The water-saving rate**

Following the implementation of the well-facilitated farmland construction project, the vigorous promotion of comprehensive water-saving measures, including channel seepage control, pipeline water transmission, micro-irrigation, sprinkler irrigation, film mulching and raised-bed cultivation, and water-suitable planting, has been undertaken. These measures have the potential to enhance the efficiency of irrigation water use within the project area. The statistical content is the percentage of irrigation water saved per mu through the implementation of water-saving measures after the implementation of the well-facilitated farmland construction project, compared with the irrigation water consumption per mu before the implementation of the project (%).

3. **The fertilizer-saving rate**

Following the implementation of the well-facilitated farmland construction project, the control of fertilizer application can effectively reduce the water, soil and gas pollution caused by chemical fertilizers. The indicator is defined as the percentage of the fertilizer saved per mu through the implementation of fertilizer-saving measures after the implementation of the well-facilitated farmland construction project, compared with the amount of fertilizer application per mu before the implementation of the project (%).

4. **The pesticide-saving rate**

Following the implementation of the high-standard farmland construction project, the control of pesticide application can effectively reduce the water, soil and gas pollution caused by pesticides. The indicator is defined as the percentage of the pesticide saved per mu through the implementation of pesticide-saving measures after the implementation of the well-facilitated farmland construction project, compared with the amount of pesticide application per mu before the implementation of the project (%).

5. **The improvement rate of farmers' net income per mu**

The well-facilitated farmland construction project has been shown to directly contribute to the growth of farmers' net income per mu. This indicator reflects the project's capacity to enhance farmers' property income per mu. The indicator is defined as the percentage of farmers' net income increased per mu after the implementation of high-standard farmland construction project, compared with farmers' net income per mu before the implementation of the project (%).

- **Results and Analysis**

1. **Improvement of comprehensive benefits**

Since the construction of the well-facilitated farmland project, the resource use efficiencies of fertilizers, pesticides, irrigation water, and land have increased by 8.8–24.3%, and the incomes of farmers have improved significantly with an average rate of 56.4%. In addition, the levels

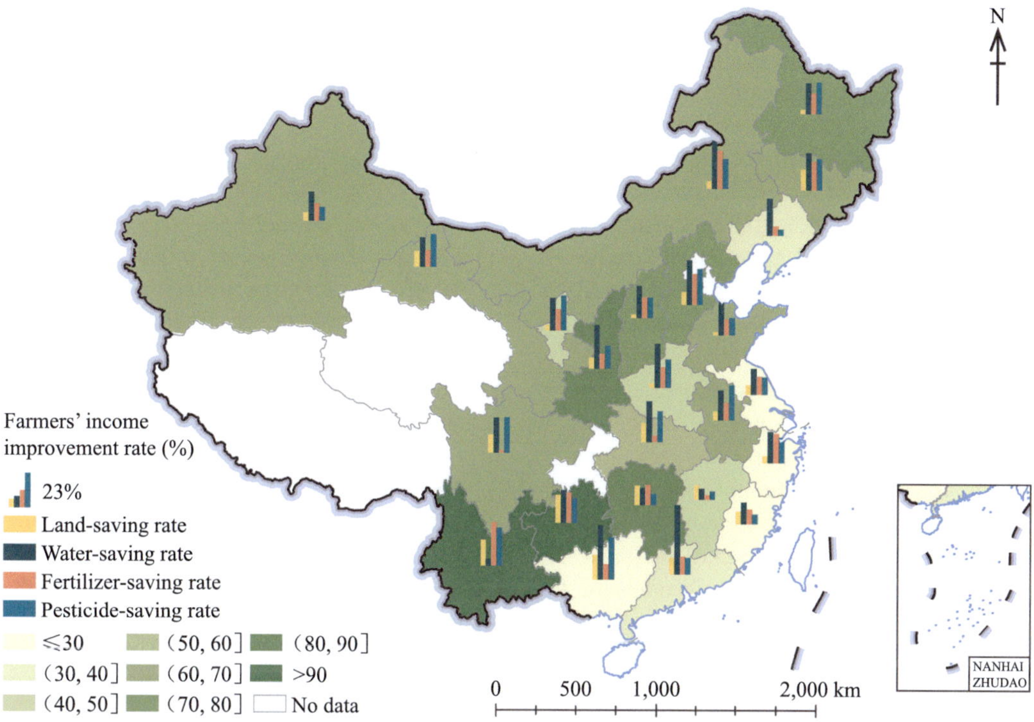

Fig. 2.13 Provincially assessed efficiency gains in high-standard farmland in China's key agricultural regions in land-saving rate, water-saving rate, fertilizer-saving rate, pesticide-saving rate, and farmers' income improvement rate

of scientific planting and the comprehensive benefits have been significantly improved. The costs of resource inputs and agricultural non-point source pollution have been reduced, and the average saving rates of land, water, fertilizer, and pesticide per mu are 8.8%, 24.3%, 13.8%, and 19.1%, respectively. The eco-environment of farmland has evolved toward sustainable agriculture with lower energy consumption, higher resource use efficiency, a healthier ecosystem, and less pollution in the environment. Moreover, the well-facilitated farmland construction has turned the main bodies in agricultural production management into large-scale farming, and the average incomes of farmers have increased by 7,464 yuan/hm^2, with an average improvement rate of 56.4%; 77.3% of the project areas achieved the economic target of increasing the average incomes of farmers by 3,000 yuan/hm^2 (Fig. 2.13). There are three main reasons for the increase in farmers' incomes. Firstly, local farmers contribute their labor and time, and are paid in the construction processes of well-facilitated farmland. Secondly, the production value of crops has increased after the construction of well-facilitated farmland. For example, traffic convenience has been significantly improved, which has led to an increase in the production value as a result of market expansion. Thirdly, the incomes from the construction of well-facilitated farmland increase through land transfers or land shares, and the local laborers can do more other things to earn money.

2. **Significant regional difference**

The comprehensive benefits of the construction of well-facilitated farmland vary significantly in different regions. Due to the high degree of fragmentation and low degree of flatness in farmland, the land-saving rate in the southwestern region increases significantly. The land-saving rate in Guizhou Province increased by 18.5%, and the farmland rent value in Yunnan Province increased by 7,080 yuan/hm^2 after the implementation of the project, with an increase rate of 63.0%. The saving rates of fertilizers and pesticides are 18–24%, which are more effective in the southeastern and southwestern regions with higher slopes and more fragmented land parcels. In contrast, the Middle and Lower Reaches of the Yangtze River region has the lowest fertilizer-saving rate of 8.0%, and the northeastern region has the lowest pesticide-saving rate of 12.8%. The water-saving rate in the northwestern region reaches 33.6%, with an average reduction of 1,621.5 m^3/hm^2 in irrigation water consumption; the irrigation water-saving rate per hectare in Guangdong Province in the southeastern region reaches 45.9%. The land-saving rate and the improvement rate in farmers' incomes in the southwestern region are the most significant, reaching 17.9% and 95.8%, respectively. The net incomes of farmers in most project areas (96.2%) increased by more than 3,000 yuan/hm^2.

Highlights

- Based on the questionnaire survey data on well-facilitated farmland in China's major agricultural areas, take the lead in carrying out a comprehensive assessment of the benefits of well-facilitated farmland construction on a national scale.
- After the implementation of the well-facilitated farmland construction project, the application of chemical fertilizers and pesticides in agriculture has been effectively controlled; the use efficiency of land and water resources has been significantly improved, and farmers' incomes have been significantly raised. The average saving rates of land, water, fertilizer, and pesticide per mu are 8.8%, 24.3%, 13.8%, and 19.1%, respectively. The average incomes of farmers have increased by 7,464 yuan/hm^2, with an average improvement rate of 56.4%.

- **Discussion and Outlook**

The following recommendations are proposed to address the risks and challenges associated with the construction of well-facilitated farmland in the future.

(1) The quantitative and empirical analysis of the effectiveness of well-facilitated farmland construction should be strengthened.
(2) A differentiated spatial layout should be implemented, and targeted development and protection strategies should be formulated.
(3) According to the characteristics of the project and the region, a highly targeted evaluation index system should be formulated.
(4) Effectively implement the management and protection responsibilities, establish a long-term management and operation mechanism, and ensure that the construction projects realize the expected benefits in a long-term and stable manner.

As an important measure for the development of productive and sustainable agriculture in China, the construction of well-facilitated farmland, while aiming to enhance grain production capacity as its primary goal, has promoted the income increase for farmers, improved resource utilization efficiency and environmental emission reduction, and has provided a typical example for realizing sustainable and high-quality agricultural development.

2.4 Summary

In China, big data and remote sensing technologies are ushering in a new era of sustainable agriculture and food security. This chapter integrates five research cases to demonstrate how these technologies support the achievement of the zero hunger goal. By conducting precise mapping of terraced fields and evaluating their soil conservation benefits, this report not only reveals the spatial heterogeneity of terraces in soil conservation, but also provides policymakers with decision-making recommendations that integrate monitoring, incentives, and regulation. Additionally, long-term land satellite data have revealed the distribution and change patterns of irrigated dryland in China, providing foundational data for the sustainable development of food, water, and environment.

In Northeast China, multi-source remote sensing data have been applied to monitoring CRC, demonstrating the impact of policy guidance and geographical factors on the spatio-temporal evolution of corn residue coverage. Future efforts can further enhance the spatio-temporal resolution of data, achieving coupling between residue coverage and crop planting distribution, and enabling nationwide monitoring and assessment of residue coverage.

The construction of well-facilitated farmland in China has not only increased grain production capacity, but also promoted farmers' income, resource efficiency, and environmental sustainability, serving as a model for sustainable and high-quality agricultural development. An analysis of over 70 years of cropland protection policies since the founding of the People's Republic of China indicates that more than half of the farmland has been converted to well-facilitated farmland. Future land protection will focus more comprehensively on quantity, quality, ecological sustainability, and functional capacity, thereby implementing a multifunctional collaborative management mechanism.

Overall, China's agricultural development is moving toward a direction that combines productivity and sustainability. Through scientific planning and comprehensive goal-oriented guidance, efforts will be made to further improve the quality of cropland and agricultural technology services, thereby promoting the integrated benefits of multi-objective synergistic land use. These efforts will contribute to the achievement of SDG 2.4. Not only do these initiatives ensure China's food security, but also they provide valuable experience for global food security and sustainable agricultural development.

References

Dong Y, Xuan F, Huang XD et al (2024) A 30-m annual corn residue coverage dataset from 2013 to 2021 in Northeast China. Sci Data 11:216

Dong Y, Xuan F, Li ZQ et al (2023) Modeling the corn residue coverage after harvesting and before sowing in northeast China by Random Forest and soil texture zoning. Remote Sens 15(8):2179

Food and Agriculture Organization of the United Nations (2021) The state of the world's land and water resources for food and agriculture: systems at breaking point: synthesis report 2021. https://openknowledge.fao.org/server/api/core/bitstreams/bc8810ae-2a13-4cfe-b019-339158c7e608/content/cb7654en.html [2024-08-01]

Food and Agriculture Organization of the United Nations (2022) Food and agriculture organization statistics: food and agriculture data. https://www.fao.org/faostat/en/#data/RL [2024-08-01]

Gabr SS, Farg EF, Habeebullah TM et al (2020) Irrigation water consumption and its impact on the groundwater aquifer of Wadi Uranah, Makkah, Saudi Arabia using remote sensing techniques. Egypt J Remote Sens Space Sci 23(2):167–180

Glesson T, Wada Y, Bierkens MFP et al (2012) Water balance of global aquifers revealed by groundwater footprint. Nature 488(7410):197–200

Han Y (2022) The policy evolution, vision goal and realization path of China's cultivated land protection and utilization. J Manag World 38(11):121–131 (In Chinese)

Jia MG (2010) Forecast and prevention of the soil erosion in typical irrigation district of Kunes River basin. Water Sci Eng Technol 6:42–44 (In Chinese)

Jia RN, Zhang WP, Hou S (2018) Resource and environment impact assessment of winter wheat production in dry land and irrigated land. J Shanxi Agric Sci 46(6):928–933, 961 (In Chinese)

Kastner T, Rivas MJI, Koch W et al (2012) Global changes in diets and the consequences for land requirements for food. Proc Natl Acad Sci USA 109(18):6868–6872

Kuang WH, Zhang SW, Du GM et al (2022) Remotely sensed mapping and analysis of spatio-temporal patterns of land use change across China in 2015–2020. Acta Geogr Sin 77(5):1056–1071 (In Chinese)

Li ZQ, Xuan F, Dong Y et al (2024) Performance of the global ecosystem dynamics investigation data combined with Sentinel-2 images for automatic labeling of wall-to-wall corn mapping. Int J Appl Earth Obs Geoinf 127:103643

Liu L, Huang JF, Xiong QX et al (2020) Optimal MODIS data processing for accurate multi- year paddy rice area mapping in China. Gisci Remote Sens 57(5):687–703

Lu YX, Wang BY (2022) The evolution and logic of farmland protection system and policy. China Land 2:4–8 (In Chinese)

Martínez-Ferrer L, Moreno-Martínez Á, Campos-Taberner M et al (2022) Quantifying uncertainty in high resolution biophysical variable retrieval with machine learning. Remote Sens Environ 280:113199

National Bureau of Statistics (2023) China rural statistical yearbook 2023. China Statistics Press, Beijing (In Chinese)

Niu SD, Fang B (2019) Cultivated land protection system in China from 1949 to 2019: Historical evolution, realistic origin exploration and path optimization. China Land Sci 33(10):1–12 (In Chinese)

Shan CH, Wang YP (2015) Effect of excessive irrigation on the soil and fertilizer loss in farmland cultivation area. Chin J Agric Resour Reg Plan 36(6):11–16 (In Chinese)

Sun CL, Yang H, Han D et al (2022) Situation and countermeasures of well-facilitated farmland in China. J Agric Sci Technol 24(7):15–22 (In Chinese)

Sun DL, Kafatos M (2007) Note on the NDVI-LST relationship and the use of temperature-related drought indices over North America. Geophys Res Lett 34(24):L24406

United Nations (2015) Transforming our world: the 2030 agenda for sustainable development. https://sdgs.un.org/2030agenda [2024-07-24]

Wang WX, Cao YG, Su RQ et al (2020a) Cultivated land protection policy in China: background, effect and future trends. Chin J Agric Resour Reg Plan 41(10):40–51 (In Chinese)

Wang XQ, Shi WJ, Sun XF et al (2020b) Comprehensive benefits evaluation and its spatial simulation for well-facilitated farmland projects in the Huang-Huai-Hai Region of China. Land Degrad Dev 31(14):1837–1850

Wang XQ, Shi WJ, Xing XX et al (2019) High-standard farmland: a review of methods on construction suitability evaluation, benefit evaluation and influencing factors. Chin Agric Sci Bull 35(19):131–142 (In Chinese)

You NS, Dong JW, Li J et al (2023) Rapid early-season maize mapping without crop labels. Remote Sens Environ 290:113496

Yue JB, Tian QJ, Dong XY et al (2020) Using broadband crop residue angle index to estimate the fractional cover of vegetation, crop residue, and bare soil in cropland systems. Remote Sens Environ 237:111538

Zhang ZX, Wang X, Zhao XL et al (2014) A 2010 update of national land use/cover database of China at 1:100 000 scale using medium spatial resolution satellite images. Remote Sens Environ 149:142–154

Zhu ZJ, Zhang ZX, Zuo LJ et al (2022) Study on the classification and change detection methods of drylands in arid and semi-arid regions. Remote Sens 14(5):1256

3 SDG 6 Clean Water and Sanitation

3.1 Background

Since the adoption of the United Nations, 2030 Agenda in 2015, there has been no fundamental change in the issues of extensive water resource utilization, poor management, and the over-extraction and pollution of freshwater and groundwater resources. Water-related ecosystems continue to degrade at an alarming rate. The failure to advance the clean water and sanitation goals (SDG 6) along the intended trajectory has become a global consensus (UN-Water 2021).

The Sustainable Development Goals Report 2022 of United Nations indicates that due to a lack of monitoring, the quality of water resources upon which at least 3 billion people depend remains unknown. Additionally, 730 million people live in countries experiencing severe water scarcity. At the current rate of progress, by 2030, 1.6 billion people will still lack safely managed drinking water, 2.8 billion people will lack basic sanitation facilities, and 1.9 billion people will lack basic hand hygiene services (Economic and Social Council of United Nations 2022). To meet the 2030 targets, the progress rate for improving drinking water, sanitation, and hygiene facilities needs to increase fourfold (United Nations 2022).

While global progress conclusions of SDG 6 have been drawn from national statistical data, such data are insufficient to effectively support policy-making and decision-making at national and sub-national levels. In recent years, non-traditional data sources such as satellite remote sensing, mobile phone data, and crowd-sourced data have increasingly complemented traditional statistical data. For example, advancements in Big Earth Data technologies have significantly enhanced monitoring and evaluation capabilities for SDG 6. These technologies enable high spatio-temporal resolution monitoring of relevant indicators through remote sensing, periodic revisits, and rapid information extraction. They not only save time and resources, but also provide more accurate and comprehensive evaluation results (Lu et al. 2021).

Over the past four years, China has conducted a series of national and provincial monitoring and evaluation case studies for SDG 6 progress. These studies have laid a solid foundation for mid-term progress evaluations of SDG 6 at various scales. However, due to the lack of foundational data, effective monitoring and evaluation of the sanitation target (SDG 6.2) have been limited. This report presents a case study evaluating the population coverage rate of China's public toilets from 2015 to 2020. It demonstrates the feasibility of using Big Earth Data technologies to address the lack of sanitation indicator data and provides progress assessments for SDG 6.2.1a (population using safely managed sanitation services) at the national level and for typical cities in China.

H. Guo, *Big Earth Data in Support of the Sustainable Development Goals (2023)—China*, Sustainable Development Goals Series, https://doi.org/10.1007/978-981-95-2636-9_3

Table 3.1 Case studies and their main contributions

Target	Tier	Case study	Contribution
SDG 6.2 By 2030, achieve access to adequate and equitable sanitation and hygiene for all and end open defecation, paying special attention to the needs of women and girls and those in vulnerable situations	Tier I	Population coverage of urban public toilets in China from 2015 to 2020	Method and model: develop a public toilet population coverage rate evaluation model integrating multi-source data, including POI, remote sensing, and statistical data Decision support: provide support for policy formulation on public toilet construction and renovation at national, provincial, and municipal levels

3.2 Main Contributions

The primary contributions of the case study in this chapter include: (1) developing a population coverage rate evaluation model for public toilets by integrating multi-source data; (2) supporting the formulation of construction and renovation policies for public toilets at national, provincial, and municipal levels (Table 3.1).

3.3 Case Study

3.3.1 Population Coverage of Urban Public Toilets in China from 2015 to 2020

Target: SDG 6.2 By 2030, achieve access to adequate and equitable sanitation and hygiene for all and end open defecation, paying special attention to the needs of women and girls and those in vulnerable situations.

Indicator: SDG 6.2.1a Proportion of the population using safely managed sanitation services.

- **Background**

The General Assembly and the Human Rights Council of the United Nations have recognized clean water and sanitation as a fundamental right that applies to all people (United Nations General Assembly 2010). Sanitation is included in the target from SDG 7.c of the Millennium Development Goals (MDGs) to SDG 6.2 of the 2030 Agenda. Halfway through the 2030 Agenda, only 57% of public sanitation services are safely managed globally, and progress on sanitation is off-track, with a fivefold acceleration of progress on sanitation construction needed to achieve universal coverage on time (United Nations 2023). The proportion of the population with access to safe management services, basic services, limited services, unimproved sanitation facilities, and open defecation sanitation facilities in the country in 2022 is 67.22%, 28.67%, 2.06%, 1.95%, and 0.1%, respectively (Journal of Manufacturing Processes 2023). Public sanitation facilities are defined as "limited sanitation services", and the proportion of the population covered by public sanitation facilities needs to be further considered to accurately assess the status of SDG 6.2.1.

Although the progress of public toilet construction in China is based on national statistics, it still does not effectively support policy-making and decision-making at the subnational level. In recent years, non-traditional data sources, such as satellite remote sensing data, mobile phone data, and public-generated data, have formed a useful complement to traditional statistics. This chapter focuses on sanitation facilities (SDG 6.2), and comprehensively adopts multi-source data such as POI data, remote sensing data, and statistical data to construct an urban public toilet population coverage rate assessment model, and analyzes the population coverage rate of urban public toilets in 2015 and 2020 in provinces and typical cities.

- **Data**

- Land use data and population data from the Resource and Environmental Science Data Platform (2015, 2020), with spatial resolutions of 100 m and 1 km, respectively.
- POI data from Autonavi (2015, 2020) and is presented in vector point format.
- DEM data from NASA (2020), with a spatial resolution of 30 m.
- Statistical data from China Statistical Yearbook (2015, 2020) and China Urban–Rural Construction Statistical Yearbook (2015, 2020).

- **Methods**

Based on the data from the China Statistical Yearbook and the China Urban–Rural Construction Statistical Yearbook, an analysis was conducted on the change in the number of urban public toilets and the number of public toilets per 10,000 people in China for the years 2015 and 2020. Seven typical cities were selected from the seven major administrative regions of China. Utilizing data on land use, DEM, POI, high-resolution population distribution, etc., the study developed a monitoring model to assess the proportion of the population with access to public sanitation facilities in these cities. The model calculated the proportions in these seven cities in 2015 and 2020, and further validated the statistical results of public sanitation facility services (Fig. 3.1).

In the specific calculation of the population coverage rate of urban public toilets, firstly, according to the type of functional area, the commercial road points, life road points and urban public toilet points are identified from POI. According to the Standard for Setting of Environmental Sanitation Facilities (CJJ 27-2012), the walking time to reach the public toilets is 3 min and 5 min, respectively, for the commercial sections and living sections, assuming a person walking 5 km/h. However, slope affects people's walking speed. Slope data are derived from DEM data to identify the slope of public toilets on commercial and living roads. With slope as the weight, the buffer radius of public toilets in commercial and residential sections is set to obtain the accessible range of public toilets. Combined with the population distribution data, the number of population covered by public toilets is counted and divided by the resident population in the region, that is, the population coverage rate of public toilets is obtained. In the setting of slope, according to the slope degree, it is divided into six degrees, namely $<5°$, $[5°, 15°)$, $[15°, 25°)$, $[25°, 35°)$, $[35°, 45°)$, and $\geq 45°$. The corresponding walking speed is set as 5 km/h, 4 km/h, 3 km/h, 2 km/h, 1 km/h, and 0.5 km/h, respectively (He and Liu 2022).

- **Results and Analysis**

1. **The overall number of urban public sanitation facilities in China's provincial administrative regions**

According to the China Urban–Rural Construction Statistical Yearbook, the number of urban public toilets in China has increased significantly, with the number of urban public toilets rising from 324,949 in 2015 to 396,617 in 2020, with a growth rate of 22.06%. Among them, the numbers of urban public toilets in Zhejiang and Jiangsu were both high-value areas during the study period, and the corresponding numbers of urban public toilets in the two provinces in 2020 were 28,033 and 26,893, respectively. The provinces (autonomous region and municipality directly under the Central Government) with the lowest number of public toilets were different during the study period. The provinces (autonomous region and municipality directly under the Central Government) with the lowest number of public toilets in 2015 was Hainan (1,516), and the provinces (autonomous region and municipality directly under the Central Government) with the lowest number of public toilets in 2020 was Ningxia Hui Autonomous Region (1,999).

From the perspective of spatial distribution, the number of urban public toilets did not show

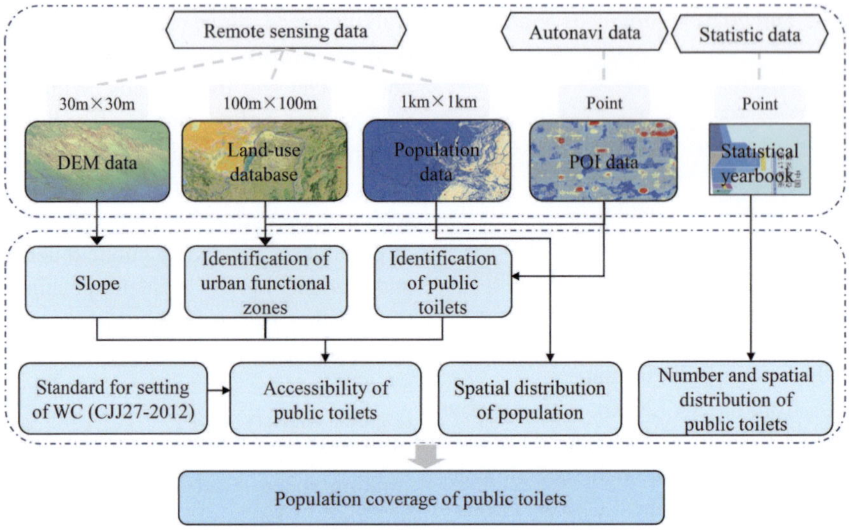

Fig. 3.1 The map of the technical roadmap

an obvious spatial distribution pattern in 2015. The overall spatial distribution was mainly in the middle-and low-value areas, while the high-value areas were mainly distributed in the eastern coastal areas. In 2020, the number of urban public toilets presents an obvious spatial distribution pattern of "high in the east and low in the west", which is consistent with the population distribution pattern of the "Hu's line". The high-value areas are widely distributed in the eastern coastal areas such as Zhejiang, Jiangsu, Guangdong, and Shandong, while the medium- and low-value areas are mainly located in the northwestern and northeastern regions such as Ningxia, Qinghai, Xizang, and Jilin.

From 2015 to 2020, the number of urban public toilets in provincial administrative regions mostly showed an upward trend, and the number of urban public toilets in the west showed an obvious upward trend, which was related to the small cardinal number in 2015 in the region, such as Xizang (129.70%), which had the highest increase, and the number of urban public toilets in 2015 was only 2017. The slight decline in the number of urban public toilets in Beijing, Heilongjiang, Jilin, Jiangsu, Liaoning, Ningxia, Shandong, and Shanghai may be related to the removal of old toilets during urban construction and renovation, and the decrease in the number of urban public toilets in Northeast China may be related to the large outflow of regional population (Fig. 3.2).

2. **The number of urban public toilets per 10,000 residents in China's provincial administrative regions**

The number of public toilets per 10,000 urban residents has also increased significantly from 4.02 in 2015 to 4.47 in 2020. Among them, urban residents in Xizang have the highest number of public toilets per 10,000 residents, with 12.13 in 2015 and 20.81 in 2020, Nei Mongol followed with 8.33 and 9.50, and Guangxi had the lowest number, with 2.26 and 2.37, respectively.

From the perspective of spatial distribution, the spatial distribution of the number of public toilets per 10,000 urban residents in 2015 was mainly in low-value areas, while the number in high-value areas was relatively low, distributed in Zhejiang with developed economy and sound infrastructure construction, as well as Xizang and Nei Mongol with continuously improved infrastructure and other regions. In 2020, the spatial distribution of the number of public toilets per

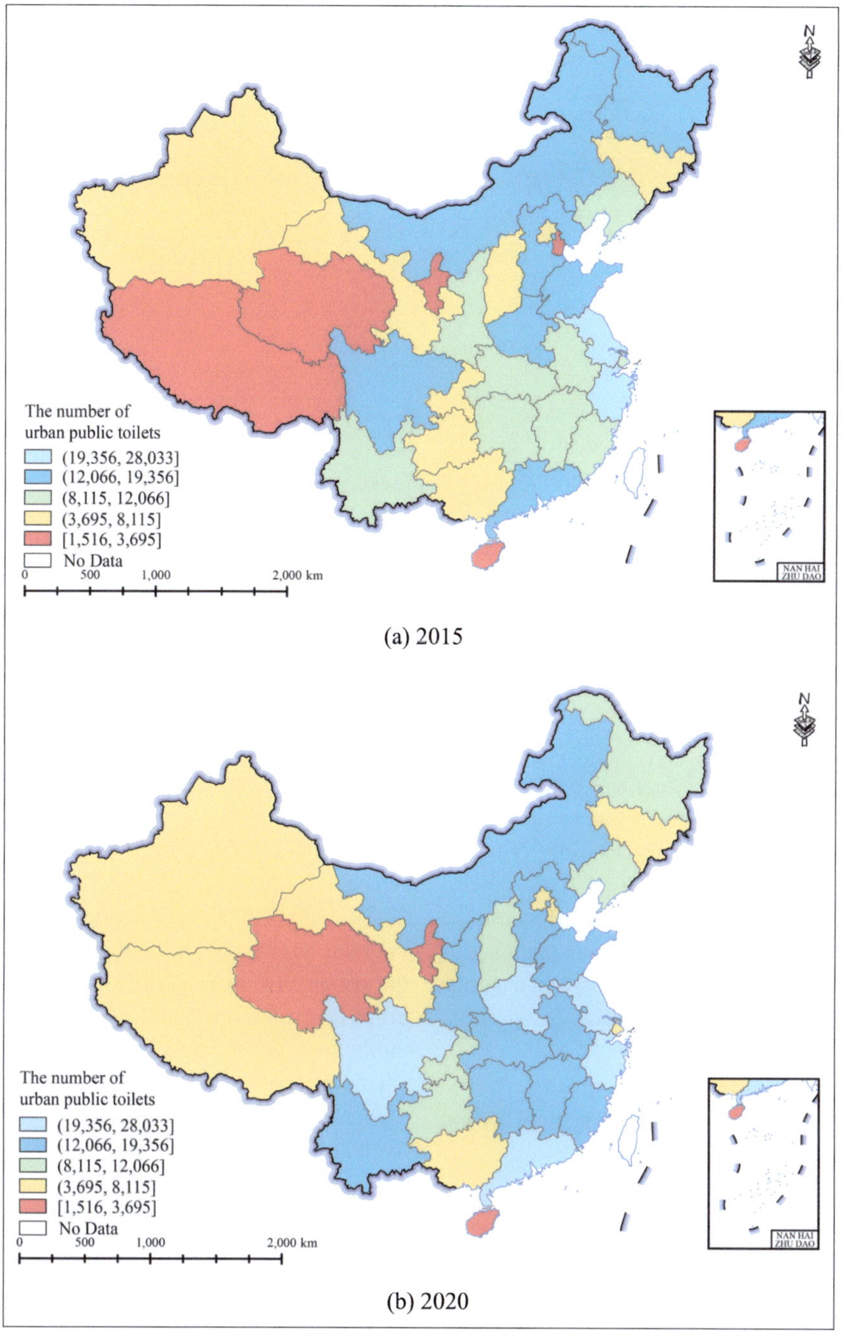

Fig. 3.2 The number and change rate of urban public toilets in China from 2015 to 2020

10,000 residents residents is mainly in middle- and high-value areas, while the scope of low-value areas is significantly reduced, mainly in the northeastern, southern, and southwestern regions.

With the construction of public toilets entering the standardized track, the number of public toilets per 10,000 urban residents also showed an increasing trend, with a growth rate of 11.16%. Among them, the number of public toilets per 10,000 residents in Xizang increased the most, reaching 71.58%, which was consistent with

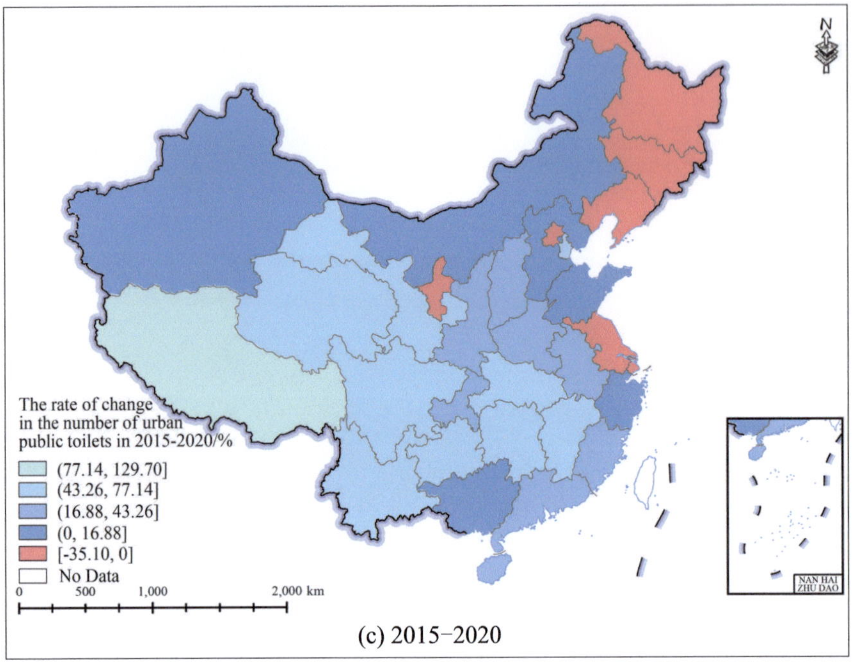

Fig. 3.2 (continued)

the substantial growth of regional public toilets. The numbers of urban public toilets in Beijing, Heilongjiang, Jilin, Jiangsu, Liaoning, Ningxia, Shandong, and Shanghai decreased, so the change rates in these regions were negative. Because the growth rate of urban residents in Zhejiang is higher than that of the number of urban public toilets, its change rate is also negative (Fig. 3.3).

3. **Population coverage rate of typical urban public toilets based on POI**

In accordance with the geographical divisions of the seven geographical regions and the four economic regions, and with reference to the Notice of the State Council on Adjusting the Standards for Dividing the Size of Cities, seven provincial capitals, including Shijiazhuang in Hebei Province, Guangzhou in Guangdong Province, Xi'an in Shaanxi Province, Nanjing in Jiangsu Province, Chengdu in Sichuan Province, Zhengzhou in Henan Province and Dalian in Liaoning Province, were selected to carry out a study on the population coverage rate of typical urban public toilets based on POI.

The total number of public toilets in the seven typical cities in 2015 and 2020 was 9,047 and 27,342, respectively, accounting for nearly 10% of the total number of public toilets in China. From the perspective of the urban level, the higher the urban level of the area is, the greater the number of public toilets is. Two megacities, Guangzhou and Chengdu, lead the list in terms of the number of public toilets, which was 6,026 and 5,560, respectively, in 2020. In addition to Dalian, the number of public toilets in the three super-large cities of Nanjing, Xi'an and Zhengzhou was also relatively large, reaching 5,082, 3,731, and 3,646, respectively, in 2020. Dalian had the lowest number of public toilets during the study period, with 380 in 2015 and 1,399 in 2020. Shijiazhuang is a large city that consistently ranks second to last in terms of the number of public toilets with 1,898 in 2020.

In terms of the population coverage rate of public toilets, the average population coverage rate of public toilets in 2015 and 2020 in seven

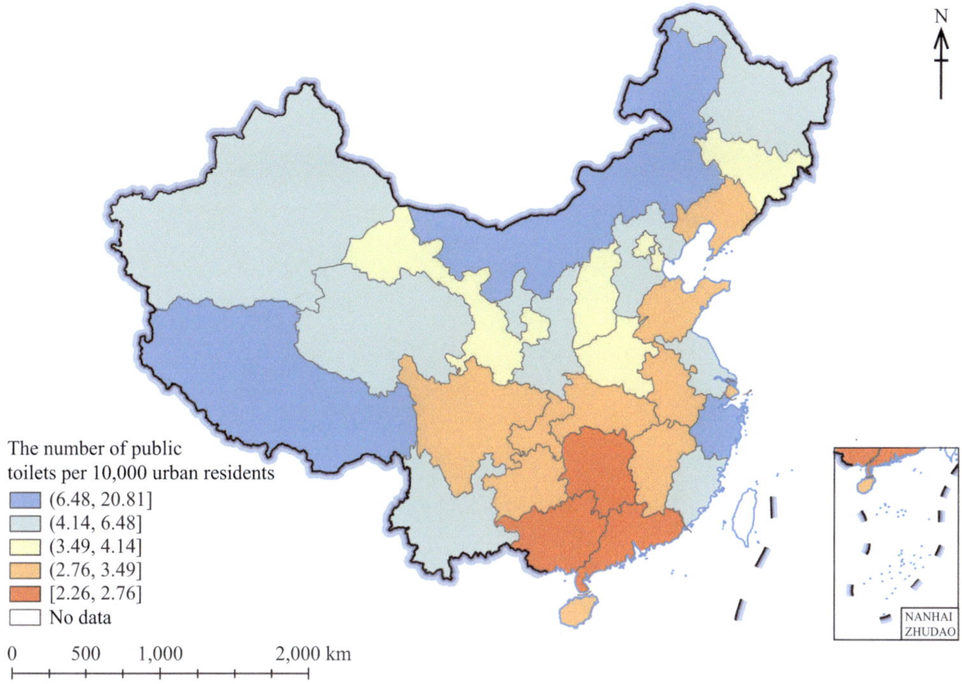

(a) 2015

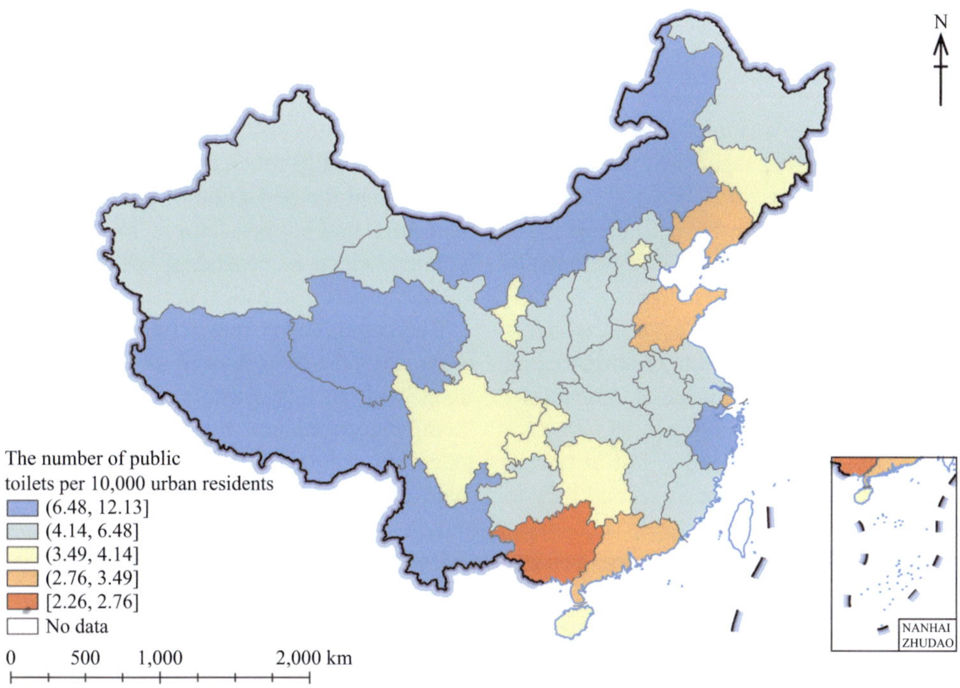

(b) 2020

Fig. 3.3 The number and change rate of public toilets per 10,000 urban residents in China from 2015 to 2020

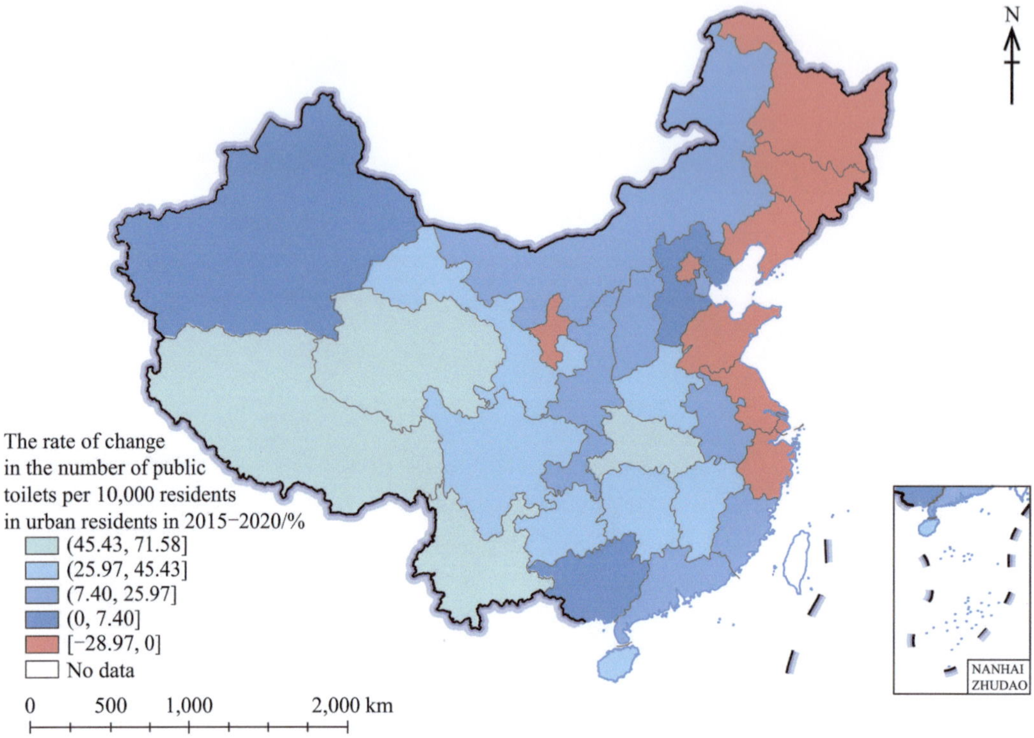

(c) 2015–2020

Fig. 3.3 (continued)

typical cities was 18.96% and 25.55%, respectively. Among the two megacities, the coverage rate of Guangzhou is the highest, reaching 37.78%, while the coverage rate of Chengdu ranks fourth, only 25.61%, which is closely related to the slope of the region. The coverage rate of Shijiazhuang only reached 14.03% in 2020, which is related to the large population base in the region and the failure to expand the scale of public toilet construction. The scale of public toilet construction in the region ranks second to last of typical cities, and the total population is second only to Chengdu and Guangzhou.

For the spatial distribution of the population coverage rate of public toilets, Guangzhou, Nanjing, Xi'an, and Chengdu ranked the top four during the study period, while Zhengzhou, Dalian, and Shijiazhuang ranked the bottom three, and Guangzhou and Nanjing reached the highest values in 2020 and 2015, respectively. It shows that the construction of public toilets in the eastern cities with rapid economic development is generally better, followed by the western regions, and the perfection of public facilities in the central and northeast ern regions is relatively lagging, which is consistent with the analysis results based on statistical data.

The population coverage rate of public toilets in typical cities also showed an obvious upward trend, with an increase rate of 34.81%. Among them, the population coverage rate of public toilets in Shijiazhuang and Dalian increased relatively high, with an increased rate of 77.54% and 51.41%, respectively. The number of public toilets in the two cities had a lower base value, but a larger increase rate, both exceeding 220%. In addition, the population growth of Dalian is relatively low, with the total permanent population increasing from 6,644,300 in 2015 to 6,881,600 in 2020, with an increase rate of only 3.57%. The lowest increase is in Nanjing, but its public facilities are relatively sound.

The coverage rate of public toilets in Nanjing increased from 28.26% in 2015 to 32.08% in 2020, which is significantly higher than that in Dalian and Shijiazhuang, where the coverage rates are both less than 15% (Fig. 3.4).

Highlights

- Combined with POI, remote sensing, and statistical data, the assessment model of population coverage rate of public toilets was constructed to conduct a comprehensive assessment of the progress of SDG 6.2.1a. The results show that the number of public toilets in Chinese cities and towns increased by 22.06%, the number of public toilets per 10,000 people increased by 11.16%, and the population coverage rate of public toilets in seven typical cities increased by 34.81% from 2015 to 2020.
- The number of public toilets in most provinces (autonomous regions and municipalities directly under the Central Government) has increased to varying degrees, but there is an obvious spatial imbalance, showing a spatial pattern along the "Hu's line" as a whole. In 2020, the number of public toilets in southeast coastal areas (Jiangsu, Shanghai, Fujian, Zhejiang, Guangdong, and Guangxi) reached 107,808. Typical cities in the east (Guangzhou and Nanjing) also had a high population coverage rate of public toilets, 37.78% and 32.08%, respectively.

- **Discussion and Outlook**

From 2015 to 2020, the number of public toilets in Chinese cities and towns increased by 22.06%, and the distribution of public toilets was more consistent with the distribution of residents along the "Hu's line", showing a spatial distribution pattern of high in the east and low in the west, while the number of public toilets in the west showed an obvious upward trend. Further analysis shows that the eastern region has a developed economy, a high degree of urbanization, and perfect infrastructure including public toilets. The economic foundation of the western region is relatively weak, but with the acceleration of the process of urbanization, the degree of attention to public toilets is deepening, and the numbers of new and rebuilt public toilets are increasing. From the perspective of urban level, the higher the

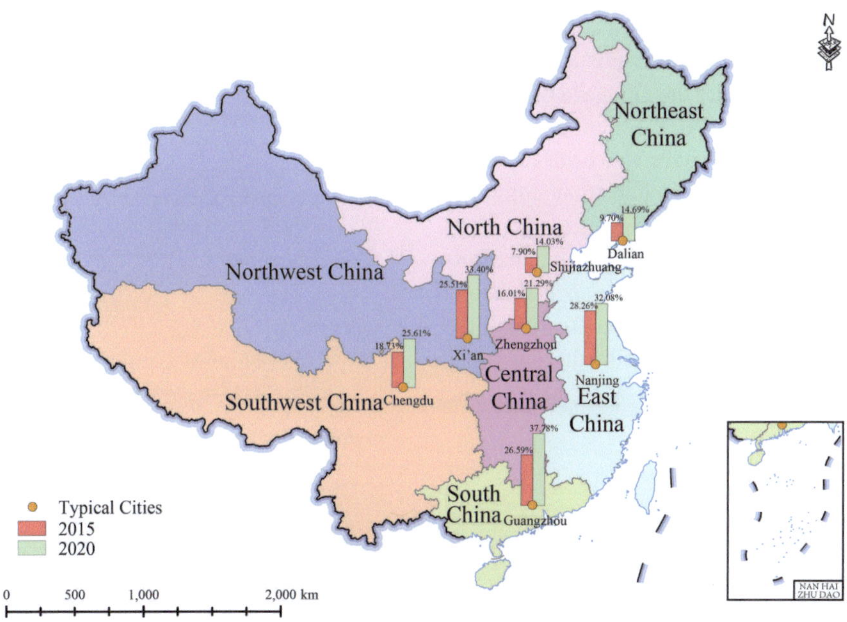

Fig. 3.4 Population coverage rate of public toilets in typical cities in 2015 and 2020

urban level is, the higher the degree of urbanization is, the more sound the construction of public facilities, but the cities with a higher slope, such as Chongqing, need to appropriately increase the number of public toilets to achieve a reasonable number of public toilets and layout.

The significant improvement in China's environmental health is mainly due to the following two aspects. On the one hand, it benefits from the government's great attention and inter-departmental collaboration, since the full implementation of the "Toilet Revolution" in 2015, The State Council, the Ministry of Culture and Tourism, the Office of the Central Leading Group for Rural Affairs of the Communist Party of China, the Ministry of Agriculture and Rural Affairs, the National Health Commission, the Ministry of Housing and Urban–Rural Development, the National Development and Reform Commission, the Ministry of Finance, the Ministry of Ecology and Environment and other departments have formulated and implemented a series of planning and policy documents, which have comprehensively promoted the construction and renovation of urban public toilets, rural public toilets, tourist public toilets and rural household toilets, and formulated corresponding standardized documents for the construction and reconstruction of toilets, organized relevant technical training, and allocated long-term and stable financial support. On the other hand, under the guidance of national policy documents, local "Toilet Revolution" action plans and models have also been explored, such as the "Changshu Model" that combines rural toilet renovation and sewage treatment, the "self-use and self-management" of retail investors in Ningde City and the "farmer specialzation" management model in centralized areas, and the "Toilet Revolution" model in Guilin that combines public toilets with commercial facade sales.

3.4 Summary

Based on the population coverage data of urban public toilets in China obtained from POI, remote sensing, and statistical data, the progress assessment of China's SDG 6.2.1a was carried out, and the following two understandings were formed.

(1) From 2015 to 2020, the number of public toilets in Chinese cities and towns has increased significantly. Since the implementation of the "Toilet Revolution" in China, the capacity of public sanitation services has been greatly improved.

(2) The number of public toilets in most provinces (autonomous regions and municipalities directly under the Central Government) has increased to varying degrees but shows an obvious spatial imbalance. Public sanitation infrastructure in eastern China is relatively perfect. The number of urban public toilets in western China shows an obvious upward trend. With the acceleration of urbanization, the demand for new construction and reconstruction of public toilets is increasing in these areas.

References

Economic and Social Council of United Nations (2022) Progress towards the sustainable development goals report of the secretary-general. https://unstats.un.org/sdgs/files/report/2022/secretary-general-sdg-report-2022--EN.pdf [2024-09-01]

He XQ, Liu C (2022) Spatial accessibility of national nature reserves in the Yangtze River economic belt and its influencing factors. J Yangtze River Sci Res Inst 39(11):9–16 (In Chinese)

Journal of Manufacturing Processes (2023) Progress on household drinking water, sanitation and hygiene in East Asia and Pacific region. https://washdata.org/reports/progress-household-drinking-water-sanitation-and-hygiene-east-asia-and-pacific-region-2023 [2024-07-24]

Lu SL, Jia L, Jiang YZ et al (2021) Progress and prospect on monitoring and evaluation of United Nations SDG 6 (clean water and sanitation) target. Bull Chin Acad Sci 36(8):904–913 (In Chinese)

United Nations (2022) The sustainable development goals report 2022. https://unstats.un.org/sdgs/report/2022/ [2024-08-01]

United Nations (2023) The sustainable development goals report 2023: special edition. https://unstats.un.org/sdgs/report/2023/ [2024-07-24]

United Nations General Assembly (2010) The human right to water and sanitation. https://www.un.org/waterforlifedecade/human_right_to_water.shtml [2024-07-24]

UN-Water (2021) Summary progress update 2021: SDG 6—water and sanitation for all. https://www.unwater.org/publications/summary-progress-update-2021-sdg-6-water-and-sanitation-all [2024-08-01]

SDG 7 Affordable and Clean Energy

4

4.1 Background

Energy, being the driving force behind human civilization, is derived from various resources that can be converted into usable forms such as thermal, electrical, light, mechanical, and chemical energy. Clean energy, also known as green energy, refers to energy sources that produce minimal or no pollution, and can be directly utilized for production and daily life, which includes nuclear energy and renewable energy sources such as hydropower, wind energy, solar energy, and tidal energy. The primary objective of utilizing clean energy is to mitigate the environmental impact of human activities while providing economically viable energy solutions. Clean energy is not only environmentally friendly with low carbon emissions and minimal pollution, but also plays a pivotal role in driving the green and low-carbon transformation of the economy and society. The development of clean energy is the cornerstone of the fourth Industrial Revolution, and is indispensable for the advancement of human civilization. Promoting clean energy is not only a global consensus for addressing climate change but also a crucial means of strengthening ecological civilization.

In recent years, China has made significant strides in the clean energy sector. For instance, the installed capacity of wind power and photovoltaic power in China has increased dramatically in the past few years. Moreover, China has demonstrated outstanding performance in terms of solar photovoltaic power generation and lithium battery production. Additionally, China is the world's largest wind energy market and hydrogen producer, ranking among the global leaders in the development and utilization of clean energy. As global attention shifts toward environmental protection and sustainable development, the development and utilization of clean energy have gained increasing prominence. To this end, this chapter conducts research on the application of clean energy in cooking in China and the ultra-high voltage (UHV) technology in the long-distance transmission of renewable energy sources. Specific case studies include the proportion of the Chinese population relying primarily on clean fuels and technologies for cooking, and the UHV technology supporting long-distance utilization of renewable energy in China.

In the case study of "The Proportion of Chinese Population Relying on Clean Cooking Energy and Technologies", we employ Big Earth Data to estimate and statistically analyze the proportion and quantity of the population in each region that relies on these clean energy sources for cooking at the national scale. This enables us to grasp the utilization patterns of various types of clean energy in different regions and among different groups, providing data support for the formulation of relevant policies, contributing to the realization of universal clean cooking energy coverage

H. Guo, *Big Earth Data in Support of the Sustainable Development Goals (2023)—China*,
Sustainable Development Goals Series, https://doi.org/10.1007/978-981-95-2636-9_4

Table 4.1 Case studies and key contributions

Targets	Tiers	Case studies	Contributions
SDG 7.1 By 2030, ensure universal access to affordable, reliable, and modern energy services	Tier I	The proportion of Chinese population relying on clean cooking energy and technologies	Data product: dataset of the proportion of the Chinese population using clean cooking energy in 2022
SDG 7.2 By 2030, greatly increase the share of renewable energy in the global energy mix	Tier I	UHV technology supporting long-distance utilization of renewable energy in China	Data product: dataset of the scale and structure of renewable energy transmitted through China's UHV lines from 2016 to 2022

by 2030, and promoting the widespread adoption and application of clean cooking energy globally.

In the case study of "UHV Technology Supporting Long-distance Utilization of Renewable Energy in China", based on the mismatch between the distribution of renewable energy resources and demand centers in China and the advantages of UHV technology in long-distance transmission with large capacity and low losses for renewable energy integration, we comprehensively combine the existing research findings and relevant survey data to compile data on the UHV transmission of renewable energy. Furthermore, we analyze data on the scale of direct transmission of renewable energy through UHV lines, the promotion of renewable energy consumption by UHV technology, and the impacts of UHV technology on the "curtailment rate of wind and solar power". This presents the supporting role of UHV technology in the development of renewable energy, and summarizes the Chinese approach to promote renewable energy consumption through UHV technology.

4.2 Main Contributions

To address the challenges posed by the energy crisis, global climate change, and SDGs, such as increasing the share of renewable energy in the global energy mix and doubling the global rate of improvement in energy efficiency, the report fully leverages the advantages of multi-source remote sensing technologies and geographical information system (GIS), among other Big Earth Data technologies. It provides monitoring and evaluation methods to support the achievement of SDG 7. Focusing on indicators SDG 7.1.2 and SDG 7.2.1, the report contributes to the global efforts by providing data products for monitoring SDG 7, showcasing China's achievements in this area (Table 4.1).

4.3 Case Study

4.3.1 The Proportion of Chinese Population Relying on Clean Cooking Energy and Technologies

Target: SDG 7.1 By 2030, ensure universal access to affordable, reliable, and modern energy services.

Indicator: SDG 7.1.2 The proportion of the population relies primarily on cleaner fuels and technologies.

- **Background**

Clean cooking energy is a concept proposed after considering whether the cooking activities in the lives of residents who have gradually entered the modernization stage are environmentally friendly. The pollutant emissions, energy utilization efficiency, and convenience are mainly analyzed. Clean cooking energy is of great significance to the life and health of residents in modern society. The type of energy used greatly impacts the environment and climate, and non-clean energy emits large amounts of carbon dioxide, and produces harmful gases such as sulfur oxide and nitrogen oxide, seriously affecting the ecological environment. There are still 2.6 billion people using wood, charcoal, dung, and coal to cook and get warm, resulting in more than 4 million premature deaths each year due to indoor air pollution (International Energy Agency 2017). At the same time, the decentralized and private nature of energy consumption for cooking activities makes it difficult to analyze and investigate the type of energy used and the cleanliness of the technology utilized through large-scale data collection and statistics. The existing surveys, studies, and statistics on the types of energy and how they are utilized have been carried out only from a macro perspective, focusing on the distribution of energy types, total utilization and industries, and regional studies on the proportion of the population relying on clean cooking energy and technologies have not yet emerged.

- **Data**
  - Vector data of China's provincial administrative divisions.
  - Date from the National Catalogue Service for Geographic Information (https://www.webmap.cn/main.do?method=index).
  - Land use data: The 30 m annual land cover dataset and its dynamics in China from 1990 to 2019 (http://irsip.whu.edu.cn/resources/CLCD.php).
  - LJ1-01 nighttime light data (http://59.175.109.173:8888/app/login.html).
  - Per capita GDP data of provinces, cities, and counties in the 2022 provincial statistical yearbooks regional GDP
  - Internet data (Internet maps, place names), field survey sampling data.

- **Methods**

The research area of the project is the representative regions from 34 provincial administrative regions (including 23 provinces, 5 autonomous regions, 4 municipalities directly under the Central Government, and 2 special administrative regions) for targeted research. Based on the Big Earth Data (including the vector data, GDP, nighttime light remote sensing image, and land use data, etc., of the provincial administrative regions), using the methods of spatial analysis, questionnaire survey and statistical analysis, and taking "selection of representative provinces (autonomous regions, municipalities directly under the Central Government)-survey sample area identification-type division of the residential area-select the sampling points of the survey sample area" as the main line, we scientifically analyze and calculate the proportion of the population relying on clean cooking energy and technology in various regions of China. In this case, a "pyramid" approach is adopted to tackle the difficult issues, and the technical route is shown in Fig. 4.1.

According to the technical route, the specific implementation steps and methods are as follows.

(1) Selection of representative provinces (autonomous regions, municipalities directly under the Central Government): Based on the similarity of economic development levels and energy policies in different regions of China, several provinces (autonomous regions, municipalities directly under the Central Government) were selected as the study area for the questionnaire survey.
(2) Survey sample area identification: Taking the district and county administrative divisions of the surveyed provinces (autonomous regions, municipalities directly under the Central Government) as units, the surveyed provinces (autonomous regions, municipalities directly under the Central Government) were categorized into four grades: high, relatively high, medium, and low, using per capita GDP statistics. Each survey sample area was formed based on the range of grades (the highest levels excluded from the sampling). Within the survey sample areas formed by the relatively high, middle, and low tertiary areas, GIS software was used to count the proportion of towns and villages land in each survey sample area for the entire surveyed area. The survey sample area was determined based on the order of occupancy in conjunction with the actual number of sampling points.

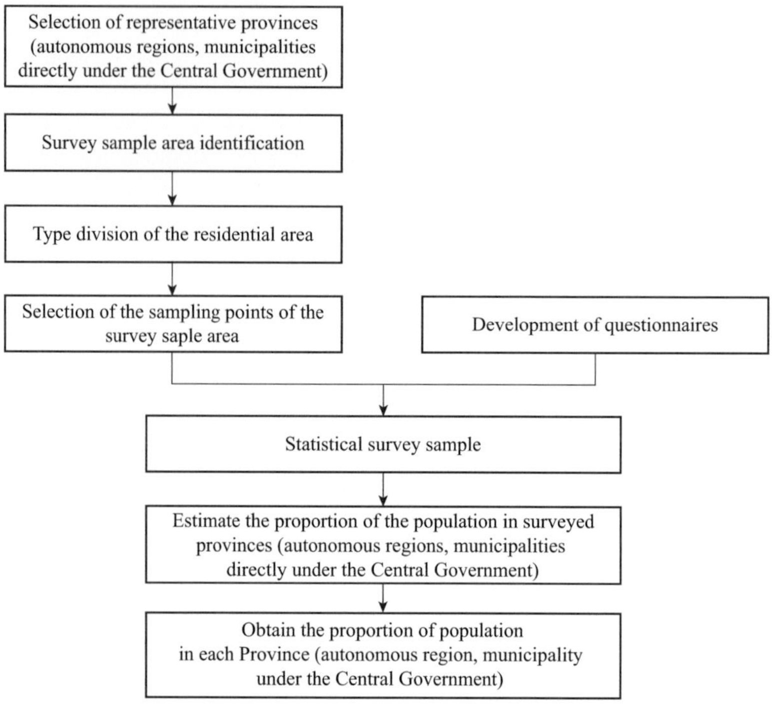

Fig. 4.1 Technical route

(3) Type division of the residential area: Based on nighttime light remote sensing data, land use data, and administrative zoning maps, the lighting index is obtained using the software. Based on town land, townland is further divided into urban areas (towns) and rural areas. Combined with village land in land use data, the classification of urban residential areas, town residential areas, and village residential areas is realized.

(4) Selection of the sampling points of the survey sample area: In the survey sample area, a county was randomly selected, and sampling points in several townland areas and village land areas were randomly selected within the selected county to conduct a field questionnaire survey (the default ratio of urban residential areas to the population using clean cooking energy is 100%).

(5) Development of questionnaires: According to the economic conditions, terrain characteristics, national policies on new energy construction, and living habits of the provinces (autonomous regions, municipalities directly under the Central Government) investigated, questionnaires were formulated, and energy availability and residents' income levels were concerned to help analyze the results in the later stage. Statistics of residents' intentions and needs provide ideas for the promotion of new energy in the future.

(6) Statistical survey sample: All valid questionnaires in town and village survey areas were counted, and the proportion of residents in each sample area relying on clean cooking energy and technology B_i and C_i were calculated (B_i represents the proportion of town residents using clean cooking energy and technology, C_i represents the proportion of village residents using clean cooking energy and technology, i represents the number of samples).

(7) Estimate the proportion of the population in surveyed provinces (autonomous regions, municipalities directly under the Central Government): According to the ratios B_i and C_i, the weighted average of the sample proportion of all towns and village areas

in the provinces (autonomous regions, municipalities directly under the Central Government) is obtained. The ratios D_i and E_i obtained (D_i and E_i are the proportions of urban and rural residents using clean cooking energy and technology after the weighted average of the sample area, respectively) are the proportions of the population relying on clean cooking energy and technology in the surveyed area.

(8) Obtain the proportion of population in each province (autonomous region, municipality directly under the Central Government): After calculating the proportions of the population in the surveyed provinces (autonomous regions, municipalities directly under the Central Government), according to the proportion of using clean energy in each region (from the regional statistical yearbooks) and the composition of village households using various cooking and heating fuels in each region (from the Communique of the Third National Agricultural Census), the proportion of town and village population using clean cooking energy and technology in similar provinces (autonomous regions, municipalities directly under the Central Government) was calculated respectively.

- **Results and Analysis**

1. **The popularization of clean energy in China has achieved full development**

This case adopts the forms of questionnaire survey and statistical analysis to conduct a statistical survey on the proportion of people who rely on clean cooking energy and technology in various regions of China. The proportion in town residents is 89.38%, the proportion in village residents is 69.77%, and the total proportion of the country is 83.55%. It can be estimated that the population who do not use clean cooking energy and technology is about 150 million people. The data are debatable, but goes some way to correct the conclusions of international organizations such as the World Health Organization (WHO)—"a total of 2.6 billion people worldwide do not have access to clean cooking energy, of which 61% are in Asia and 36% are in Africa" (International Energy Agency 2021).

Figures 4.2, 4.3 and 4.4 present a hierarchical color chart of the proportion of the population relying on clean cooking energy and technology, a bar chart of the proportion of the population, and a bar chart of the population in the country in 2022, respectively.

2. **Gas and electricity are the main cooking energy in China, and economy and convenience are the main considerations in the choice of energy for town and village residents in China**

The proportions of gas and electricity used in household cooking energy in town and village areas are higher than those of other energy sources, of which 86% of town residents use gas and electricity, and 71% of village residents use gas and electricity. With the further improvement of the proportion of renewable energy in China and the continuous improvement of power infrastructure, electric energy will become the main choice of clean cooking energy in rural areas. Due to the relative perfection of town gas facilities, the proportions of gas and electricity in town household cooking are significantly higher than those of in village households. In village areas without natural gas pipelines, liquefied gas has become the main cooking energy. In the choice of energy use, town and village residents mainly consider economy, accounting for 50% and 53%, respectively, while 30% of town residents and 20% of village residents give priority to the convenience of energy use.

Highlights

- Based on Big Earth Data (including vector maps of administrative divisions, GDP, nighttime light remote sensing images, land use remote sensing classification results, etc.)

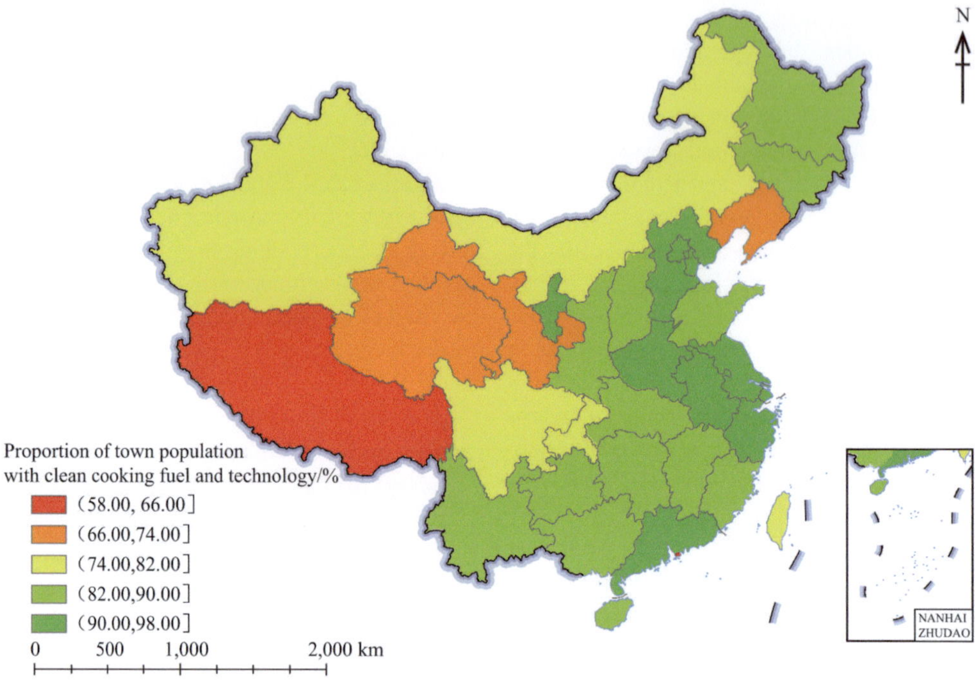

(a) Towns

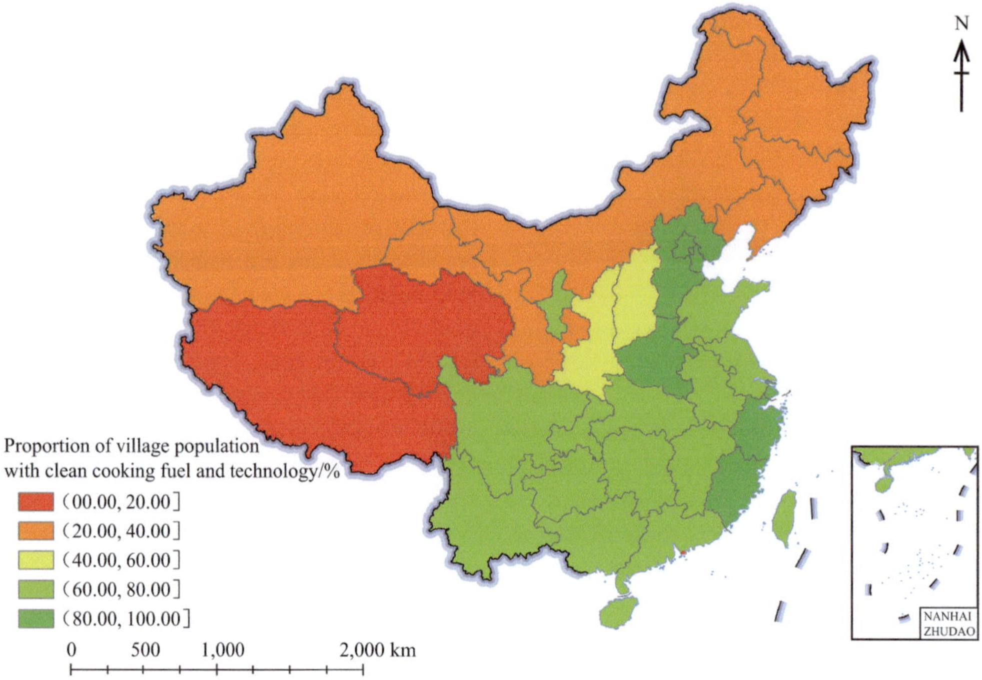

(b) Villages

Fig. 4.2 Color chart of the proportion of the population relying on clean cooking energy and technology in provinces (autonomous regions, municipalities directly under the Central Government) in 2022

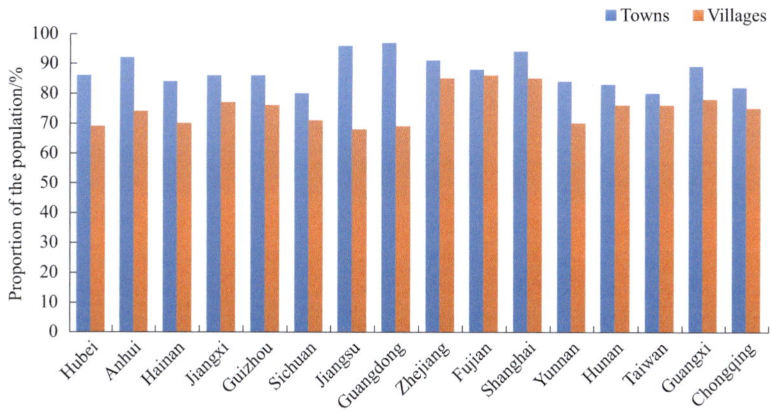

(a) The southern region of the Yangtze River

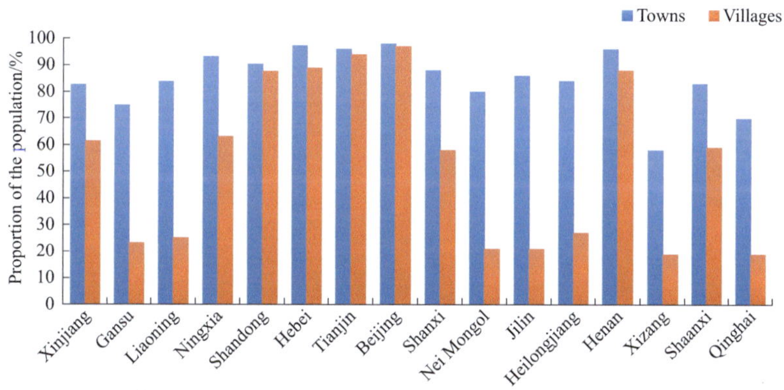

(b) The northern region of the Yangtze River

Fig. 4.3 Bar chart of the proportion of the population relying on clean cooking energy and technology in provinces (autonomous regions, municipalities directly under the Central Government) in 2022

and field investigations, conduct quantitative inversion of the population proportion of large-scale clean cooking energy and technology. The proportion and number of people who rely on clean cooking energy and technology in all regions of China were calculated (100% in urban areas, 89.38% in town areas, 69.77% in village areas, 83.55% in the whole country, with a total population of 1.26 billion).

– The results of the case can correct WHO's previous conclusions, promote the upgrading of clean energy use, and provide a data basis for relevant decision-making and policy-making.

- **Discussion and Outlook**

According to the data from the International Energy Agency (IEA) at the end of 2020, the global proportion of the population with clean cooking energy supply is 65%. In Asia, 66%, 65%, and 82% of the population in China, India, and Indonesia use clean cooking energy, 59% in other Southeast Asian countries and only 39% in other developing Asian countries use clean cooking energy. However, according to the statistics of this case, the proportion of the population using clean cooking energy in China in 2022 has exceeded 80%, which corrects the previous

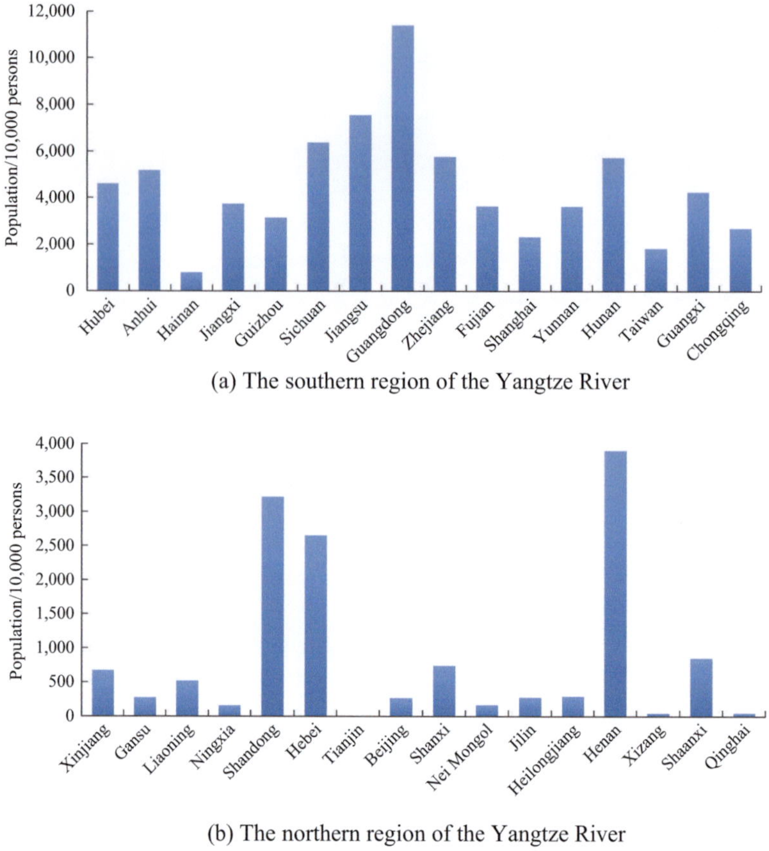

Fig. 4.4 Bar chart of the population relying on clean cooking energy and technology in each province (autonomous regions, municipalities directly under the Central Government) in 2022

conclusion of international organizations such as WHO. In China, 500 million people lack access to clean cooking energy (International Energy Agency 2021). Using household surveys, WHO estimated that 500 million people in China lacked access to clean cooking energy, probably because that count only counts electricity as clean energy. The reason for adopting this evaluation method is that the United States and other countries mainly use electricity for cooking activities, so the previous survey only regarded electricity as clean energy, and did not consider other types of clean energy. In the investigation of this case, electricity, solar stoves, biogas, and liquefied gas were used as clean cooking energy, which may be the main reason for the excessive data gap. Knowing the proportion of people who mainly rely on clean cooking energy and technology in each region of China can promote the upgrading of clean energy utilization, and provide a data basis for relevant decision-making and policy-making. Later, further surveys can be carried out in various regions, and nationwide data can be continuously updated to provide a data basis for the formulation of relevant targeted policies, help to achieve the full coverage of clean cooking energy by 2030, then promote the popularization and application of clean cooking energy worldwide, and better assist the realization of SDG goals.

4.3.2 UHV Technology Supporting Long-Distance Utilization of Renewable Energy in China

Target: SDG 7.2 By 2030, greatly increase the share of renewable energy in the global energy mix.
Indicator: SDG 7.2.1 Share of renewable energy in final energy consumption

- **Background**

China experiences a significant spatial mismatch between renewable energy-rich regions and electricity load centers. Wind and solar power, along with coal resources, are primarily concentrated in the northern and northwestern regions, while hydropower is abundant in the southwestern part. In contrast, electricity demand is concentrated in the central and eastern regions. To effectively utilize renewable energy, it is necessary to implement large-scale, "West-to-East Power Transmission", and "North-to-South Power Transmission". Developing highly efficient, low-loss, and cost-effective, long-distance, and large-capacity power transmission technology has become an imperative.

UHV technology, referring to direct current voltage levels of ±800 kV and above and alternating current voltage levels of 1,000 kV and above, offers advantages such as large transmission capacity, long transmission distances, high efficiency, low losses, and economic viability. According to relevant calculations, the transmission capacity and distance of UHV lines are 4–5 times that of UHV lines, with transmission losses being only about 25% and saving approximately 60% of land resources (Yang 2011). Therefore, the construction of UHV grids enables the transmission of large amounts of electricity to remote load centers, reduces losses, and saves resources.

- **Data**
  - National Energy Administration of China, Monitoring and Evaluation Report on the Development of Renewable Energy in China (2016–2023).
  - Editorial Board of China Electric Power Yearbook, China Electric Power Yearbook (2016–2019).
  - China Electricity Council, China Power Statistics Yearbook (2020–2023).

- **Methods**

A combination of qualitative and quantitative research methods was employed in this case study. Leveraging the research findings accumulated by Beijing Jipeng Investment Information & Consultant Ltd. over the years, data on UHV transmission of renewable energy was compiled through literature review and expert interviews. By applying multi-dimensional data analysis methods such as comparative analysis and data classification, the scale of renewable energy directly transmitted through UHV lines, the promotion of renewable energy consumption by UHV, and the impact of UHV on the "curtailment rate of wind and solar power" were analyzed. This presentation demonstrates the supporting role of UHV technology in the development of renewable energy.

- **Results and Analysis**

The UHV project in China has undergone four major development stages. Research on UHV transmission technology in China began in 1986. After more than 20 years of practical exploration and scientific research, the period from 2006 to 2010 marked the exploration phase of UHV engineering in China, with the first commercial UHV line being completed in 2009. From 2011 to 2013, China experienced its first peak of UHV development, with 5 UHV projects initiated, including 2 alternating current (AC) and 3 direct current (DC) projects. The second peak of development occurred from 2014 to 2017, during which 13 UHV projects were launched, consisting of 5 AC and 8 DC projects. The third peak of development spanned from 2018 to 2022. By the end of 2022, China had completed nearly 40 UHV transmission projects, including 20 DC UHV transmission

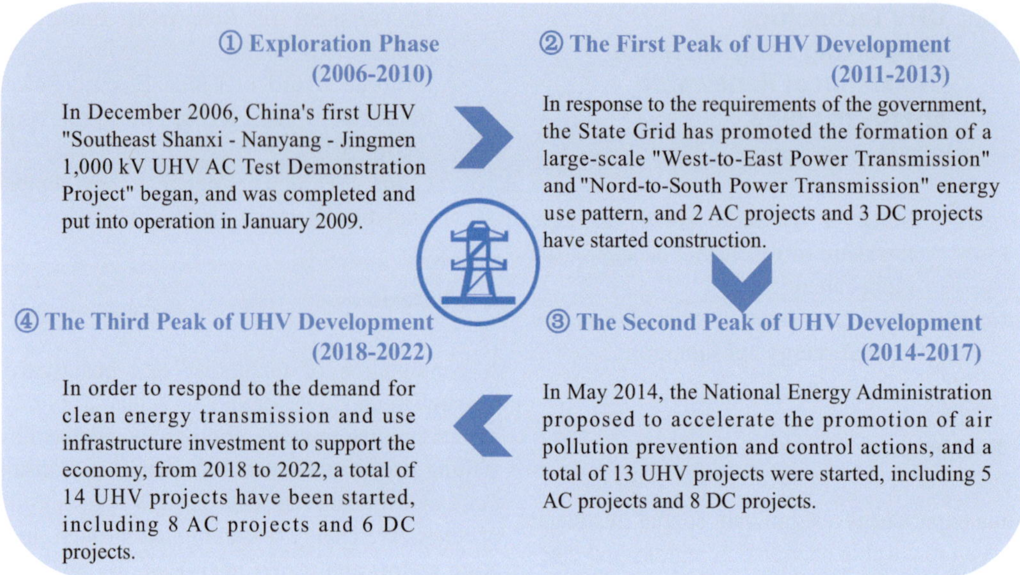

Fig. 4.5 Development of China's UHV grid

channels, 3 AC UHV ring networks in East China, North China (partially completed), and Central China (partially completed), and multiple cross-regional UHV interconnection lines in North China-Central China, Northwest China-North China, Northeast China-North China, West Nei Mongol-North China, and Zhejiang-Fujian. The total length of completed and under-construction lines exceeded 50,000 km, with a substation (converter station) capacity exceeding 500 million kW and a transmission capacity approaching 20 million kW. The cumulative transmitted electricity exceeded 3 trillion kW h (Fig. 4.5).

1. **The rapid development of renewable energy in China has driven the demand for UHV line construction, while the construction of UHV lines has in turn promoted the development of renewable energy**

From the perspective of power generation, as of the end of 2022, China's cumulative installed capacity of renewable energy power generation reached 1.213 billion kW, accounting for 47.30% of the total installed power capacity. Renewable energy installed capacity in 2022 was 2.53 times that of 2015. In 2022, China's renewable energy power generation reached 2.7 trillion kW h, accounting for 30.80% of the total power generation, nearly double the level of 2015 (Fig. 4.6a). Among them, hydropower bases in the southwest and wind and photovoltaic bases in the "Three Northern" regions accounted for over 50% of the total renewable energy power generation in China, requiring large-scale and long-distance transmission to load centers in the central and eastern regions. From the perspective of power consumption, the actual consumption of renewable energy electricity in China in 2022 reached 2681 billion kW h, accounting for 31.6% of the total social electricity consumption, increasing by 7.1 percentage points compared with that of 2015 (Fig. 4.6b).

In 2022, renewable energy accounted for 56.14% of the total electricity transmitted through all DC UHV lines (Fig. 4.7). The annual electricity transmitted through the 20 DC UHV lines reached 563.9 billion kW h, 2.42 times that of 2016. Among them, renewable energy electricity was 316.6 billion kW h, 1.84 times that of 2016. Therefore, UHV lines are a significant driving force for the development of renewable energy.

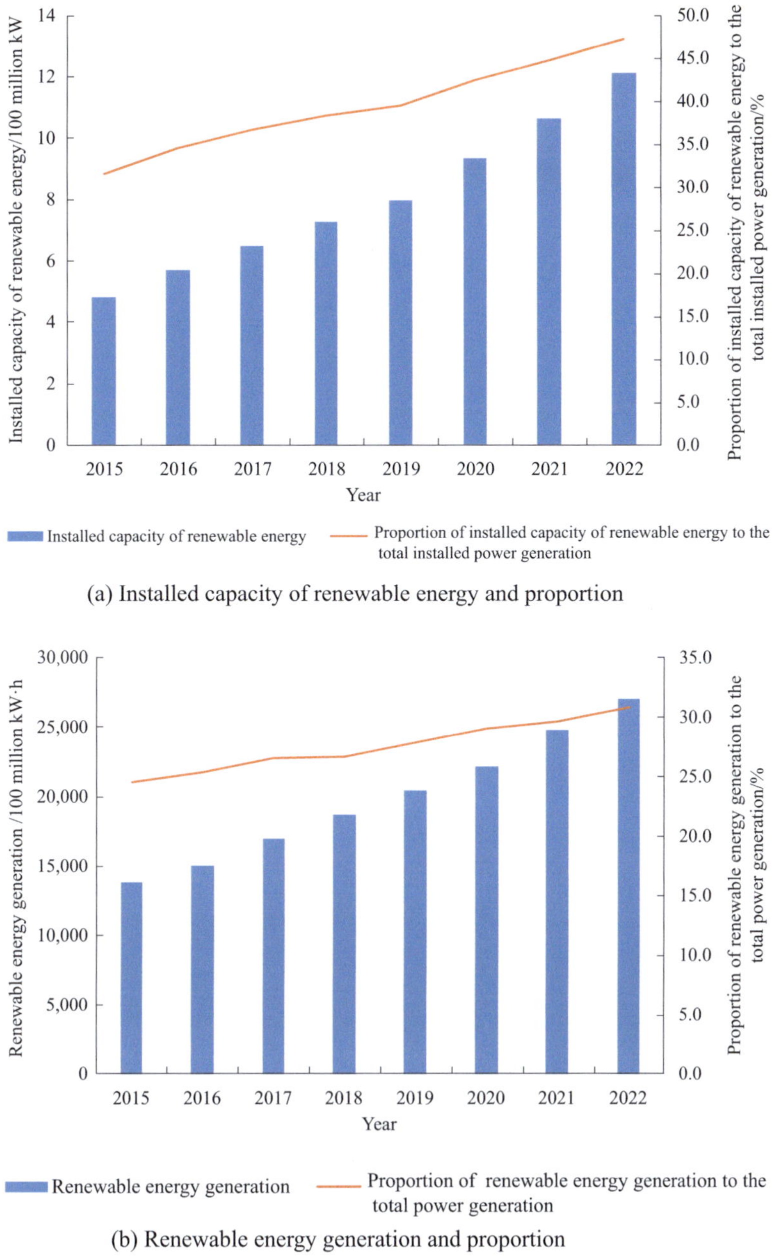

Fig. 4.6 Installed capacity and power generation of renewable energy in China from 2015 to 2022

2. **UHV-transmitted renewable energy is a crucial guarantee for renewable energy consumption**

China's renewable energy resources are relatively concentrated and far from load centers. By transmitting electricity from renewable energy bases to distant load centers through UHV, the consumption of renewable energy has been effectively promoted. Data show that from 2016 to 2022, renewable energy transmitted through 20 DC UHV lines accounted for 10.2–11.7% of the total renewable energy electricity in China, promoting renewable energy consumption (Fig. 4.8).

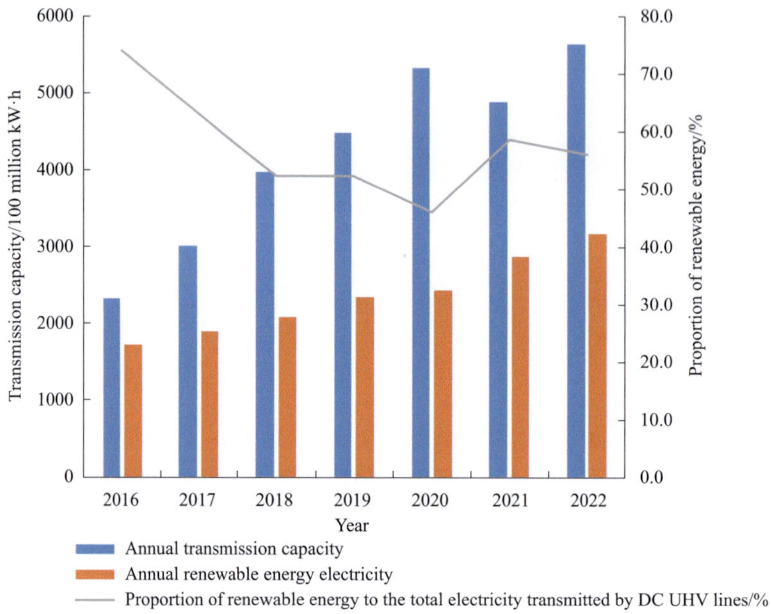

Fig. 4.7 Electricity transmitted through China's UHV lines from 2016 to 2022

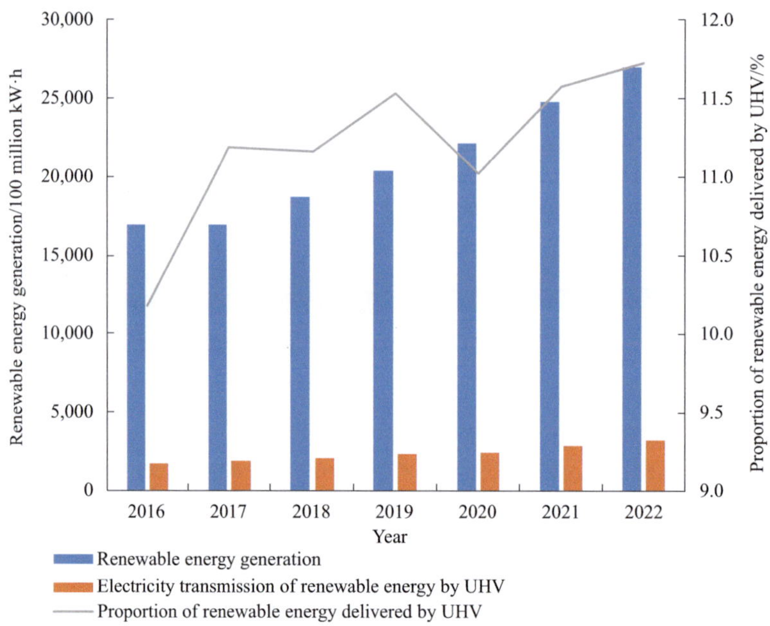

Fig. 4.8 Proportion of renewable energy transmitted through UHV to the total renewable energy from 2016 to 2022

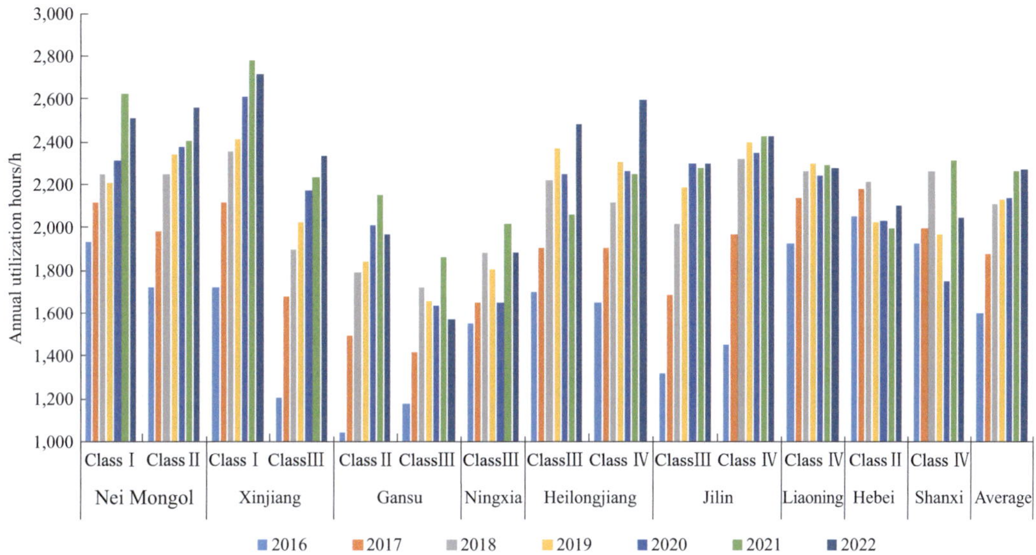

Fig. 4.9 Annual utilization hours of wind power in key regions of China from 2016 to 2022

3. **UHV promotes the solution of "curtailment of wind and solar power"**

During the "12th Five-Year Plan" period, China's northwestern, northern, and northeastern regions experienced severe "curtailment of wind power" and "curtailment of solar power" problems, mainly due to insufficient transmission lines. Conventional transmission lines were insufficient to address the consumption of renewable energy. Data show that after 2016, with the completion of UHV lines in the "Three Northern" regions, the annual utilization hours of wind and photovoltaic power have steadily increased. In 2022, the curtailment of wind and solar power in China was basically controlled within a reasonable range, and the problem of "curtailment of wind and solar power" has been gradually solved (Figs. 4.9 and 4.10).

4. **UHV has solved the problem of hydropower consumption in Southwest China**

China's hydropower is mainly concentrated in Yunnan and Sichuan provinces. To alleviate the tight power supply situation, China has accelerated the construction of hydropower bases in the southwestern region since the 21st century. However, the large-scale centralized transmission of hydropower has also brought many challenges. From 2011 to 2022, China has built 10 DC UHV hydropower transmission lines. In 2022, the hydropower transmitted through UHV accounted for nearly 30% of the total hydropower generation in Sichuan and Yunnan provinces, effectively ensuring the consumption of clean hydropower (Fig. 4.11).

Highlights

- Constructing a new power system is crucial for promoting clean energy transition. China has long been committed to the development of renewable energy, ranking first in the world in terms of installed capacity of hydropower, wind power, and photovoltaic power. The country is vigorously building a clean, low-carbon, safe, and efficient energy system and a new power system, accelerating the development of renewable energy and energy infrastructure. With its advantages of large transmission capacity, long transmission distance, low energy consumption, small land occupation, and significant economic benefits, UHV technology can

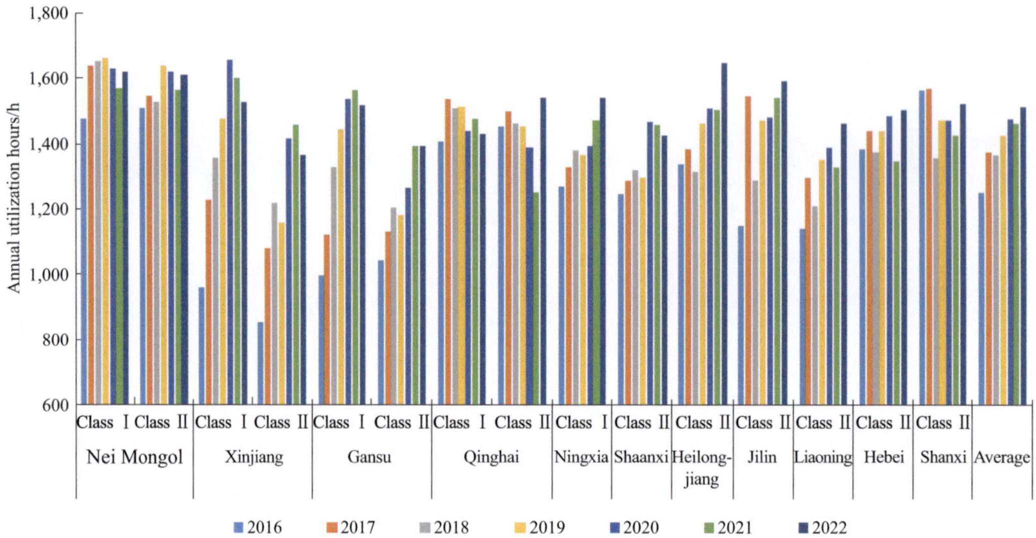

Fig. 4.10 Annual utilization hours of photovoltaic power in key regions of China from 2016 to 2022

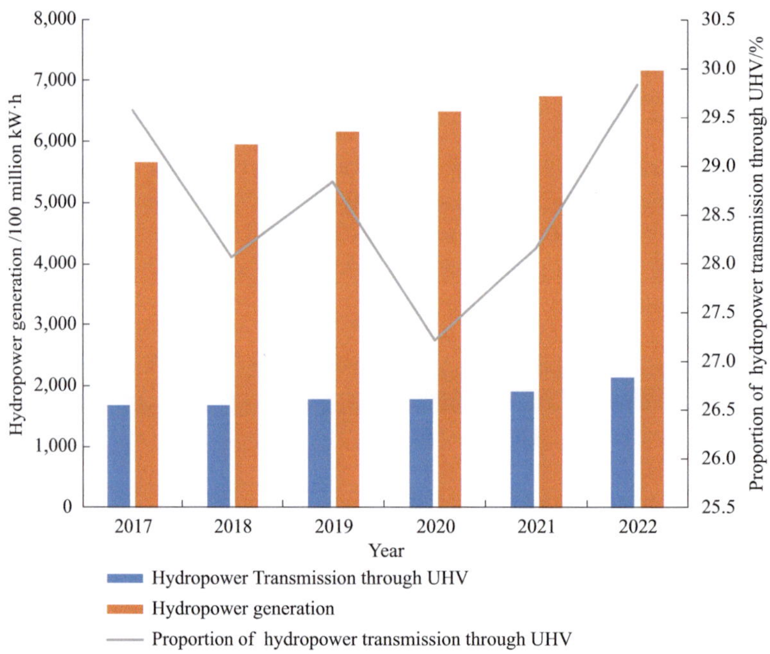

Fig. 4.11 Hydropower transmission through UHV lines from 2017 to 2022

facilitate China's "West-to-East Power Transmission", "North-to-South Power Transmission", "hydrothermal coordination", and "wind-solar complementarity", enabling a fundamental shift from local energy balance to large-scale resource allocation and strongly promoting energy transition. Based on the Monitoring and Evaluation Report on the Development of Renewable Energy in China (National Energy Administration of China 2017–2023) and the installed capacity and generation data of renewable energy in China from 2015 to 2022 (Editorial Board of China Electric Power Yearbook 2016–2019;

China Electricity Council 2020–2023), this chapter analyzes the scale, structure, and growth trend of renewable energy transmitted through UHV DC lines, and examines the supporting role of UHV in the development of China's renewable energy. It summarizes China's approach to promoting renewable energy consumption through UHV.

- **Discussion and Outlook**

Renewable energy sources such as wind, solar, and hydropower in China are primarily distributed in areas far from load centers. In the past, some projects faced the issue of curtailed electricity due to the lagging development of transmission channels and limited local consumption capacity. Considering the continued rapid development of large-scale wind farms in the north and northwest, large-scale hydropower bases in the southwest, and ultra-large-scale photovoltaic power stations in the west and north, the transmission of renewable energy electricity remains a significant challenge that needs to be addressed. With its advantages of large transmission capacity, long transmission distance, and high efficiency, UHV transmission technology can enable large-scale cross-regional transmission of renewable energy, effectively solving the problem of renewable energy consumption in remote areas, and promoting the optimization and sustainable development of China's energy structure.

Looking ahead, China will continue to strengthen the construction of a digital and resilient power grid, promoting the green and low-carbon transformation of energy. UHV engineering will remain a top priority, and the UHV industry will become an important pillar of China's economic development. As the proportion of renewable energy increases, on the one hand, China will need UHV transmission systems to ensure the stable supply of energy and optimize its allocation. On the other hand, technological advancements and intelligent development of the power grid have opened up more possibilities for UHV development. In the future, UHV transmission capacity is expected to continue growing, the level of intelligence will continuously improve, the coverage of UHV line networks will be expanded, and the application of UHV technology will be further enhanced. The upstream development of the UHV industry chain is expected to receive more support. The development of China's UHV will also pay more attention to the integration with renewable energy, the construction of smart grids, and cross-border interconnection. Through continuous innovation and technological upgrades, China's UHV is expected to play an even more important role in the global energy transition, and make greater contributions to promoting green and sustainable energy development.

4.4 Summary

This chapter focuses on SDG 7, utilizing remote sensing, GIS, and other Big Earth Data technologies to acquire and comprehensively analyze multi-source clean energy data in China. By focusing on two key indicators, namely clean cooking energy and renewable energy, this study explores the methods for monitoring and analyzing SDG 7 using Big Earth Data. The research provides scientific data support, and contributes to the development of Chinese solutions for achieving the global SDG 7 target.

In terms of data products, China has independently developed a dataset of the proportion of the population utilizing clean cooking energy and technology in 2022, as well as the data on the scale and structure of renewable energy transmitted through China's UHV lines from 2016 to 2021.

Regarding decision support, the findings of this study provide important scientific evidence and analytical methods to support the clean energy utilization, the construction of UHV lines, and the development of renewable energy in China. The research offers data foundation and reference for relevant decision-making and policy formulation. It provides independent scientific data and capacity support for China to effectively address global climate change, actively engage in energy transition, ensure energy security, improve the atmospheric environment, and promote the global sustainable development process and the building of a community with a shared future for mankind.

References

China Electricity Council (2020) China power statistics yearbook of 2020. China Statistics Press, Beijing (In Chinese)

China Electricity Council (2021) China power statistics yearbook of 2021. China Statistics Press, Beijing (In Chinese)

China Electricity Council (2022) China power statistics yearbook of 2022. China Statistics Press, Beijing (In Chinese)

China Electricity Council (2023) China power statistics yearbook of 2023. China Statistics Press, Beijing (In Chinese)

Editorial Board of China Electric Power Yearbook (2016) China electric power yearbook of 2016. China Electric Power Press, Beijing (In Chinese)

Editorial Board of China Electric Power Yearbook (2017) China electric power yearbook of 2017. China Electric Power Press, Beijing (In Chinese)

Editorial Board of China Electric Power Yearbook (2018) China electric power yearbook of 2018. China Electric Power Press, Beijing (In Chinese)

Editorial Board of China Electric Power Yearbook (2019) China electric power yearbook of 2019. China Electric Power Press, Beijing (In Chinese)

International Energy Agency (2017) Energy access outlook 2017. https://www.iea.org/reports/energy-access-outlook-2017 [2024-07-24]

International Energy Agency (2021) Tracking SDG7: the energy progress report. https://www.iea.org/reports/tracking-sdg7-the-energy-progress-report-2021 [2024-07-24]

National Energy Administration of China (2017) Monitoring and evaluation report on the development of renewable energy in China for 2016. http://zfxxgk.nea.gov.cn/auto87/201704/t20170418_2773.htm [2024-07-28] (In Chinese)

National Energy Administration of China (2018) Monitoring and evaluation report on the development of renewable energy in China for 2017. https://www.gov.cn/zhengce/zhengceku/2018-12/31/content_5433577.htm [2024-07-28] (In Chinese)

National Energy Administration of China (2019) Monitoring and evaluation report on the development of renewable energy in China for 2018. https://www.gov.cn/zhengce/zhengceku/2019-09/29/content_5434697.htm [2024-07-28] (In Chinese)

National Energy Administration of China (2020) Monitoring and evaluation report on the development of renewable energy in China for 2019. https://zfxxgk.nea.gov.cn/2020-05/06/c_139059627.htm [2024-07-28] (In Chinese)

National Energy Administration of China (2021) Monitoring and evaluation report on the development of renewable energy in China for 2020. http://zfxxgk.nea.gov.cn/2021-06/20/c_1310039970.htm [2024-07-28] (In Chinese)

National Energy Administration of China (2022) Monitoring and evaluation report on the development of renewable energy in China for 2021. https://www.nea.gov.cn/2022-09/16/c_1310663387.htm [2024-07-28] (In Chinese)

National Energy Administration of China (2023) Monitoring and evaluation report on the development of renewable energy in China for 2022. http://zfxxgk.nea.gov.cn/2023-09/07/c_1310741874.htm [2024-07-27] (In Chinese)

Yang L (2011) Ultra-high voltage transmission technology. China Water & Power Press, Beijing (In Chinese)

SDG11 Sustainable Cities and Communities

5

5.1 Background

Achieving "Sustainable Cities and Communities" (SDG 11) is crucial in addressing climate change, promoting economic growth, and reducing poverty and pollution. By 2030, it is projected that 60% of the global population will reside in urban areas (Chen et al. 2022a, b), and this proportion is expected to reach nearly 70% by 2050 (Sun et al. 2020). However, despite contributing 75% of the global GDP, urban areas also generate 75% of waste and carbon emissions worldwide (Chen et al. 2022a, b). Rapid urbanization has led to challenges such as housing shortages, traffic congestion, air pollution, waste management issues, and inadequate infrastructure and services. Unplanned urban expansion makes cities more susceptible to the impacts of climate change and natural disasters. Currently, over 4 billion urban residents still face serious problems such as air pollution, inadequate infrastructure, and uncontrolled urban development (Sun et al. 2020).

SDG 11 aims to build inclusive, safe, resilient, and sustainable cities and human settlements, comprising seven technical targets and three policy targets. Among the 17 SDGs included in the 2030 Agenda, SDG 11 is interconnected with multiple other SDGs, and around one-third of the more than 230 indicators can be measured at the city level. However, many indicators lack data support despite having evaluation methods, presenting a significant challenge in monitoring and assessing progress toward the targets and indicators. Big Earth Data technologies, such as satellite remote sensing and GIS, can regularly obtain global data, uncover drivers of progress toward SDG 11, and serve as important data sources and analysis tools for global sustainable development research.

Over the past four years, progress on SDG 11 has been monitored and assessed at both global and Chinese scales, providing strong support and guidance for urban sustainable development. The report will focus on targets, such as urban land use efficiency (SDG 11.3), urban air pollution (SDG 11.6), urban public open spaces (UPOS) (SDG 11.7), and rural–urban disparity (SDG 11.a). The report will reveal the quantified impact of urban vitality on carbon emissions, analyze changes in population exposure to air pollution in China, extract information on urban public open spaces in China's five major urban clusters, construct indicators for urban–rural economic development differences in China, and carry out a comprehensive assessment of SDG 11 in typical Chinese cities. This chapter will focus on the progress and comprehensive assessment of SDG 11 at the Chinese scale, providing scientific foundation and experiential reference to reveal the current status of the Chinese implementation of SDG 11 and to support the achievement of SDG 11.

H. Guo, *Big Earth Data in Support of the Sustainable Development Goals (2023)—China*, Sustainable Development Goals Series, https://doi.org/10.1007/978-981-95-2636-9_5

5.2 Main Contributions

This chapter evaluates four specific targets, namely SDG 11.3, SDG 11.6, SDG 11.7, and SDG 11.a as well as the comprehensive SDG 11 target through six case studies, with the main contributions shown in Table 5.1.

5.3 Case Study

5.3.1 The Quantitative Impact of Urban Vitality on Carbon Emissions in Cities in China

Target: SDG 11.3 By 2030, enhance inclusive and sustainable urbanization and capacity for participatory, integrated, and sustainable human settlement planning and management in all countries.

- **Background**

In the context of SDG 11.3, there is a global need to strengthen inclusive and sustainable urban development, improving the sustainability of human settlement planning and management. The process of urbanization has a significant and far-reaching impact on China's transformation and development. In addition to exacerbating the urban greenhouse effect, China's current growth model, characterized by high consumption and high emissions, has also led to varying levels of urbanization quality. In some regions, prioritizing the cultivation of urban vitality over environmental protection has resulted in poor sustainability of urban development. In the face of challenges such as prioritizing development over protection during urbanization (Wang et al. 2016), studying the relationship between urban vitality and carbon emissions becomes crucial. The exploration is essential for establishing vibrant and sustainable urban environment.

Localized emissions of pollutants may affect the air quality of neighboring areas. The diffusion of pollutants varies with changes in wind direction and speed, and may accelerate with an increase in wind speed (Shan et al. 2023). Xia et al. (2022) found that fiscal imbalances reduce carbon emissions, while discrepancies between expenditures and revenues hinder pollution control. The Central Government's fiscal transfer alleviates the effects of fiscal imbalance. Both the economy and fiscal policies influence carbon emissions and air pollution, making the study of the impact of urban vitality on carbon emissions essential. Urban vitality refers to the ability of a city to foster economic activity, population growth, and social interaction, encompassing economic vitality, population vitality, and social vitality. In urban development, the ability and potential for economic growth are referred to as economic vitality (Jin 2007). Currently, the studies on urban vitality and carbon emissions mainly focus on the economic aspect (Jiang et al. 2023), such as how GDP growth and per capita GDP influence carbon emissions. However, it is more important to recognize that the economic aspect is only one dimension of urban vitality (He et al. 2018). Few studies explore how urban vitality as a whole impacts carbon emissions. From a broader perspective, Zhu et al. (2022) studied the relationship between urban vitality and carbon emissions. Although cities are major contributors to carbon emissions, previous research has typically not explored this relationship at the urban level. As a primary source of emissions, achieving low-carbon and sustainable urban development is crucial for the achievement of low-carbon SDGs across society (Tan et al. 2017; Pan et al. 2022). Furthermore, the study went beyond short-term impacts, and used time series data to examine the long-term effects of urban vitality on carbon emissions. Analyzing historical data helps identify evolving trends, and provides valuable insights into future decision-making.

This case study explores how urban vitality in different dimensions affects carbon emissions, and provides the latest information on the mechanisms at play. The research covers 222 prefecture-level cities, and constructs a multi-dimensional urban vitality index. Additionally, the case study

Table 5.1 Case studies and their major contributions

Targets	Tiers	Case studies	Contributions
SDG 11.3 By 2030, enhance inclusive and sustainable urbanization and capacity for participatory, integrated, and sustainable human settlement planning and management in all countries	Tier I	The quantitative impact of urban vitality on carbon emissions in prefecture-level cities in China	Data product: urban vitality assessment of 222 prefecture-level cities Method and model: spatial durbin model Decision support: a more suitable quantitative vitality framework at the urban scale, using long-term data to analyze the quantitative impact of urban vitality on carbon emissions to enhance the persuasiveness and representativeness of the results
SDG 11.6 By 2030, reduce the adverse per capita environmental impact of cities, including by paying special attention to air quality and municipal and other waste management	Tier I	Comprehensive assessment of population exposure to air pollution in China	Data product: comprehensive air pollution index dataset for China from 2013 to 2020, and dataset of population-weighted annual average exposure concentrations for six air pollutants from 2013 to 2020 Decision support: provide reference for formulating future air pollution policies in China
		Long-term high-resolution $PM_{2.5}$ retrieval and trends in China	Data product: gridded $PM_{2.5}$ data product at a 0.1° daily scale from 2010 to 2022 Method and model: machine learning methods based on multi-source datasets Decision support: provide critical data support for assessing progress on SDG 11.6, evaluating the effectiveness of clean air action plans, and conducting health impact studies
SDG 11.7 By 2030, provide universal access to safe, inclusive, accessible, and green public spaces, in particular for women and children, older persons, and persons with disabilities	Tier I	Extraction of urban public open space information in China's five major urban agglomerations	Data product: high-resolution (1 m) urban public open space information extraction products for China's five major urban agglomerations Method and model: multi-attention network (MANet) Decision support: provide universal access to safe, inclusive, accessible, green and public spaces
SDG 11.a Support positive economic, social, and environmental links between urban, peri-urban, and rural areas by strengthening national and regional development planning	Tier I	Metrics of urban–rural economic development disparities in China	Data product: indicators of urban–rural economic development disparities in China Method and model: nighttime light landscape metrics Decision support: balance traditional economic dimensions with new economic dimensions, better reflecting equilibrium across economic levels, social security, and resource efficiency in urban and rural areas, and providing a comprehensive measurement framework for reducing urban–rural disparities in China
Comprehensive assessment of SDG 11	–	Comprehensive assessment of SDG 11 in typical Chinese cities	Data product: localized comprehensive index for SDG 11 in 59 typical Chinese cities in 2022 Method and model: comprehensive index synthesis based on the united nations sustainable development solutions network (UNSDSN) Decision support: develop a localized SDG 11 indicator system tailored to China's conditions, advancing the collection and integration of multi-source data, and promoting quantitative assessment of development goals

takes into account both the direct and spillover effects of urban vitality on carbon emissions. The study aims to address the following questions.

(1) To what extent do different aspects of urban vitality—economy, population, and society—affect carbon emissions?
(2) Does the urban vitality of one city (i.e., population, economy, and society) influence the carbon emissions of neighboring cities?
(3) How significant is the impact of carbon emissions in one city on the carbon emissions of neighboring cities?

This case study makes the following contributions.

(1) A more comprehensive assessment framework: Current research on the impact of carbon emissions often focuses primarily on the influence of economic vitality, neglecting the importance of a holistic exploration from the perspective of urban vitality. This case study proposes a more comprehensive evaluation framework.
(2) Filling the research gap at the urban level: Existing studies mainly focus on macro (national) or micro (rural) levels, with limited research at the meso (city) level. By taking cities as the unit of analysis, this case study holds a significant exploratory value.
(3) Quantitative analysis using long-term data: Although some studies attempt to explore how urban vitality affects carbon emissions, they are usually qualitative. This case study employs a quantitative approach using long-term data, which enhances the persuasiveness and representativeness of the findings.

- **Data**
  - Carbon emission data of China Carbon Emission Accounting Databases (CEADs) from 2010 to 2019.
  - Data from the China City Statistical Yearbook in 2010–2019, including GDP, fixed assets, disposable income, general public budget, education expenditure, population, urbanization, built-up area.
  - Nighttime light data from 2010 to 2019, including 1,000 m spatial resolution data from the U.S. Defense Meteorological Satellite Program (DMSP in 2011–2013, and 500 m spatial resolution data from the National Polar-orbiting Partnership (NPP) satellite in 2014–2019.

- **Methods**

To address the scientific questions outlined above, this case study employs a range of methods, including global spatial autocorrelation, ordinary least squares (OLS) regression, and the Spatial Durbin Model (SDM). The SDM is particularly effective in capturing the interactions between the dependent variables and neighboring units due to spatial dependencies, providing a more accurate modeling of the data (Anselin 2013). The SDM allows for the inclusion of spatial lag terms of the dependent variables, helping to capture spatial autocorrelation (Elhorst 2014). When spatial correlation is present, estimating with the SDM is more accurate than using models that ignore spatial dependencies, enhancing the efficacy and significance of parameter estimates.

The SDM is particularly suited for fields like geography and economics, where data are significantly influenced by spatial factors, including data distribution, spatial dependence, spatial heterogeneity, and spatial autocorrelation. Moreover, this case study investigates the impact of urban vitality on carbon emissions and its spatial spillover effects. As Goodchild et al. (1992) argued, spatial autocorrelation is an inherent characteristic of nearly all data. It measures the degree of correlation between the attribute values of a given area and the same attribute values in surrounding areas.

Global spatial autocorrelation studies the overall pattern of geographical data distribution across the system, revealing typical spatial correlations of processes or entities. In contrast, local spatial autocorrelation delves into the distribution characteristics within local subsystems. It distinguishes high-value and low-value areas, and identifies the most influential regions in the overall spatial autocorrelation. The spatial correlation analysis considers the geographical distance matrix, which forms the foundation of spatial econometric analysis.

Urban vitality includes economic vitality, population vitality, and social vitality. Economic vitality consists of GDP, fixed assets, and per capita disposable income; population vitality includes population size and education expenditure; social vitality encompasses the public budget and the Gini coefficient. Although each vitality dimension has its own set of indicators, there are commonalities among the indicators within each dimension. To simplify the model and reduce the dimensionality of each vitality dimension, this case study employs the principal component analysis (PCA). The PCA is a method that transforms the original variables into a smaller number of uncorrelated variables (principal components) while retaining as much information as possible. This approach allows for a more efficient and manageable representation of the different dimensions of urban vitality, facilitating a clearer analysis of their impact on carbon emissions.

In this case study, GDP, fixed assets, and per capita disposable income are used as sub-indicators of economic vitality. Population vitality in this context refers to the potential that the population provides for urban development. Some scholars use population size and education expenditure as indicators (Li et al. 2021b), so this case study adopts population size and education expenditure as sub-indicators of population vitality. Social vitality refers to the ability to ensure social progress and fairness. The general public budget is primarily composed of fiscal revenue, which is used to ensure and improve people's livelihoods, promote economic and social development, maintain national security, and ensure the normal operation of national institutions. The Gini coefficient, which reveals social vitality from the perspective of social fairness, is therefore included as a sub-indicator of social vitality. By using these indicators, this case study explores how to balance various aspects of urban vitality with environmental protection. All of the information can become an important resource for urban planning and sustainable development, helping local governments formulate more effective policies and initiatives to reduce carbon emissions and enhance urban sustainability.

This case study focuses on 222 prefecture-level cities in China, and uses the SDM to explore the impact of urban vitality on carbon emissions. The study follows the following steps. First, the PCA method is used to summarize and calculate the urban vitality index. Second, the study examines both global and local spatial autocorrelation of carbon emissions, assessing the spatial dependencies and distribution patterns of carbon emissions across the cities. Third, the OLS model is used for correlation testing, and the most appropriate model is selected based on the results. This step helps to determine the relationship between urban vitality and carbon emissions. By following these steps, the study seeks to uncover the influence of urban vitality on carbon emissions, and to identify the spatial spillover effects of carbon emissions across neighboring cities.

According to the theory of the SDM, when certain conditions are met, the spatial autoregressive model can be transformed into spatial lag models (SLM) and spatial error models (SEM). The SDM posits that the values of neighboring independent variables, as well as the neighboring values of the dependent variables, can influence the observed dependent variable values. This helps to gain a more comprehensive understanding of the endogeneity of the dependent variables and the external influencing factors. The SDM will degenerate into the SLM under certain conditions ($\theta = 0$), and into the SEM under other conditions ($\theta + \rho\beta = 0$). The general form of the SDM is represented as

$$Y_{it} = \rho \sum_{j=1}^{N} W_{ij} Y_{it} + \beta X_{it} + \sum_{j=1}^{N} W_{ij} X_{it} \theta + \lambda_i + \mu_t + \varepsilon_{it} \tag{5.1}$$

where Y_{it} represents the dependent variable for individual i at time t. ρ is the spatial autoregressive coefficient, indicating the spatial lag effect of the dependent variable Y. W_{ij} represents the elements of the spatial weight matrix, indicating the spatial relationship between the observations of individual i and individual j. X_{it} represents the independent variables for individual i at time t. β is the vector of regression coefficients for the independent variables X. θ is the vector of coefficients for the spatially lagged independent variables, indicating the spatial lag effect of the independent variables. λ_i represents the individual fixed effects, capturing the heterogeneity between different individuals. μ_t represents the time-fixed effects, capturing the heterogeneity across different time points. ε_{it} is the term error, representing random disturbances. The model helps to account for both direct and indirect (spillover) effects of urban vitality on carbon emissions, incorporating spatial dependencies and interactions between cities.

In addition, this case study conducts the necessary tests on the data. Firstly, the Hausman test is used to determine whether the fixed effects model or the random effects model should be selected. Secondly, the likelihood ratio (LR) test is used to assess the significance of spatial effects, determining whether spatial lag terms (ρ and θ) are required and testing the significance of fixed effects (both individual fixed effects and time fixed effects). Thirdly, to select the best model, the LR test and Wald's test are used to determine whether the SDM can be simplified to the SLM or SEM (Liu and Song 2020). This process helps selecting the optimal spatial model. Finally, the SDM is constructed.

- **Results and Analyses**

1. **PCA Results**

The principal components of economic vitality, population vitality, and social vitality are extracted as shown in Fig. 5.1. Economic vitality is derived from the first principal component extracted from GDP, fixed assets, and per capita disposable income, with the proportion of the first principal component exceeding 80%. Population vitality is determined by extracting the first principal component from the population size and education expenditure, with the proportion of the first principal component being 88%. Social vitality is identified by extracting a principal component from the public budget and the Gini coefficient, with the proportion of the first principal component being 70%.

2. **Urban Vitality Performance**

From 2011 to 2019, there were significant changes in the economic vitality, population vitality, and social vitality of China's prefecture-level cities, as shown in Fig. 5.2a. During this period, the economic vitality of most cities continued to grow. Notably, Fuyang City experienced the most significant growth in economic vitality, increasing from −2.234 in 2011 to 0.871 in 2019, while Anshan City saw the smallest increase, rising from −0.315 in 2011 to −0.274 in 2019 (Fig. 5.2a, b).

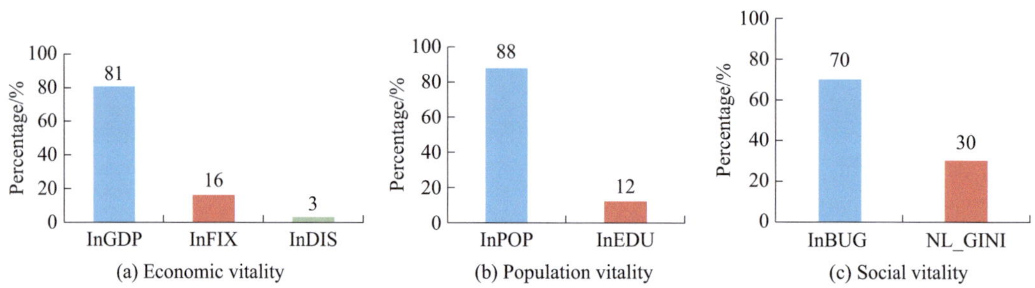

Fig. 5.1 PCA results. *Note* FIX refers to fixed assets; DIS refers to per capita disposable income; POP refers to population size; EDU refers to education expenditure; BUG refers to public budget; NL_GINI refers to the Gini coefficient

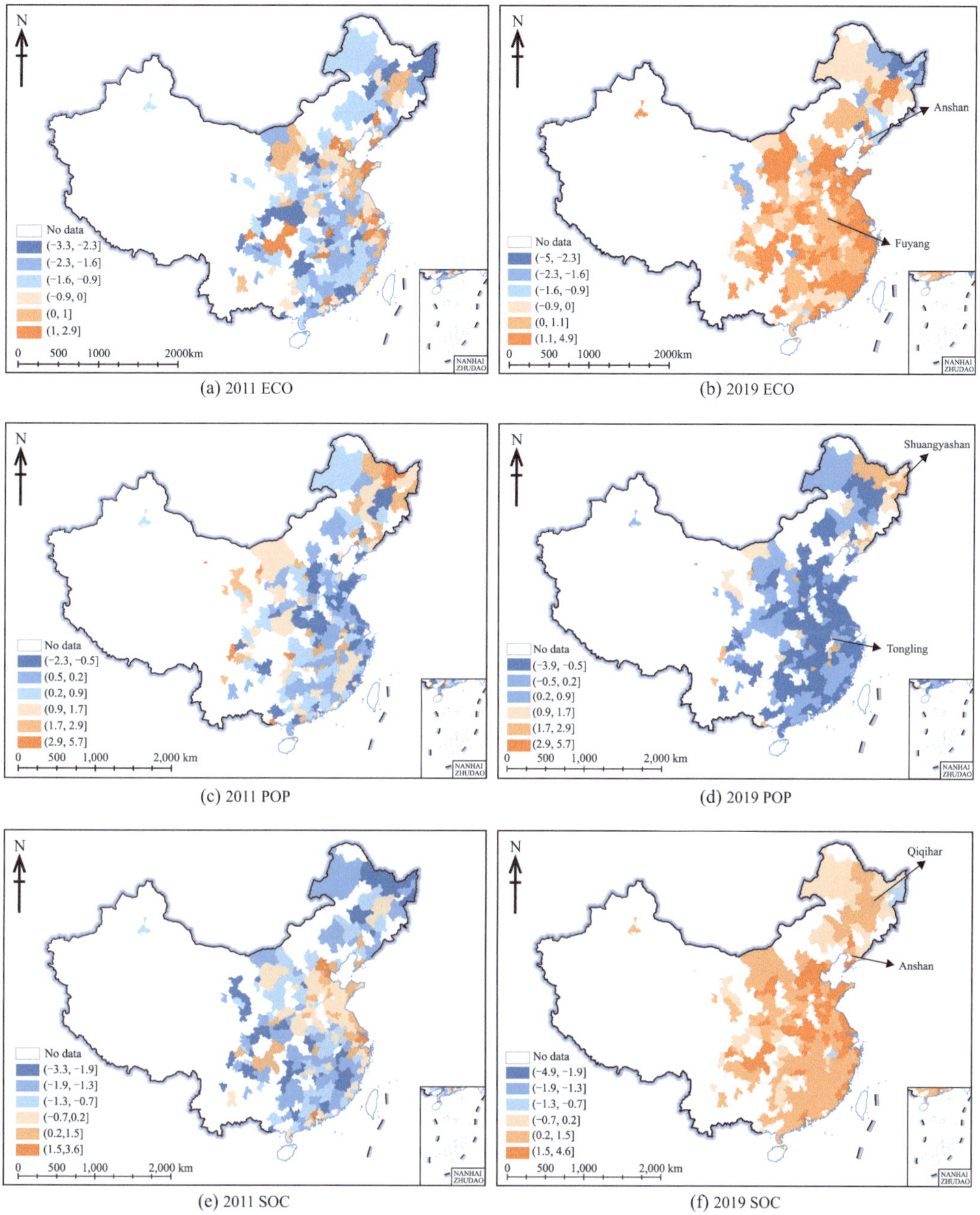

Fig. 5.2 Spatial distribution of urban vitality of 222 prefecture-level cities in China in 2011 and 2019. *Note* ECO refers to the economic vitality; POP refers to the population vitality; SOC refers to the society vitality

During the same period, the population vitality of these cities showed different development trends, with both increases and decreases. Tongling City experienced the largest decrease in the population vitality, falling from 4.360 in 2011 to 1.352 in 2019. In contrast, Shuangyashan City saw the highest increase, rising from 1.470 in 2011 to 2.011 in 2019 (Fig. 5.2c, d). From 2011 to 2019, the social vitality of each city continued to grow. Notably, Qiqihar City saw the most

Table 5.2 Moran's *I* result

Year	Carbon emissions		
	Moran's *I*	*Z* value	*p* value
2011	0.053***	8.945	0.000
2012	0.058***	9.731	0.000
2013	0.056***	9.329	0.000
2014	0.06***	9.958	0.000
2015	0.061***	10.101	0.000
2016	0.063***	10.378	0.000
2017	0.068***	11.205	0.000
2018	0.055***	9.176	0.000
2019	0.058***	9.743	0.000

*** indicates significance at the 0.01 level

significant growth, increasing from −2.234 in 2011 to 0.871 in 2019, while Anshan City had the smallest growth, rising from −4.165 in 2011 to 0.090 in 2019 (Fig. 5.2e, f).

The economic vitality in South China and Southwest China is generally higher, indicating that these areas are relatively economically developed. The population vitality is mainly thriving in Northeast China, while the social vitality is more prominent in East China and Central China.

3. **SDM Results**

Table 5.2 shows the global Moran's *I*. Spatial autocorrelation is considered significant when the *Z* value exceeds 1.65 or the *p* value of Moran's *I* is less than 0.05. A value of Moran's *I* close to 1 indicates stronger geographical connectivity. As shown in Table 5.2, between 2011 and 2019, there was spatial dependence in the explanatory factors for the 222 prefecture-level cities.

To understand the changes in annual carbon emissions across different cities, this study uses local indicators of spatial association (LISA) to identify clustering patterns, as shown in Fig. 5.3. The distribution pattern remained largely unchanged in 2016 and 2017. Cities that are close to each other and have similar carbon emission levels tend to cluster in specific regions, with high-high clusters predominantly located in northern China. This suggests that these cities share similar carbon emission patterns or are influenced by each other's carbon emissions. However, there are also areas with significant differences in carbon emissions, such as Huanggang, Quanzhou, Guilin, and Guangzhou. These areas have high carbon emission levels, but the surrounding regions do not exhibit similarly high emissions. Additionally, there are notable differences in carbon emission levels between neighboring cities. For example, in northern China, some areas with high carbon emissions are adjacent to areas with lower emissions. This highlights the differences in carbon emission levels between neighboring regions. Furthermore, in central China, there are regions where

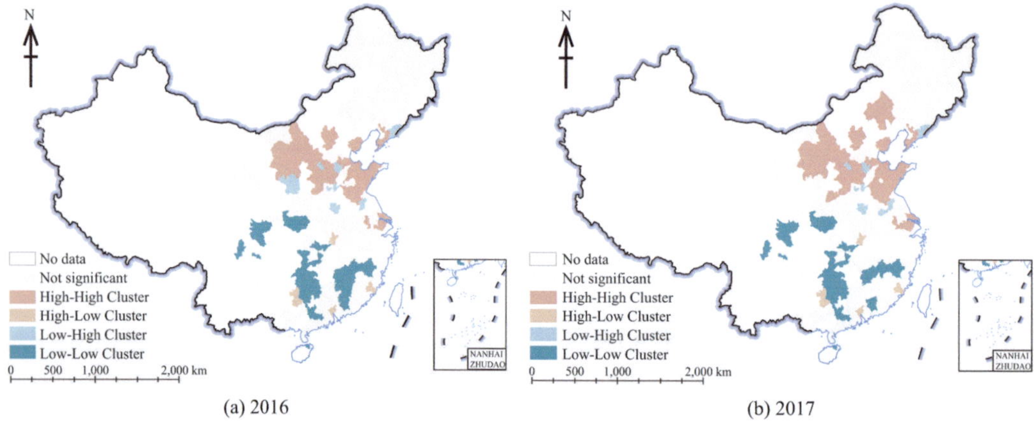

Fig. 5.3 LISA results of carbon emissions in 2016 and 2017

Table 5.3 Decomposition of spatial effects based on the prefecture-level cities

Variables	Direct effect	Spillover effect	Total effect
Economic vitality	−0.02	0.35^{***}	0.33^{**}
p value	0.35	0.01	0.01
Population vitality	0.05^{**}	0.12	0.17
p value	0.03	0.36	0.20
Social vitality	0.01	0.04	0.06^{**}
p value	0.31	0.19	0.06
lnURB	0.29^{***}	-2.16^{**}	-1.89^{**}
p value	0	0.02	0.047
lnBUI	−0.01	−0.58	−0.59
p value	0.54	0.16	0.16

*** and ** indicate significance at the 1% and 5% levels, respectively. URB refers to urbanization, BUI refers to built-up area

both the local and surrounding cities' carbon emissions were relatively low in both years.

The research results (Table 5.3) indicate that economic vitality has a direct negative impact on carbon emissions, but there is a significant positive spillover effect, and the overall effect is also positive. The spillover effect of economic vitality on the carbon emissions of neighboring cities is significant, with a value of 0.35, which is significant at the 1% level. An increase in economic vitality is associated with an increase in carbon emissions in neighboring cities, but has no significant impact on the emissions of the local city. Improving economic vitality can reduce the local city's carbon emissions while increasing the carbon emissions of neighboring cities. The decrease in local carbon emissions may be attributed to environmentally friendly consumption habits, such as purchasing energy-efficient appliances (Kelly 2012) and using low-carbon vehicles (Bonsu 2020; Yao et al. 2011). Economic prosperity may also promote the optimization of industrial structure and the development of low-carbon industries, such as clean energy, technology, and services. In contrast, the increase in carbon emissions in neighboring cities may be related to the relocation of high-energy-consuming and high-polluting industries from the local city. According to studies by Shafik and Bandyopadhyay (1992), Grossman and Krueger (1995), and Selden and Song (1994), income has a profound impact on various environmental quality indicators. As income increases, environmental indicators initially decline, but as the overall national prosperity improves, environmental indicators tend to improve as well. Considering the overall effect of economic vitality, its total impact on carbon emissions in both the local and neighboring regions is 0.33, which is significant at the 1% level, reflecting the important role of economic vitality in promoting carbon emissions in both local and neighboring areas.

Population vitality has positive direct and spillover effects on carbon emissions, with the direct effect being statistically significant, and the total effect is positive. Examining the direct impact, population vitality influences carbon emissions in local cities with a significant coefficient of 0.05 at the 5% level. This aligns with previous research on the association between population and carbon emissions (Zhang 2011; Wang et al. 2017). Additionally, further evidence suggests that the population expansion of prefecture-level cities significantly contributes to urban carbon emissions, consistent with earlier surveys (Liang et al. 2020; Liu and Song 2020).

The direct and spillover effects of social vitality on carbon emissions are both positive, with the total effect significantly positive statistically.

When social vitality is taken into account as a whole, the effect it has on carbon emissions in nearby and local cities is considerable at the 5% level, with an overall coefficient of 0.06. This emphasizes how important social vitality is in encouraging carbon emissions in the surrounding and local communities.

In the realm of control variables, urbanization exhibits a significant positive direct impact, while its spillover and total effects manifest as significant negative influences. Elevated urbanization contributes to increased carbon emissions in local cities, but concurrently aids in reducing carbon emissions in nearby prefecture-level cities. 1% rise in urbanization is correlated with 1.178% growth in per capita carbon emissions, which is in line with the findings of Wang and Li (2021). Contrarily, the direct, spillover, and total effects of expanding built-up areas on carbon emissions are negative, although not statistically significant. This implies that an increase in built-up areas may not only decrease carbon emissions in local urban zones, but also assist in reducing carbon emissions in adjacent urban areas. The optimization and enhancement of infrastructure can improve the public transport system and create a convenient transportation network, diminishing residents' reliance on private cars and lowering traffic-related carbon emissions (Creß et al. 2021). Furthermore, expanded built-up areas offer more cultural, entertaining, and social facilities, promoting closer mobility within the city (Li et al. 2021a), and consequently reducing traffic congestion and carbon emissions in the surrounding regions.

Highlights

– Using readily accessible data from the China City Statistical Yearbook and nighttime light data, this study examines the multidimensional impact of urban vitality on carbon emissions across 222 prefecture-level cities in China.

- **Discussion and Outlook**

This study uses data from the statistical yearbook and nighttime light data, applying the SDM to quantify the impact of various dimensions of urban vitality on carbon emissions. The main conclusions are as follows.

(1) Economic vitality exhibits a negative direct impact on local carbon emissions but a significantly positive spillover effect, resulting in an overall positive impact. The findings suggest that increased economic vitality reduces carbon emissions locally, but contributes to elevated carbon emissions in neighboring cities. The local decline in carbon emissions may be attributed to environmentally friendly consumption habits and the promotion of low-carbon industries. Conversely, the upsurge in carbon emissions in adjacent urban areas may result from the relocation of high-energy consumption industries.
(2) Regarding population vitality, both direct and spillover effects are positive, with a statistically significant direct effect. The total effect is positive, emphasizing its impact on local carbon emissions, supported by a significant coefficient of 0.06 at the 5% level. This aligns with prior research on the association between population growth and carbon emissions. Furthermore, evidence suggests that the expansion of prefecture-level cities leads to urban carbon emissions.
(3) Similarly, the direct and spillover effects of social vitality on carbon emissions are positive, with a statistically significant positive total effect. Examining the total effect of social vitality, it exerts noteworthy impact, on carbon emissions in both local and adjacent cities, with an overall coefficient of 0.06, significant at the 5% level. This underscores the substantial role of social vitality in promoting carbon emissions in both local and neighboring areas. To summarize, this research emphasizes the necessity of implementing focused policies to effectively manage and reduce environmental consequences in nearby and local areas.

This research has certain limitations: (1) The assessment of urban vitality considers the most common and fundamental dimensions, namely economic, population, and social vitality. However, the inclusion of cultural and tourism vitality, which are increasingly significant aspects, should also be taken into account.

(2) The lack of robustness testing using multiple weight matrices may result in variations in the outcomes due to data differences. (3) Economic, population, and social vitality, constrained by the limitations of yearbook data, did not consider a broader range of indicators. Future studies could supplement the results by considering additional indicators in the context of data availability.

This case offers the following policy recommendations to assist governments and policymakers in crafting effective strategies for carbon emission control and sustainable development. Governments should incentivize and support the adoption of clean energy sources such as solar, wind, and hydropower. This can be achieved through tax incentives, subsidies, and concessions to attract investors and businesses toward renewable energy. Invest in environmental education and awareness campaigns to strengthen public awareness regarding environmental issues. Foster green consumption habits and popularize low-carbon lifestyles. Foster a culture of research and innovation, especially in environment protection and energy-related fields. Invest in research projects and provide training to professionals to advance environmental technologies. Optimize urban planning, and maximize the utilization of built-up areas. Improve public transportation systems, and reduce private car usage to minimize transportation-related carbon emissions.

Recognize that carbon emissions are a global challenge. Actively participate in international cooperation efforts, engage in international climate agreements, and share experience and technologies in carbon markets to collectively reduce global carbon emissions. These policy recommendations should be tailored to the unique characteristics and requirements of each region. To ensure a sustainable future for the Earth, the governments, industry, and society should collaborate to minimize carbon emissions and foster sustainable development to the largest extent.

5.3.2 Comprehensive Assessment of Population Exposure to Air Pollution in China

Target: SDG 11.6 By 2030, reduce the adverse per capita environmental impact of cities, including by paying special attention to air quality and municipal and other waste management.

- **Background**

SDG 11.6.2: Annual mean levels of fine particulate matter (e.g., $PM_{2.5}$ and PM_{10}) in cities (population-weighted) can be used to quantify the adverse health effects caused by human exposure to air pollutants. Many studies have confirmed that there is a significant correlation between air pollution and many physiological diseases such as cardiovascular diseases, respiratory diseases, nervous system diseases, and cancers which lead to an increase in related mortality. Recently, some studies have shown that air pollution is also related to many psychological diseases such as anxiety, phobia, depression, and so on.

Previous studies have shown that although the overall mortality rate caused by air pollution in China decreased from 1990 to 2017, there are still 12 provinces showing an upward trend, and about 1.24 million deaths in China in 2017 can be attributed to air pollution. Therefore, it is of great practical significance to assess the comprehensive risk of population exposure to air pollution in China, which can provide data support for related health research and provide some reference for formulating air pollution policies in China in the future.

- **Data**
  - The China High-quality Air Pollution (CHAP) dataset in 2013–2020.
  - WorldPop population count dataset in 2013–2020.

- **Methods**

In this case, firstly, the PCA is used to construct a comprehensive air pollution index combining six air pollutants at the 1 km grid scale. Theil-Sen Median and Mann–Kendall method are used to test the trend of the index from 2013 to 2020. Theil-Sen Median is a robust nonparametric statistical trend calculation method, and the Mann–Kendall method can further test the significance of long-time series data trends.

Secondly, the population-weighted annual average exposure concentrations of six air pollutants are calculated at the prefecture-level city scale, and the formula is as follows:

$$\overline{\mathrm{AP}} = \frac{\sum_j \mathrm{AP}_j P_j}{\sum_j P_j} \tag{5.2}$$

where $\overline{\mathrm{AP}}$ indicates the population-weighted annual average exposure concentrations of air pollutants in the total spatial unit; j denotes the jth spatial unit; P_j represents the population in space unit j, and AP_j represents the annual average concentrations of air pollutants in space unit j. This value can reflect the annual average exposure concentrations of air pollutants per capita on the scale of prefecture-level cities.

Combining the Markov chain to evaluate the dynamic change process of population exposure to air pollutants. This method introduces a transition probability matrix to reflect the state of each region and its dynamic evolution process of upward or downward transfer in different periods.

- **Results and Analysis**

1. **Distribution Pattern and Change Trend of Comprehensive Air Pollution Index in China**

Figure 5.4 shows the spatial pattern of China's comprehensive air pollution index from 2013 to 2020 based on PCA. The spatial distribution of the comprehensive air pollution index from 2013 to 2020 is generally stable, and the highest value mainly appears in North China and southern Xinjiang, showing obvious agglomeration. Other high values mainly appear in the urban agglomeration in the Middle Reaches of the Yangtze River, the north of Yangtze River Delta urban agglomeration, and Chengdu-Chongqing city group, and show zonal distribution between North China and southern Xinjiang. From 2013 to 2020, the main high-value areas show a process of first decreasing, then increasing, and then decreasing, while the Pearl River Delta region shows a rising trend.

The change in the comprehensive air pollution index from 2013 to 2020 reflected by the Mann-Kendell is shown in Fig. 5.5. The areas with the most obvious downward trend are mainly concentrated in the west of Hunan Province, the north of Guangxi Zhuang Autonomous Region, the east of Sichuan Province, the south of Shaanxi Province, Guizhou Province, and Northeast China. There is also a cluster downward trend in southern Hebei Province and Zhejiang Province. At the same time, there is still an obvious upward trend in the western part of Xizang Autonomous Region, northern Xinjiang, central Hainan Province, and the intersection of Yunnan Province, Xizang Autonomous Region, and Sichuan Province.

2. **Spatial Distribution Patterns and Changes of Population-Weighted Annual Average Exposure Concentrations of Six Air Pollutants in China**

The spatial distribution of population-weighted exposure concentrations of six air pollutants at prefecture-level cities in China from 2013 to 2020 is shown in Fig. 5.6. There are some similarities in the spatial patterns of exposure concentrations of six air pollutants, among which North China is a region with high exposure concentrations of all air pollutants, which is mainly due to the high concentration of man-made emission pollution. In addition to North China, the high population exposure to $PM_{2.5}$ is also distributed in the northern part of East China and southern Xinjiang. The high population exposure value of PM_{10} also exists between Xinjiang and North China, showing an

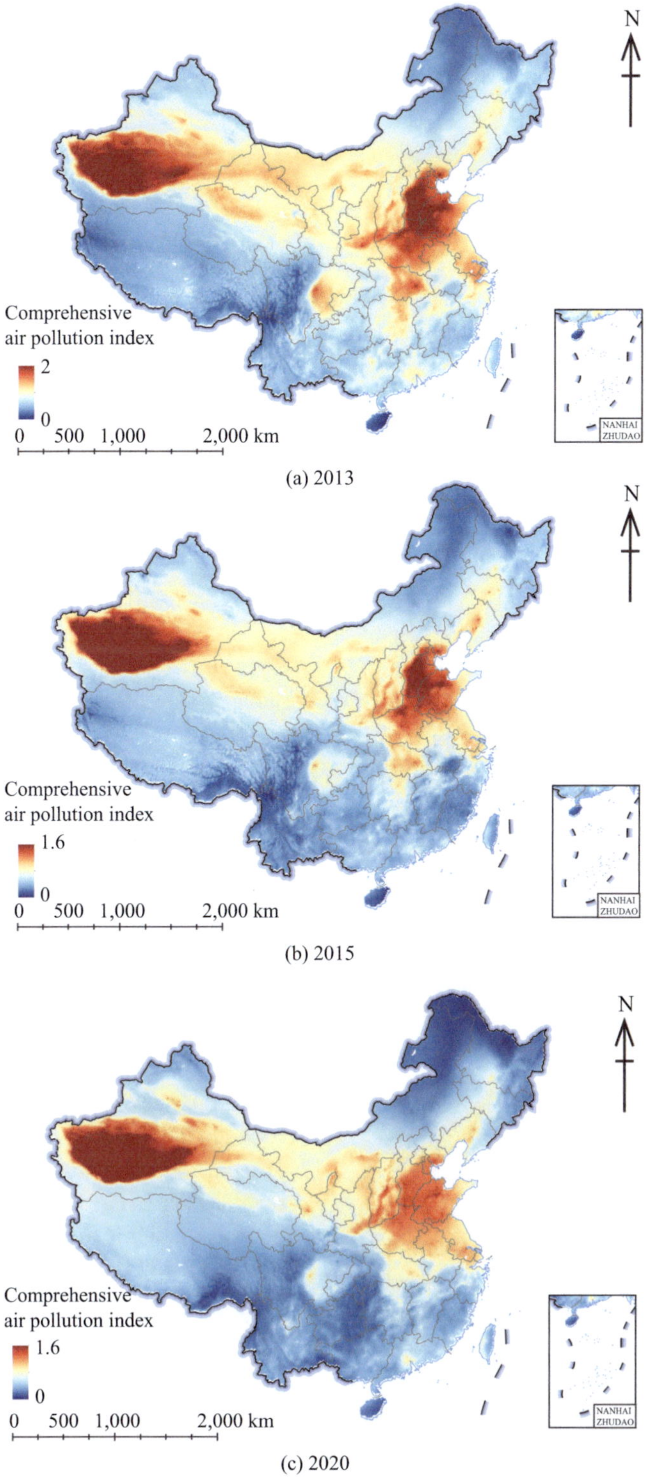

Fig. 5.4 Spatial distribution of China's comprehensive air pollution index from 2013 to 2020

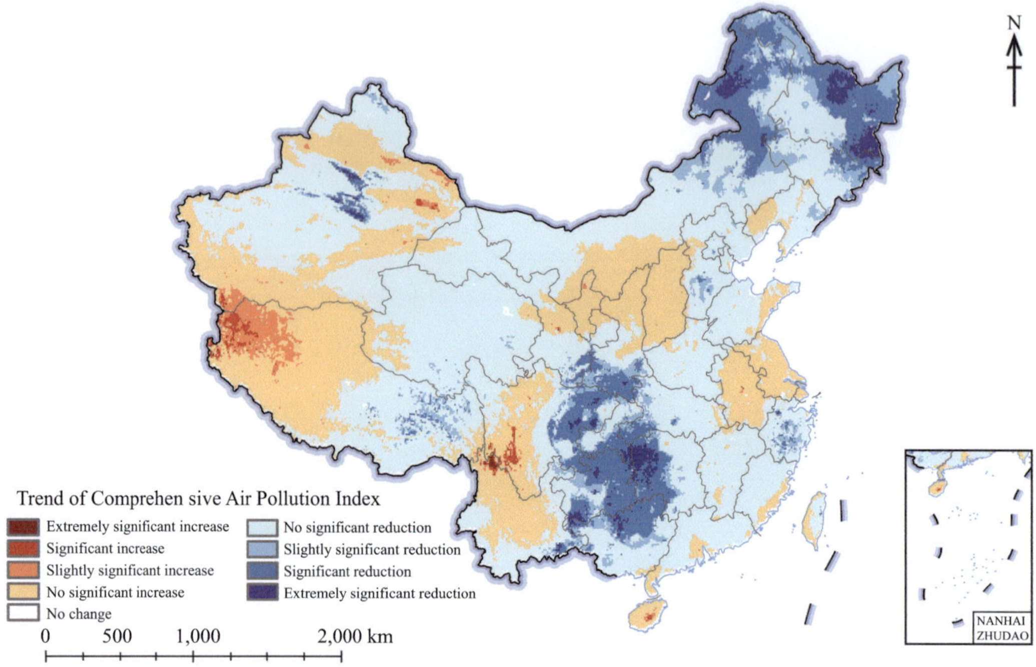

Fig. 5.5 The trend of comprehensive air pollution index in China from 2013 to 2020

east–west zonal distribution. The high value of O_3 population exposure shows a scattered distribution. The high value of SO_2 population exposure is also mainly distributed in East China and Central China. The high value of CO exposure is mainly distributed in the central region. NO_2 is mainly concentrated in the Yangtze River Delta urban agglomeration, Pearl River Delta urban agglomeration, and Chengdu-Chongqing city group.

Based on the Markov transition matrix (Fig. 5.7), the analysis of the evolution process from 2013 to 2020 reveals a significant decline in the population-weighted annual average exposure concentrations of $PM_{2.5}$, SO_2, CO, and PM_{10} in high-value regions. The average probabilities of transitioning from a higher concentration level to a lower one are 28%, 33%, 27.6%, and 27.6%, respectively. Notably, $PM_{2.5}$ and SO_2 exhibit considerable fluctuations, with variances of 0.0013 and 0.0018, respectively. NO_2 demonstrates a downward trend, but its rate of decline is relatively slower, with an average probability of transitioning to a lower concentration level of 15% (±1%), while its high-value regions largely maintain their original spatial distribution. In contrast, O_3 exhibits a noticeable upward trend in eastern regions, with an overall probability of transitioning to a higher concentration level at 17.6%, accompanied by a variance of 0.0008.

Highlights

- The high values of the comprehensive air pollution index from 2013 to 2020 are mainly found in North China and southern Xingjiang, and the main high-value areas show a process of first decreasing, then increasing, and then decreasing, while the Pearl River Delta region shows a rising trend.
- The spatial pattern of population exposure concentrations of the six air pollutants from 2013 to 2020 has some similarities. North China is a region that has high exposure concentrations of all six air pollutants. The population exposure concentration value of O_3 has a certain upward trend, while the other five air pollutants show a downward trend.

Fig. 5.6 Spatial distribution of population-weighted annual average exposure concentrations of six air pollutants in China from 2013 to 2020

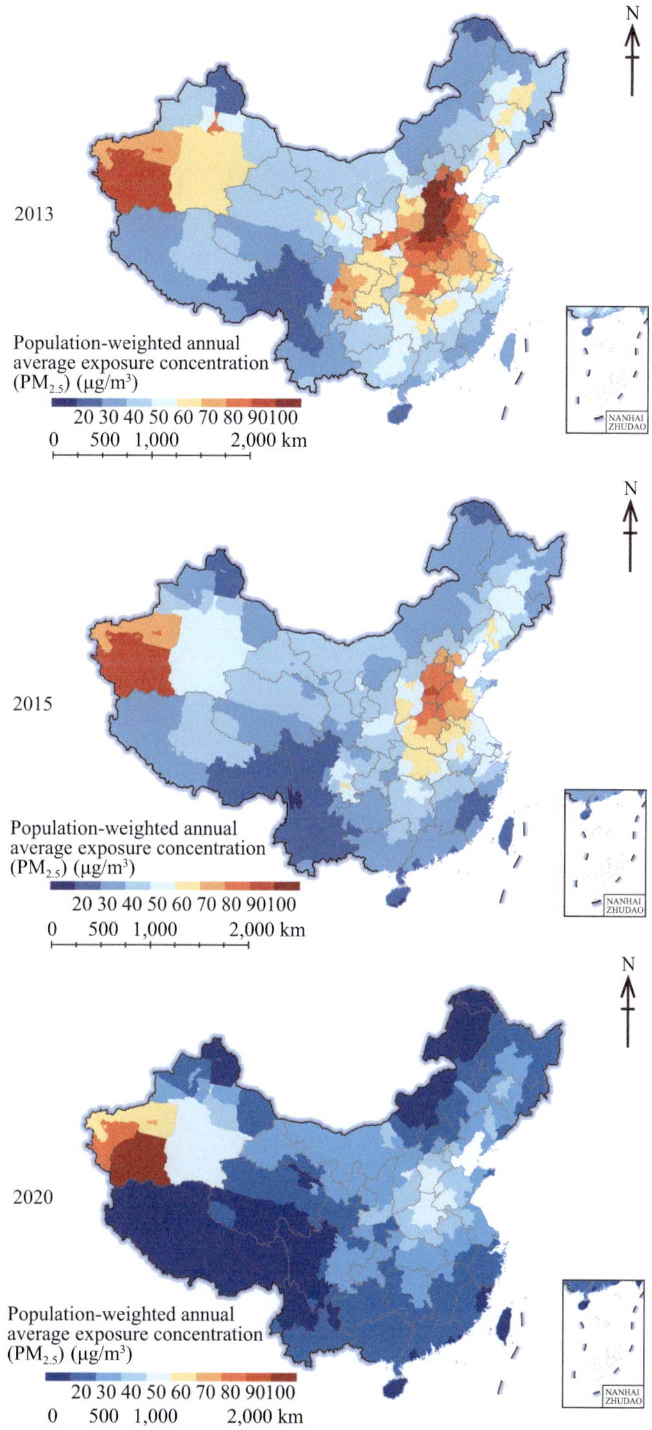

(a) $PM_{2.5}$

Fig. 5.6 (continued)

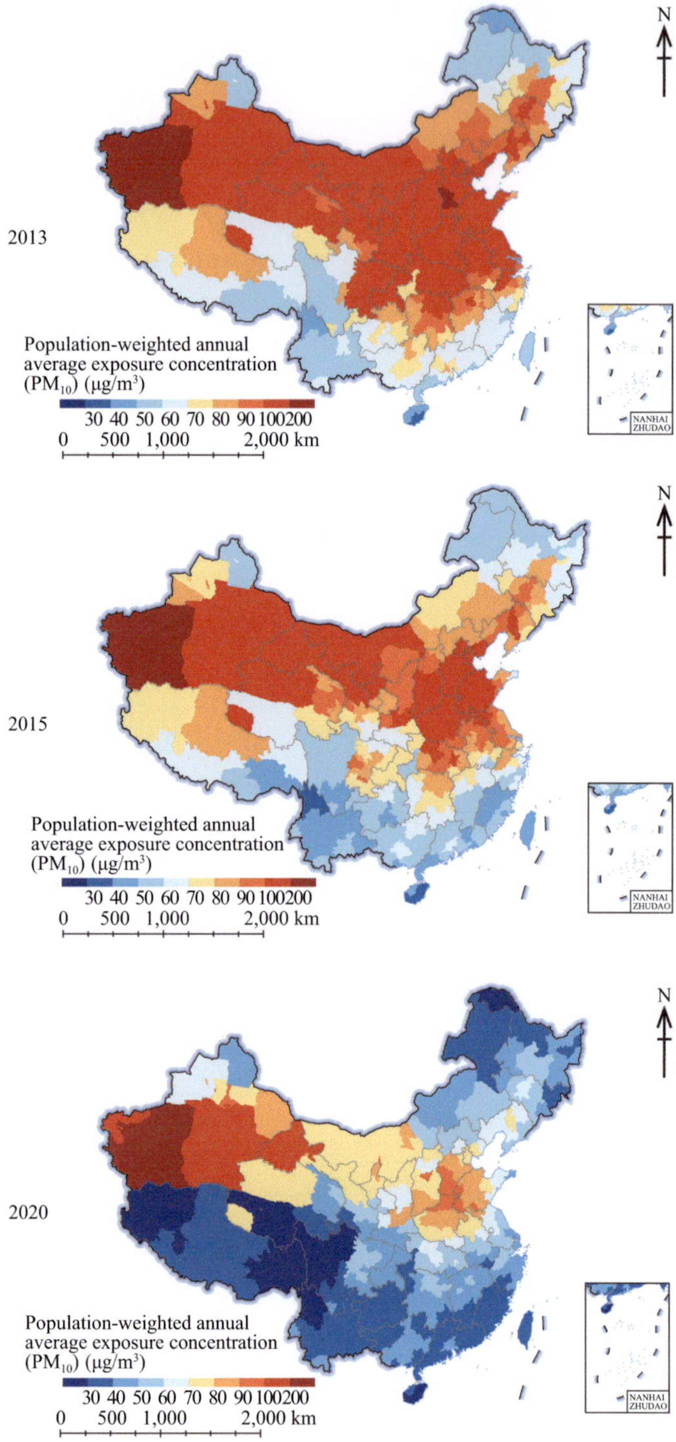

(b) PM_{10}

Fig. 5.6 (continued)

2013

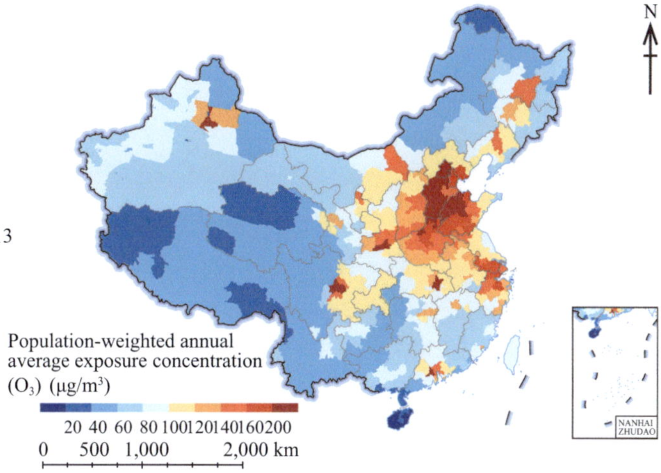

2015

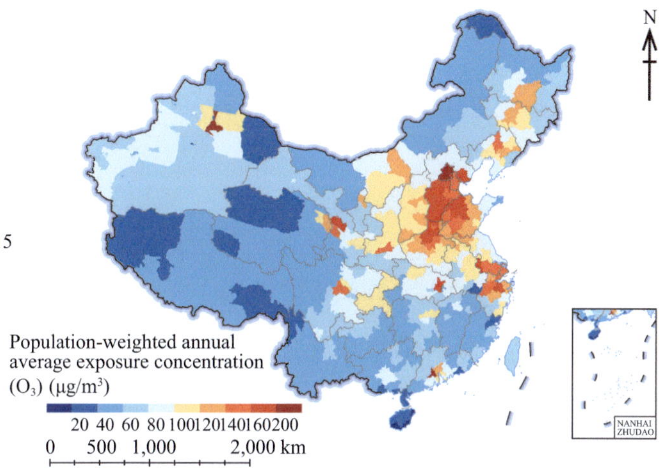

2018

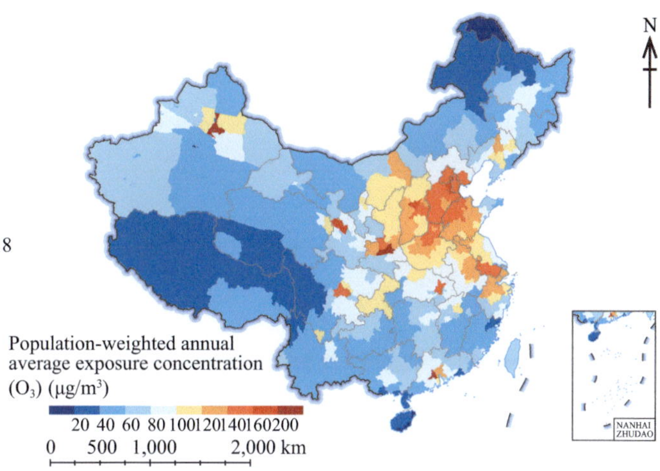

(c) O_3

Fig. 5.6 (continued)

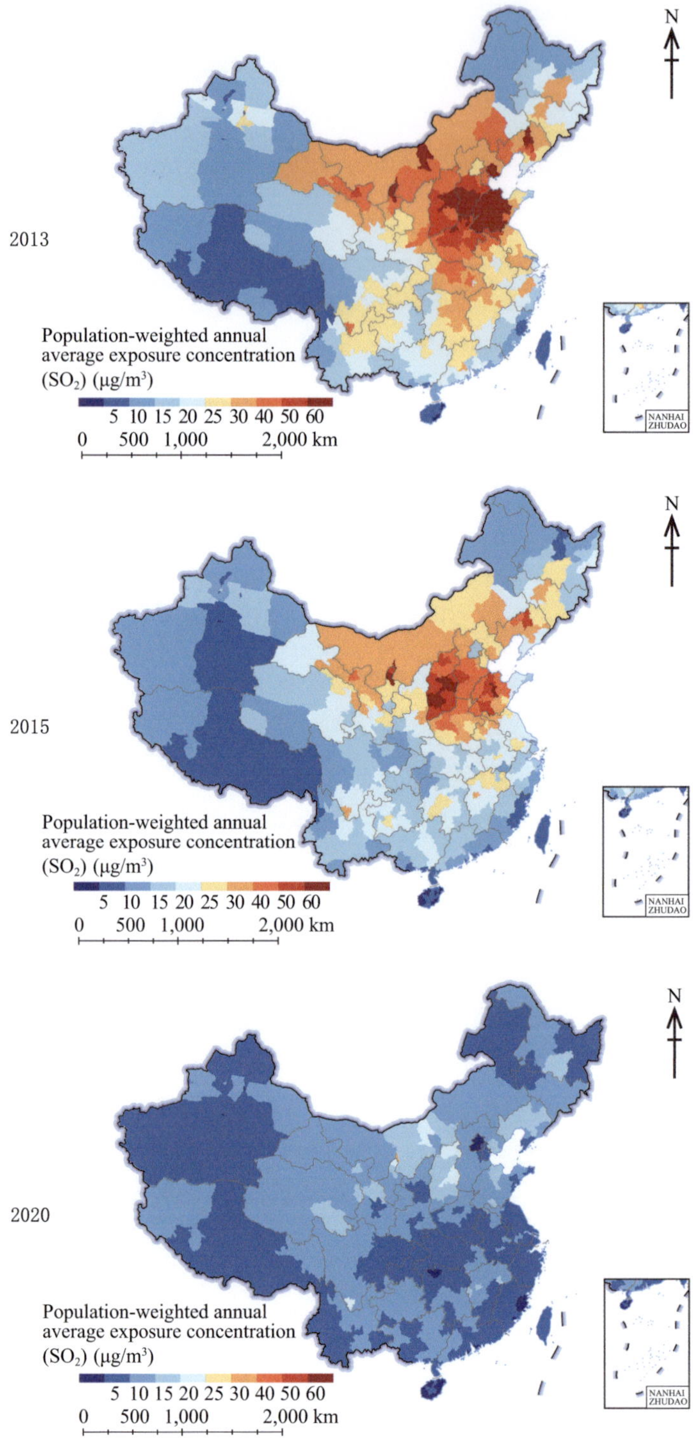

(d) SO_2

Fig. 5.6 (continued)

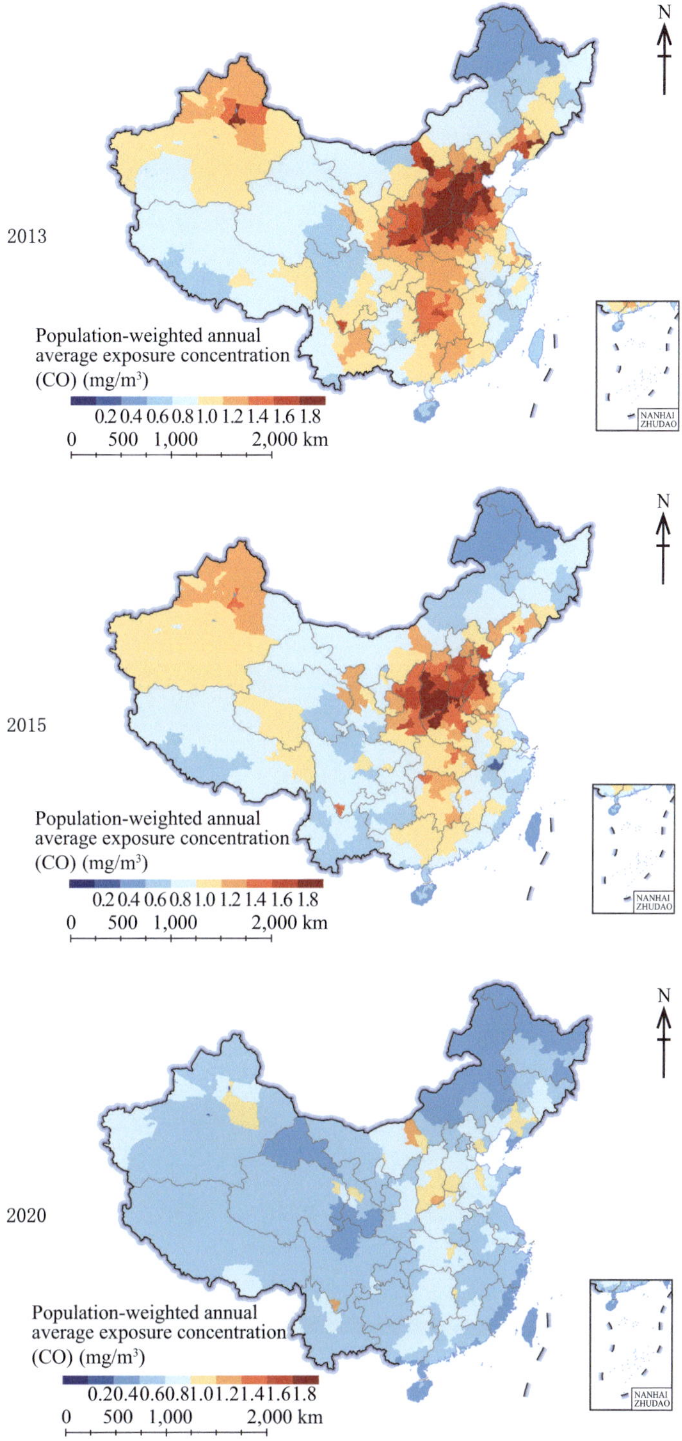

(e) CO

Fig. 5.6 (continued)

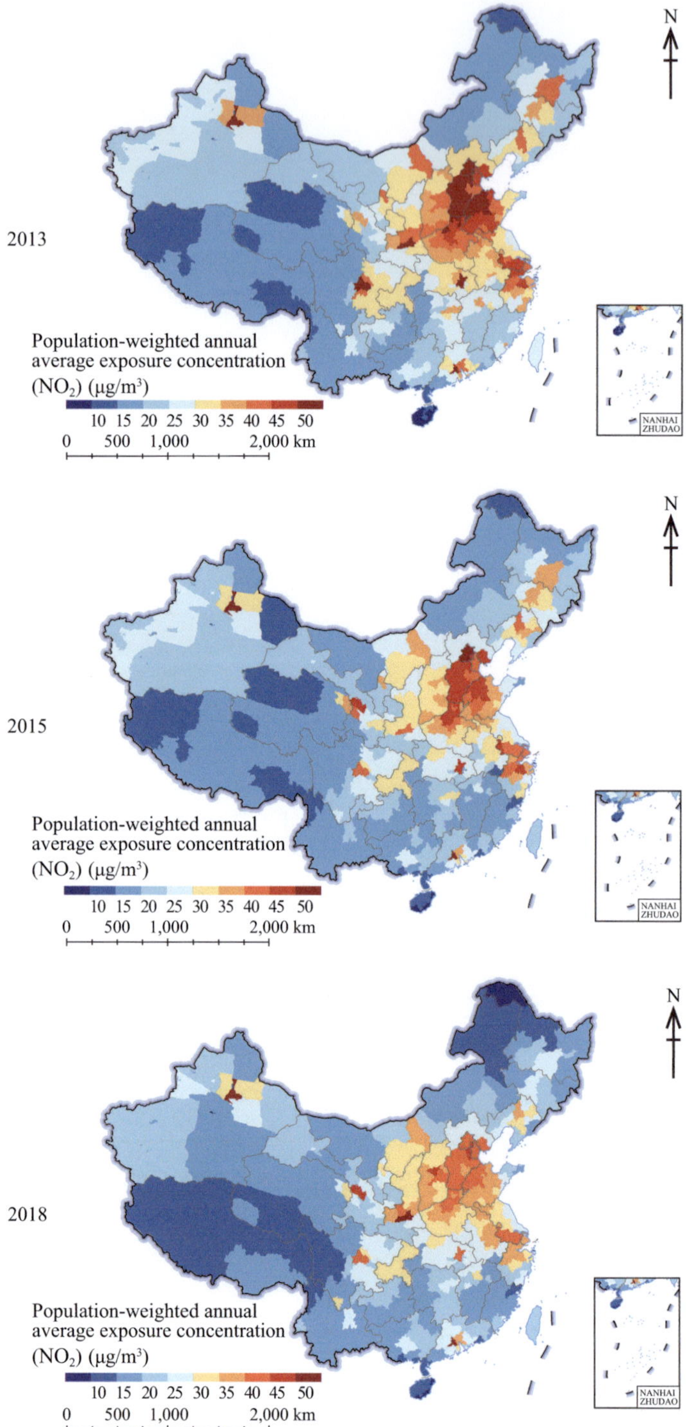

(f) NO_2

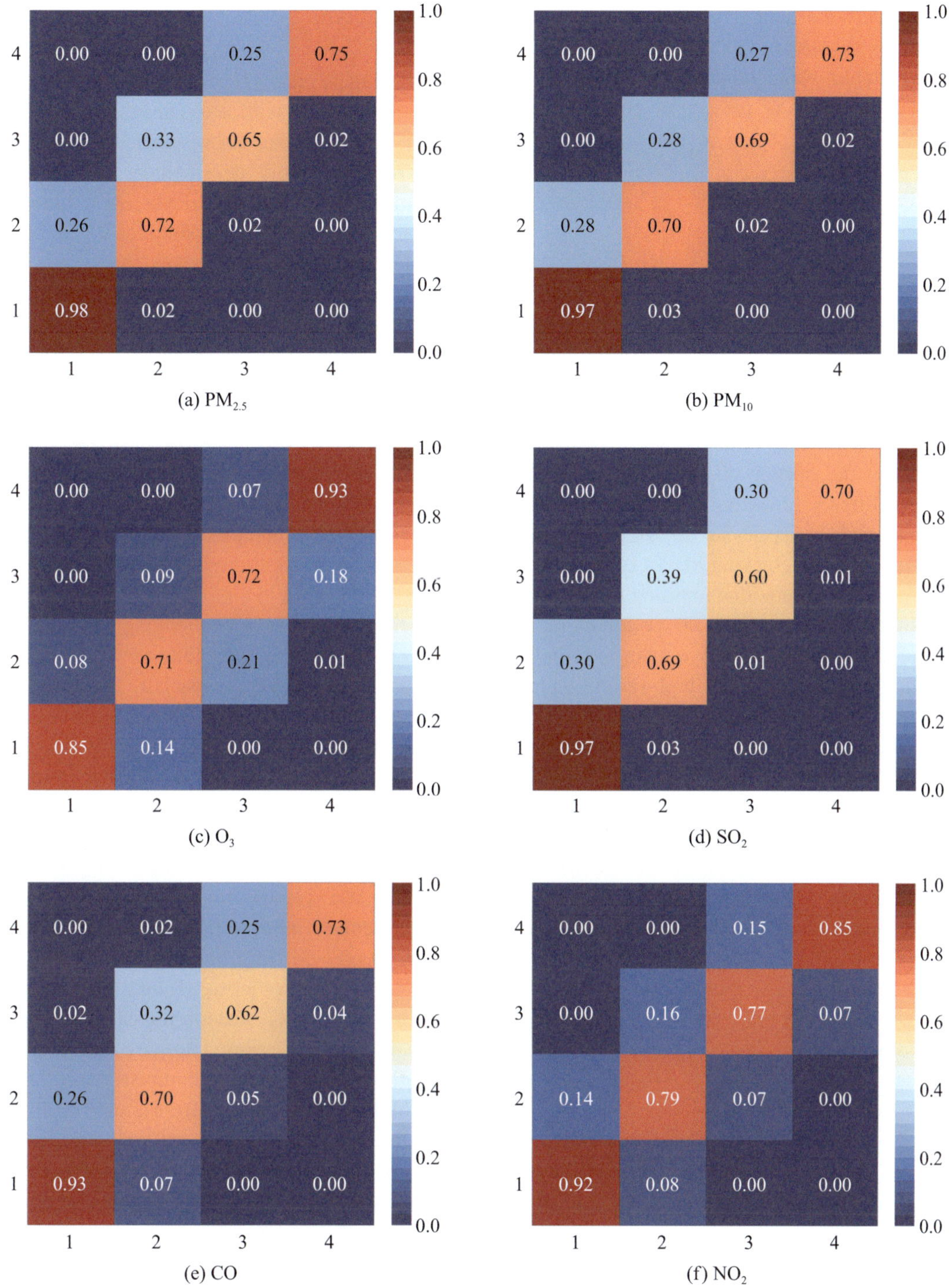

Fig. 5.7 Transition probability matrix of population exposure concentrations of six air pollutants in China from 2013 to 2020

– Compile the dataset of China's comprehensive air pollution index and the dataset of population-weighted annual average exposure concentrations of the six air pollutants from 2013 to 2020.

- **Discussion and Outlook**

This case study utilizes remotely sensed atmospheric pollutant concentration data and WorldPop population data to construct a comprehensive air pollution index incorporating six major air pollutants, along with the population-weighted annual average exposure concentrations of these pollutants. The dataset systematically evaluates the spatial distribution patterns and temporal trends of air pollution exposure across China, providing a basis for tracking the country's progress toward achieving SDG 11.6.

Due to the limitations imposed by the spatial resolution of the current air pollution dataset, future research will focus on downscaling the existing data by integrating high-resolution influencing factors. This approach aims to predict residuals at finer grid scales, thereby enhancing the resolution of the dataset by combining the original data with the predicted residuals, ultimately improving the accuracy of exposure assessments. Furthermore, given the complex interactions between air pollution and climate change, future studies should further explore the potential impacts of climate variability on the distribution of atmospheric pollutants and population exposure, enabling more refined and precise exposure assessments.

5.3.3 Long-Term High-Resolution $PM_{2.5}$ Retrieval and Trends in China

Target: SDG 11.6 By 2030, reduce the adverse per capita environmental impact of cities, including by paying special attention to air quality and municipal and other waste management.

- **Background**

Fine particulate matter ($PM_{2.5}$) refers to particles in the atmosphere with a dynamic diameter of less than 2.5 μm that are suspended and inhalable. With the rapid development of China's economy and urbanization, $PM_{2.5}$ has become a primary pollutant, which significantly affects human health and climate change. To improve air quality, our country has implemented multiple rounds of national-scale clean air action plans since 2013. Therefore, assessing the evolution of $PM_{2.5}$ spatial patterns and the effectiveness of pollution control in urban and non-urban areas is crucial for supporting the continued improvement of air quality in our country. However, the monitoring of $PM_{2.5}$ in China started relatively late, with a monitoring network primarily based on prefecture-level cities established only in 2013. Due to limited time and uneven space distribution, it is challenging to comprehensively understand the long-term evolution characteristics of $PM_{2.5}$, especially the changes in urban–rural differences. There is an urgent need to obtain long-term continuous high spatiotemporal resolution data to assess the long-term impacts of $PM_{2.5}$ on the atmospheric environment and human health, reveal the long-term evolution patterns of urban–rural differences in $PM_{2.5}$, and quantify the effectiveness of air pollution control after removing meteorological influences. By utilizing multi-source data such as satellite observations, ground monitoring visibility, and other meteorological variables, regional high-resolution $PM_{2.5}$ concentration estimation can effectively compensate for the shortcomings of site observations, and has high predictive capability. The method has been widely used, and several long-term datasets have been released (Bai et al. 2022; Wei et al. 2021). However, the accuracy of $PM_{2.5}$ concentration estimation still needs continuous improvement. This can be achieved by incorporating visibility and upper air meteorological data, which are strongly correlated with $PM_{2.5}$ as additional input datasets. Additionally, increasing the number of $PM_{2.5}$ observation stations for model training is essential. Furthermore, using more independent sites that have never been involved in modeling to perform a unified comparison of the accuracy across different datasets is crucial for clarifying the differences between them. This research can provide key data support for the progress assessment of SDG 11.6, as well as for evaluating clean air policies and studying health impacts.

- **Data**

1. **Remote Sensing Data**

   - land cover data of China with 30 m resolution from 2010 to 2021 (Yang and Huang 2022).
   - Monthly 1 km MODIS vegetation index product MOD13A3 from 2010 to 2022 (https://ladsweb.modaps.eosdis.nasa.gov/archive/allData/61/MOD13A3/).
   - Monthly 10 km resolution NO_2 tropospheric column concentration in China from 2013 to 2018, monitored by the (Ozone Monitoring Instrument, OMI) sensor of the Aura satellite (https://zenodo.org/record/7574177); monthly 1 km tropospheric column concentration of NO_2 in China from 2019 to 2020 (https://zenodo.org/record/6622056); monthly 0.125° tropospheric column concentration of NO_2 globally monitored by the OMI satellite for the years 2010 to 2012 and 2021 (https://www.temis.nl/airpollution/no2col/no2regioomimonth_qa.php).
   - The 1 km MODIS aerosol optical depth product for China from 2010 to 2021 (https://zenodo.org/record/5652257).
   - The 1 km resolution elevation data for China provided by the Resource and Environmental Science Data Platform, Chinese Academy of Sciences.

2. **Ground Observation Data**

   - $PM_{2.5}$ data from the China air quality monitoring network provided by the China National Environmental Monitoring Centre and various provincial environmental monitoring centers from 2013 to 2022.
   - $PM_{2.5}$ data from CERN network by the "Campaign on Atmospheric Aerosol Research Network of China" (CARE-China) and Beijing-Tianjin-Hebei Air Quality Monitoring Network provided by the Institute of Atmospheric Physics, Chinese Academy of Sciences from 2011 to 2014.
   - Daily visibility data from 2100 national meteorological stations in China from 2010 to 2022, provided by the Meteorological Information Comprehensive Analysis and Process System (MICAPS) of the China Meteorological Administration.

3. **Other Data**

   - 1 km LandScan population data from 2010 to 2022.
   - Surface and upper air meteorological data of the 0.1° the fifth generation of ECMWF (European Centre for Medium-Range Weather Forecasts) atmospheric reanalyses of the global climate (ECMWF ERA5) dataset in China from 2010 to 2022.

- **Methods**

In-situ $PM_{2.5}$ ground measurements in China were collected from the Ministry of Ecology and Environment of China (https://air.cnemc.cn:18007/) and the provincial environmental monitoring centers during 2013–2022. The effective number of the Minitry of Ecology and Environment sites is about 802 in 2013 and 1,296 during 2014–2022, and the average number of provincial environmental monitoring sites is about 3,110 during 2019–2021. Input data include surface visibility, satellite aerosol optical depth and NO_2 column concentrations, surface and the upper-air meteorology from

ERA5 reanalysis data (2 m temperature, relative humidity, boundary layer height, evaporation, 10 mu component of wind, 10 mv component of wind, surface pressure and the total precipitation; geopotential height, relative humidity, temperature, u component of wind, and v component of wind at 850 and 500 hPa), as well as geographic and social factor data including NDVI, land cover, elevation, and population, etc. (Bai et al. 2022; Yang and Huang 2021; Wang et al. 2023; Wei et al. 2023). The above $PM_{2.5}$ and all input factors were interpolated into 0.1° gridded data. Using RF and Light Gradient Boosting Machine (LightGBM) algorithms, daily-scale $PM_{2.5}$ retrieval models for the years 2013–2022 were established. The RF algorithm, which demonstrated the best performance based on sample and site tenfold cross-validation methods, was selected to construct the $PM_{2.5}$ dataset during 2013–2022. The historical $PM_{2.5}$ observation data for the years 2010–2012 are very limited. Using the RF algorithm, a daily-scale $PM_{2.5}$ retrieval model for China from 2013 to 2015 was established, which predicted the daily $PM_{2.5}$ concentration values for China from 2010 to 2012. A dataset of daily $PM_{2.5}$ values at the 0.1° grid in China from 2010 to 2022 was produced. The 10-fold cross-validation results of the dataset each year show R^2 (coefficient of determination) of 0.80 to 0.90, and $PM_{2.5}$ validation by external datasets exhibited R^2 ranging from 0.49 to 0.74 in different years and regions. Overall, the accuracy of this dataset is slightly better than those of other public datasets. Based on this dataset, the variation trend of $PM_{2.5}$ concentration weighted by population was evaluated, and the evolution of urban–rural differences in $PM_{2.5}$ and the effectiveness of air pollution control in key areas were analyzed. For detailed information on the construction methods and analysis results of the aforementioned dataset, please refer to the study by Liu et al. (2024).

- **Results and Analysis**

The spatial distribution of $PM_{2.5}$ concentrations in China from 2010 to 2022 is shown in Fig. 5.8, which exhibited a decreasing trend with the rate of 1.8 μg/(m^3 a). The key urban clusters, including the Beijing-Tianjin Hebei and surrounding areas, Yangtze River Delta, Fen-Wei Plain, Chengdu-Chongqing, and Pearl River Delta, experienced more prominent declining trends in $PM_{2.5}$ concentrations. This reflects the significant achievements of China's comprehensive actions in air pollution control and prevention. However, in the northwestern regions of China, such as Xinjiang, where the population density is low and natural geographical conditions such as vast deserts and Gobi exert a negative impact, $PM_{2.5}$ concentrations remain fairly high.

From 2010 to 2022, the exposure risk of Chinese residents to $PM_{2.5}$ gradually decreased, particularly in densely populated urban clusters where the risk had significantly decreased. The changes in annual average $PM_{2.5}$ concentrations, weighted by population, are noticeable, with a national decline rate of 2.6 μg/(m^3 a), representing a reduction of 41%. In the Beijing-Tianjin-Hebei region, the decline rate and reduction reached 4.8 μg/(m^3 a) and 54%, respectively. The gap in population-weighted $PM_{2.5}$ concentrations between the Beijing-Tianjin-Hebei region and the entire country has been gradually narrowing; additionally, by calculating the proportion of the population exposed to high $PM_{2.5}$ concentrations (annual average > 35 μg/m^3), the results show that by 2022, the national proportion decreased to 44.2%, while the proportion in the Beijing-Tianjin-Hebei region decreased to 66.5% (Fig. 5.9). These findings reflect the substantial reduction in $PM_{2.5}$ pollution levels in China in recent years, achieved through comprehensive air pollution control actions, leading to significant improvements in the health of Chinese residents and benefiting a wider population.

To further reveal the long-term evolution patterns of $PM_{2.5}$ in six key urban agglomerations in China with heavier pollution (Beijing-Tianjin-Hebei, Shanghai-Jiangsu-Zhejiang, Guangdong, Chongqing-Sichuan, Shandong-Henan, Shanxi-Shaanxi, as shown in Fig. 5.10), this study analyzed the evolution of urban–rural differences in $PM_{2.5}$ across different regions. At the same time, considering the combined effects

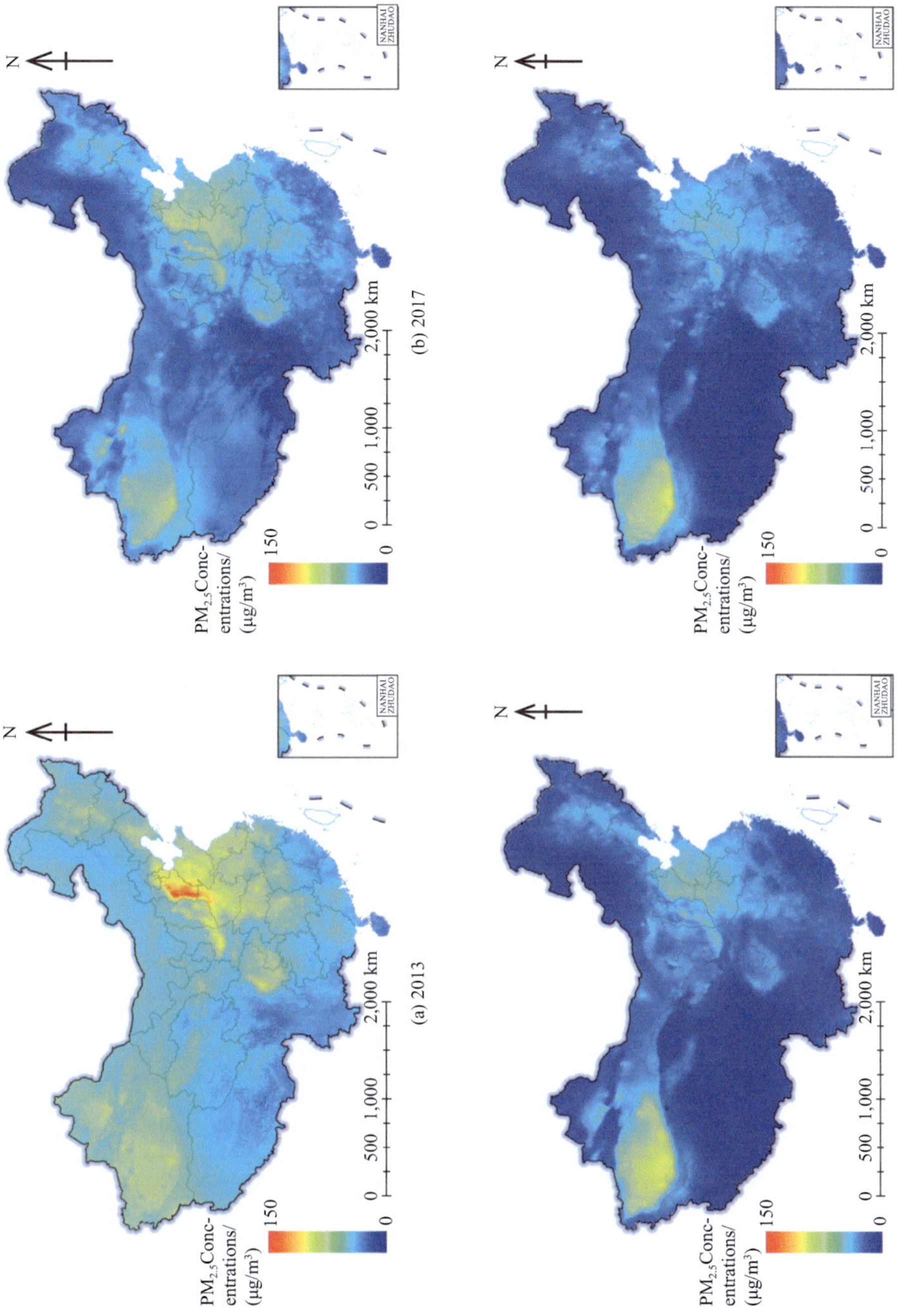

Fig. 5.8 Spatial distribution of annual average $PM_{2.5}$ concentrations in China during 2010–2022

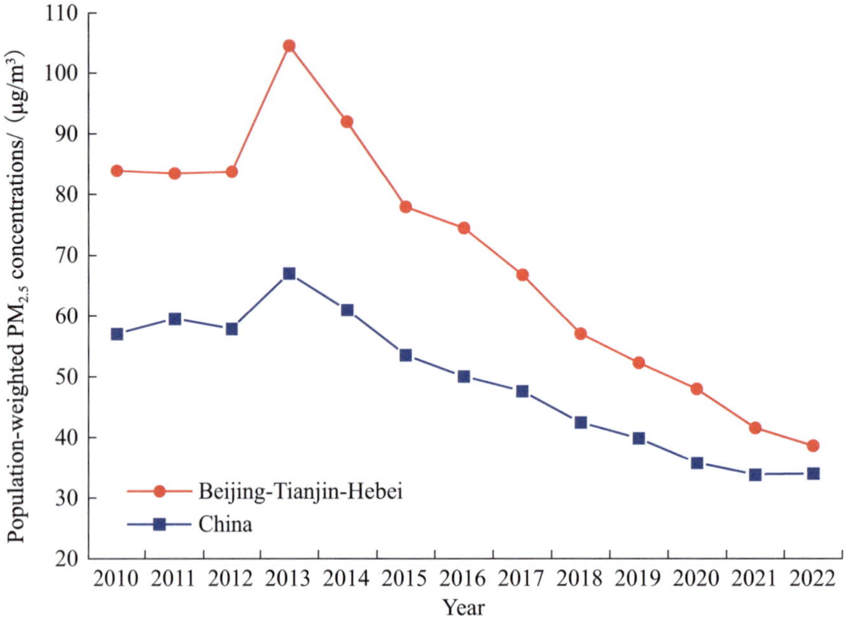

(a) Annual average population-weighted $PM_{2.5}$ concentrations in China and the Beijing-Tianjin-Hebei region

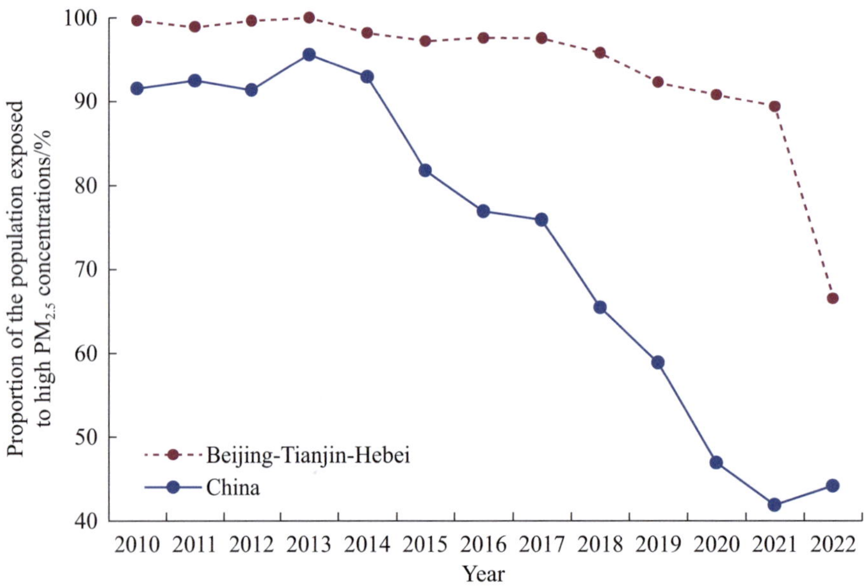

(b) The trend of the proportion of the population exposed to high $PM_{2.5}$ concentrations in China and the Beijing-Tianjin-Hebei region

Fig. 5.9 The exposure risk of Chinese residents to $PM_{2.5}$ during 2010–2022

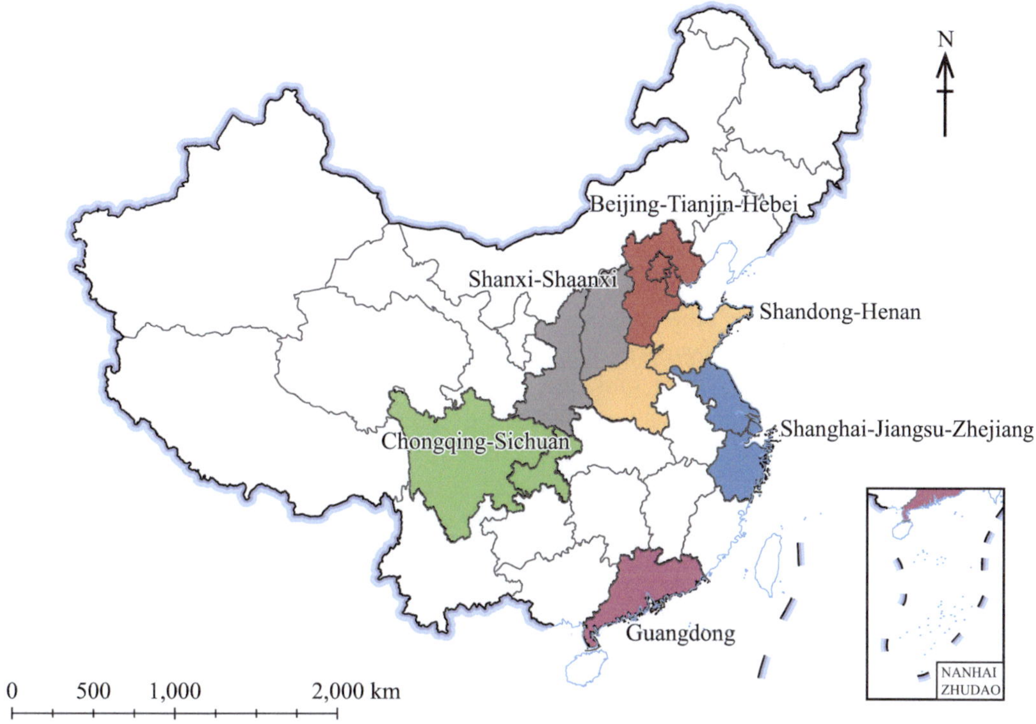

Fig. 5.10 Spatial locations of the six key urban agglomerations in China

Table 5.4 The decline rates of regional averages of $PM_{2.5}$ concentrations, urban–rural differences, and meteorological normalization concentrations in the six key urban agglomerations of China from 2010 to 2022

Key regions	Regional averages	Urban–rural differences	Meteorological normalization concentrations (urban; non-urban)
Beijing-Tianjin-Hebei	2.8	1.6	3.5;2.0
Shanghai-Jiangsu-Zhejiang	2.7	−0.2	1.9;2.0
Guangdong	1.8	0.3	1.5;1.3
Chongqing-Sichuan	2.3	1.4	2.5;1.7
Shandong-Henan	3.0	0.9	2.1;2.1
Shanxi-Shaanxi	1.9	0.6	1.7;1.3

Unit: $\mu g/(m^3\ a)$

of meteorological factors and emission sources on $PM_{2.5}$ concentrations, a deweathered $PM_{2.5}$ model was constructed based on daily data and the RF algorithm. This model provided the meteorological normalization concentrations of $PM_{2.5}$, and the changes in concentration reflect the effectiveness of human control measures on $PM_{2.5}$ (Liu et al. 2024). As shown in Table 5.4, from 2010 to 2022, the Shandong-Henan region had the fastest annual average $PM_{2.5}$ concentration decline rate among the six key urban agglomerations. With the exception of the Shanghai-Jiangsu-Zhejiang region, the urban–rural (urban-non-urban) $PM_{2.5}$ differences in each key region decreased year by year, with the Beijing-Tianjin-Hebei region experiencing the largest reduction in urban–rural differences. Combined with Fig. 5.8, it can be observed that during the Clean Air Action period (2013–2022), especially in Phase I (2013–2017), the spatial extent and pollution intensity of $PM_{2.5}$ in China significantly decreased and weakened, respectively. However, in Phase III (2021–2022), the decreasing percentage of $PM_{2.5}$ concentrations in each key region, except for Beijing-Tianjin-Hebei and Shandong-Henan, was below 1% (Table 5.5). As shown in Table 5.4, from 2010 to 2022, the $PM_{2.5}$ meteorological normalization concentrations in the six key urban agglomerations showed a significant downward trend. Human control measures played an important role in reducing $PM_{2.5}$ concentrations in each region, with the effectiveness of human control being particularly prominent in the urban areas of Beijing-Tianjin-Hebei. Overall, the contribution of control measures to the reduction of annual $PM_{2.5}$ concentrations in the six key urban agglomerations ranged from 67 to 88%, with the remainder attributed to meteorological conditions. These results indicate that the Clean Air Action in China from 2013 to 2022 significantly improved air quality; however, the challenges of $PM_{2.5}$ management have also increased. In the future, scientific governance of $PM_{2.5}$ pollution needs to place greater emphasis on controlling non-urban areas and understanding the interactions between weather/climate and air pollution.

Highlights

- Based on multi-source datasets and machine learning methods, a 0.1° daily-scale $PM_{2.5}$ gridded data product for China from 2010 to 2022 was generated.
- From 2010 to 2022, the overall $PM_{2.5}$ concentrations in the country decreased, with significant reductions in key urban agglomerations such as Beijing-Tianjin-Hebei and surrounding areas, Yangtze River Delta, Fen-Wei Plain, Chengdu-Chongqing, and Pearl River Delta. The risk of $PM_{2.5}$ exposure for residents in China has gradually decreased, and the comprehensive air pollution prevention and control actions in China have achieved significant results. The contribution of control measures to the reduction of annual $PM_{2.5}$ concentrations in the six key urban agglomerations ranged from 67 to 88%.

- **Discussion and Outlook**

Since the implementation of the Clean Air Action in 2013, China's air quality has substantially changed. Assessing the spatial patterns of $PM_{2.5}$ and the effectiveness of human control measures in urban and non-urban areas is crucial. This study employed an RF model, integrating multi-source data, including as many $PM_{2.5}$ in-situ observation data as possible, visibility, satellite retrieval data, ground and upper-air meteorological data, and other auxiliary data, to generate a high spatio-temporal resolution long-term $PM_{2.5}$ dataset for China from 2010 to 2022, which is more reliable and accurate than other public datasets. The analysis of this dataset shows that the $PM_{2.5}$ concentrations in most urban agglomerations in eastern China is decreasing at a higher rate, with reduced urban–rural differences. By using meteorological normalization methods to separate the contributions of meteorological factors and emissions to the decline of annual $PM_{2.5}$, it reveals that pollution control measures are the main driving force for the reduction of $PM_{2.5}$. This demonstrates the significant achievements made in air pollution prevention and control in China, leading to a gradual decrease in the risk of $PM_{2.5}$ exposure of residents. Based on this research, we recommend that, in addition to focusing on areas with existing monitoring networks and complete data, greater attention should be paid to regions where data are inaccurate or incomplete, such as desert areas like the Gobi Desert. Furthermore, it is essential to analyze the causes and sources of $PM_{2.5}$, clarify the interactions between weather/climate and air pollution, and strengthen control measures in non-urban areas to continuously promote deep reductions in $PM_{2.5}$. In response to the increasingly prominent issue of ozone, similar methods can be applied in the future to the construction of ozone datasets, health impact

Table 5.5 Decreasing percentages of $PM_{2.5}$ meteorological normalization concentrations for urban and non-urban areas in the six key urban agglomerations during the three phases of the clean air action from 2013 to 2022

Key regions	Urban				Non-urban			
	2013–2017	2018–2020	2021–2022	2013–2022	2013–2017	2018–2020	2021–2022	2013–2022
Beijing-Tianjin-Hebei	21.6	11.7	1.1	41.7	13.9	11.3	1.5	33.2
Shanghai-Jiangsu-Zhejiang	17.8	8.7	0.4	30.6	15.8	10.5	0.4	32.1
Guangdong	23.7	7.5	0.9	34.5	20.1	8.4	0.3	32.5
Chongqing-Sichuan	25.1	6.1	0.2	34.7	22.0	10.2	−0.7	38.2
Shandong-Henan	15.6	5.6	1.0	26.9	16.5	7.5	1.0	29.5
Shanxi-Shaanxi	9.3	10.3	0.9	26.2	8.2	12.1	0.4	27.5

Unit: %

assessments, cause diagnosis, and control effectiveness evaluations, to scientifically and accurately support the comprehensive improvement of air quality in China.

5.3.4 Extraction of Urban Public Open Space Information in China's Five Major Urban Agglomerations

Target: SDG 11.7 By 2030, provide universal access to safe, inclusive, accessible, and green public spaces, in particular for women and children, older persons, and persons with disabilities.

- **Background**

Urban public open space refers to the public external spaces in a city. These spaces are the most vivid and active part of the urban space, and have rich urban landscape elements, such as park green areas, rivers, natural landscapes, and artificial landscapes. The value of urban public open space lies in its ability to improve the ecological environment and provide a place for people to relax both physically and mentally, as well as beautify residential neighborhoods, and enhance the sense of belonging among residents and the integration and cohesion of the community. The monitoring and evaluation of urban public open space, will help promote the transformation and quality improvement of urban space, and promote the sustainable development of cities. In the United Nations' SDGs, public open space is the key to achieving several goals, such as SDG 3, SDG 5, SDG 11, SDG 8, and SDG 13 (Guo 2020; Sims et al. 2019). In the new era, the planning and renewal of urban public open space provides a viable pathway for the transformation and quality improvement of urban space. In the face of the uniqueness and complexity of urban public space, reasonable and effective compilation ideas for public space renewal planning are constantly being explored (United Nations 2015).

Current research on urban public open space relies heavily on existing global land use data products. However, these studies usually only cover a limited number of open space types, such as parks, green areas, squares, and roads, and fail to fully reflect the diversity of urban public open space (Shahtahmassebi et al. 2021; Ozdil et al. 2020). In addition, the spatial granularity of these data products is relatively low, which limits the extraction and analysis of various types of urban open space and their high-precision information. To address this issue, the case team analyzed remote sensing images with a spatial resolution of 1 m for China's five major urban agglomerations, and defined the types of urban public open space as public spaces, water bodies, sports fields, green areas, and roads. Based on the previous research results, the types of urban public open spaces such as public spaces, water bodies, sports fields, and green areas in some cities were manually annotated to form a benchmark tile dataset that can be used for deep learning (DL) model training; at the same time, the model was applied to remote sensing images with a spatial resolution of 1 m for the urban areas of China's five major urban agglomerations, and a method system for extracting urban public open space information from high-resolution remote sensing images was constructed to provide technical support for subsequent urban development decisions.

- **Data**

- The historical remote sensing image data of China's five major urban agglomerations in 2017 and 2022 were provided by World Imagery Wayback with a spatial resolution of 1 m.
- The global 30 m spatial resolution surface cover classification data in 2020 were provided by the Aerospace Information Research Institute, Chinese Academy of Sciences.
- The urban road network data of China in 2017 and 2022 were provided by OpenStreetMap (OSM).

– LandScan global population distribution data for 2017 and 2022 were provided by East View Cartographic with a spatial resolution of 1 km.

• **Methods**

The study selected China's five major urban agglomerations for research, namely the Yangtze River Delta Urban Agglomeration (27 cities, including Shanghai, Nanjing, Wuxi, Changzhou, Suzhou, Nantong, Yangzhou, Zhenjiang, Yancheng, Taizhou, Hangzhou, Ningbo, Wenzhou, Huzhou, Jiaxing, Shaoxing, Jinhua, Zhoushan, Taizhou, Hefei, Wuhu, Ma'anshan, Tongling, Anqing, Chuzhou, Chizhou, Xuancheng), the Beijing-Tianjin-Hebei Urban Agglomeration (13 cities, including Beijing, Tianjin, Shijiazhuang, Baoding, Tangshan, Handan, Xingtai, Hengshui, Cangzhou, Qinhuangdao, Langfang, Zhangjiakou, Chengde), the Guangdong-Hong Kong-Macao Greater Bay Area Urban Agglomeration (11 cities, including Hong Kong, Macao, Guangzhou, Shenzhen, Zhuhai, Foshan, Huizhou, Dongguan, Zhongshan, Jiangmen, Zhaoqing), the Middle Reaches of the Yangtze River Urban Agglomeration (31 cities, including Wuhan, Huangshi, Ezhou, Huanggang, Xiaogan, Xianning, Xiantao, Qianjiang, Tianmen, Xiangyang, Yichang, Jingzhou, Jingmen, Changsha, Zhuzhou, Xiangtan, Yueyang, Yiyang, Changde, Hengyang, Loudi, Nanchang, Jiujiang, Jingdezhen, Yingtan, Xinyu, Yichun, Pingxiang, Shangrao, Fuzhou, Ji'an), the Chengdu-Chongqing Urban Agglomeration (16 cities, including Chongqing, Chengdu, Zigong, Luzhou, Deyang, Mianyang, Suining, Neijiang, Leshan, Nanchong, Meishan, Yibin, Guang'an, Dazhou, Ya'an, Ziyang).

This method defines the extraction of urban public open space as a semantic segmentation task. The types of urban public open space are divided into five categories: public space, water body, sports field, green area, and road (Table 5.6). First, some urban areas are selected for manual sample annotation, and the annotated dataset is divided into 128-pixel × 128-pixel tiles. Then, after random cropping, random flipping, color enhancement, and other data enhancement operations, the data are expanded 10 times, and a benchmark dataset of urban public open spaces containing 231,000 tiles is constructed.

This task uses the MANet model as the open space information extraction network. As shown in Fig. 5.11, the network integrates the local feature maps extracted by ResNet-50 with their corresponding global dependencies, adaptively reweights the interdependent channel maps, and extracts contextual dependencies through multiple efficient attention modules, solving the problems of insufficient utilization of multi-scale feature information and insufficient long-range dependencies of feature maps. The ResBlock extracts features by stacking multiple convolutional layers to extract local features. The Attention performs residual connections through spatial attention and channel attention to enhance the discriminative ability of features extracted at each layer and achieve more effective feature representation. The DeBlock is an upsampling module in the decoder, which performs corresponding upsampling on the fused features so that detailed information and context information can be further fused at different levels. The cross-entropy loss function plays an important role in the semantic segmentation task. It improves the model performance by promoting the model output probability to be close to the true label distribution, so the cross-entropy loss function is selected as the loss function. The Adam optimizer helps the model converge to a better solution faster in the semantic segmentation task through its adaptability, efficiency, and robustness while reducing resource consumption during training. These characteristics make Adam an ideal choice for handling such tasks, so Adam is used as the optimizer for this task. The urban public open space benchmark dataset is divided into a training set and a test set in a ratio of 8:2. The model is trained 20 times using the training set, and the accuracy of the model is evaluated using the test set. Finally, the model with the best performance on the test set is selected for urban open space information extraction. MIoU is an evaluation indicator used to measure the overall performance of the

Table 5.6 Types of urban public open spaces

Types	Contents
Public space	Artificially constructed surfaces within the urban areas, including squares, parking lots, etc.
Water body	Areas covered by liquid water within the urban areas, including lakes, rivers, ponds, etc.
Sports field	Sports venues within a city, including basketball courts, volleyball courts, open-air gymnasiums, etc.
Green area	Vegetation-covered areas within the urban areas, including woodlands, grasslands, shrublands, etc.
Road	Roads within the urban areas, including cement roads, asphalt roads, etc.

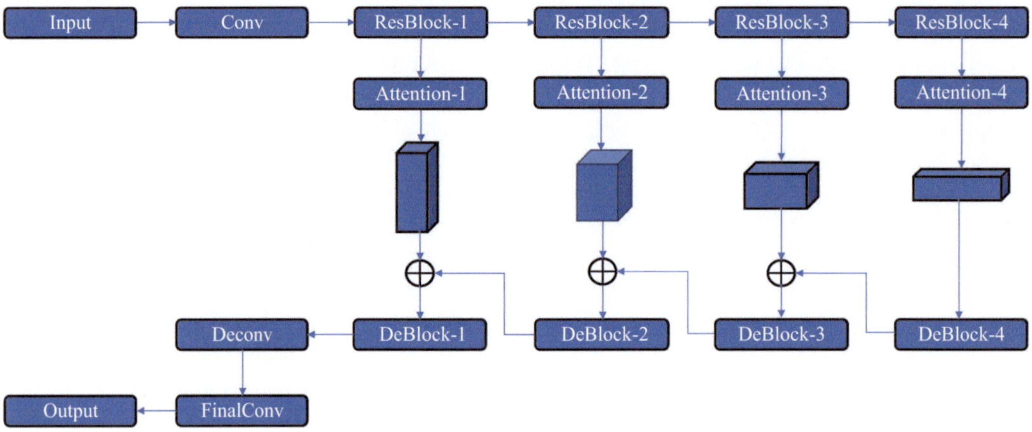

Fig. 5.11 MANet model architecture

semantic segmentation model. It reflects the segmentation accuracy of the model for different categories by calculating the weighted average of the IoU of each category. Experiments have shown that the highest MIoU reached 91.32%.

The case defines the urban area of a city according to the density of the urban built-up area. After obtaining the remote sensing image of the urban area, the trained model is used to predict the image. Based on the preliminary urban public open space results obtained using the MANet model, the water body part in the global land cover classification data is used to correct the water body extraction results, and then the urban road network vector data are used to extract road data at all levels. According to the road construction width specification, the road line data are converted into surface data, and then the urban public open space ratio is calculated and then combined with the LandScan population data, the urban per capita public open space ratio is obtained. The specific process includes the following: (1) resampling the global 30 m land cover classification data to 1 m, and correcting the water body part of the urban public open space extracted by the semantic segmentation network; (2) according to the national road construction width specifications at all levels, converting highways, provincial roads, county roads, county town roads, and other urban roads into surface data to generate grid road space data; (3) at the grid scale, superimposing the road data with the urban public open space results after water body correction to obtain the final urban public open space information; (4) dividing the area occupied by the urban public open space in the final result by the total area of the city to obtain the urban public open space ratio; (5) finally, calculating the urban public open space ratio within each 1 km $\times$ 1 km grid, and using the 1 km LandScan population data to perform a weighted average to obtain the urban per capita public open space ratio.

- **Results and Analysis**

The urban public open space extraction results of China's five major urban agglomerations in 2017 and 2022 are shown in Fig. 5.12. The urban public open space ratio and the urban per capita public open space ratio of China's five major urban agglomerations in 2017 and 2022 are shown in Table 5.7, and the proportions of different types of open space are shown in Table 5.8. From the results of Fig. 5.12, Tables 5.7, and 5.8, it can be seen that the proportions of urban public open space areas in the southern urban agglomerations of China are higher than those in the northern urban agglomerations. The open public spaces in China's five major urban agglomerations are mainly composed of water bodies, green areas, and roads, with the green areas having the largest share. In both years, the proportions of green areas exceeded 20%. The proportions of open space types in different urban agglomerations show distinct characteristics. In the Yangtze River Delta Urban Agglomeration, water bodies, green areas, and roads dominate urban public open space. In the Beijing-Tianjin-Hebei Urban Agglomeration, green areas and roads dominate urban public open space. In the Chengdu-Chongqing Urban Agglomeration, green areas dominate urban public open space. In the Middle Reaches of the Yangtze River Urban Agglomeration and the Guangdong-Hong Kong-Macao Greater Bay Area, water bodies and green areas dominate the urban open space.

In 2022, the average proportion of urban open public space in China's five major urban agglomerations was 45.38% with an increase of about 3.32 percentage points compared to 2017 (42.06%). Compared to 2017, the proportions of public spaces, sports fields, green areas, and roads all showed an upward trend in 2022. The increase in the proportion of roads was the largest, rising by about 2.88 percentage points, while the proportion of water bodies showed a downward trend, decreasing by about 1.6 percentage points. From 2017 to 2022, the proportion of urban public open space in the Yangtze River Delta Urban Agglomeration, Guangdong-Hong Kong-Macao Greater Bay Area, Middle Reaches of the Yangtze River Urban Agglomeration, and Beijing-Tianjin-Hebei Urban Agglomeration all showed an increasing trend, reflecting that China's cities, in the rapid process of urbanization, have also paid attention to the planning and improvement of public space (Fig. 5.13). However, the proportion of urban public open space in the Chengdu-Chongqing Urban Agglomeration showed a slight decline, as this region is located in the middle hill in Sichuan Basin and ridge-valley of East Sichuan, leading to a high proportion of green areas in urban public open space types. In both 2017 and 2022, the proportion of green areas in this region reached over 30%. In these cities in the Chengdu-Chongqing Urban Agglomeration, the reduction in green areas during urbanization was greater than the increase in other types of urban public open space, leading to a slight decline in the overall proportion of urban public open space.

Compared to 2017, the proportion of the per capita public open space in China's five major urban agglomerations increased by approximately 3.67% in 2022. The proportion of the per capita public open space increased in the Yangtze River Delta Urban Agglomeration, Guangdong-Hong Kong-Macao Greater Bay Area, Middle Reaches of the Yangtze River Urban Agglomeration, and Chengdu-Chongqing Urban Agglomeration, with the Middle Reaches of the Yangtze River Urban Agglomeration showing the highest increase, reaching 5.99 percentage points (Fig. 5.14). This indicates that the Middle Reaches of the Yangtze River region has placed more emphasis on the construction of public open space in urban planning and development, providing residents with a better living environment and urban experience. In contrast, the proportion of the per capita public open space in the Beijing-Tianjin-Hebei Urban Agglomeration showed a slight decrease. This is because the Beijing-Tianjin-Hebei Urban Agglomeration is one of the most densely populated areas

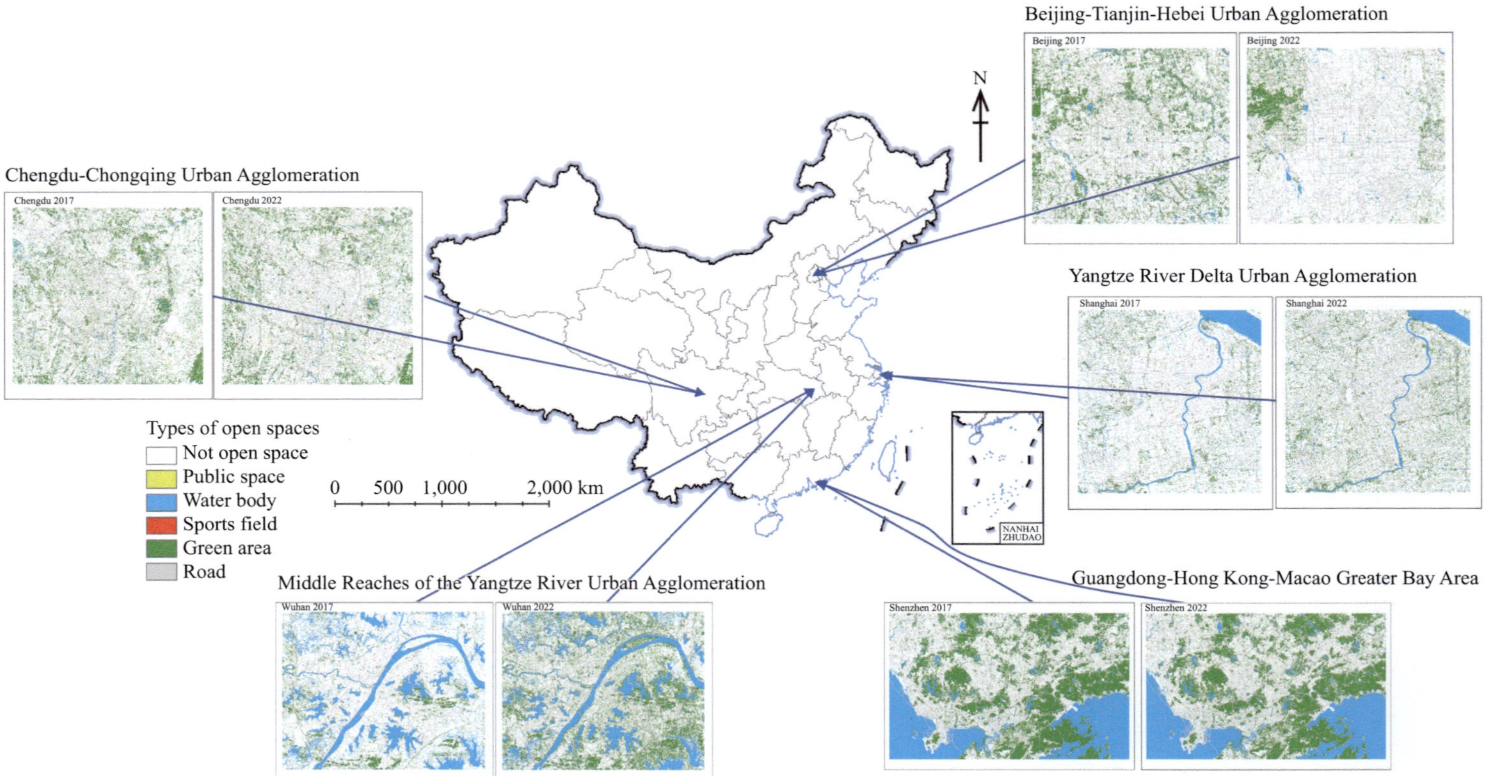

Fig. 5.12 The distribution of urban public open spaces in China's five major urban agglomerations in 2017 and 2022. *Note* Due to limited space, only one key city was selected to present the open spaces of each major urban agglomeration

Table 5.7 Extraction of public open spaces in China's five major urban agglomerations

Area	Year	Proportion of urban public open space/%	Proportion of urban per capita public open space/%
Yangtze river delta urban agglomeration	2017	35.00	27.77
	2022	40.86	33.40
Beijing-Tianjin-Hebei urban agglomeration	2017	28.82	26.82
	2022	28.88	26.56
Guangdong-Hong Kong-Macao greater bay area	2017	56.80	39.28
	2022	58.59	42.61
Middle reaches of the Yangtze river urban agglomeration	2017	40.51	30.63
	2022	46.55	36.62
Chengdu-Chongqing urban agglomeration	2017	51.04	35.95
	2022	50.96	37.44
China's five major urban agglomerations	2017	42.06	31.75
	2022	45.38	35.42

Table 5.8 Extraction of public open spaces in cities of various categories within China's five major urban agglomerations

Area	Year	Proportion of public spaces	Proportion of water bodies	Proportion of sports fields	Proportion of green areas	Proportion of roads	Proportion of urban public open space
Yangtze river delta urban agglomeration	2017	0.23	12.91	0.31	12.03	9.52	35.00
	2022	0.32	12.45	0.38	15.44	12.26	40.86
Beijing-Tianjin-Hebei urban agglomeration	2017	0.42	6.41	0.36	12.90	8.73	28.82
	2022	0.45	4.91	0.39	11.66	11.47	28.88
Guangdong-Hong Kong-Macao greater bay area	2017	0.20	20.23	0.21	28.13	8.02	56.80
	2022	0.26	16.77	0.27	31.22	10.06	58.59
Middle reaches of the Yangtze river urban agglomeration	2017	0.20	17.43	0.23	14.73	7.93	40.51
	2022	0.23	13.21	0.30	22.15	10.66	46.55
Chengdu-Chongqing urban agglomeration	2017	0.22	4.87	0.19	39.50	6.25	51.04
	2022	0.34	6.45	0.22	32.97	10.98	50.96
China's five major urban agglomerations	2017	0.24	13.13	0.26	20.09	8.33	42.06
	2022	0.31	11.53	0.32	22.00	11.21	45.38

Unit: %

in China, and the continuous growth of the urban population may have led to a decrease in the proportion of the per capita public open space. As the population increases, the demand for urban public open space also rises, but the growth of such space may lag behind the population growth, causing the proportion of the per capita public open space to decline.

Highlights

- The extraction and evaluation of urban public open space for China's five major urban agglomerations in 2017 and 2022 with a spatial resolution of 1 m were carried out.
- In this case, five types of urban public open space were defined: public space, water body, sports field, green area, and road. Urban public open space types (public spaces, water bodies,

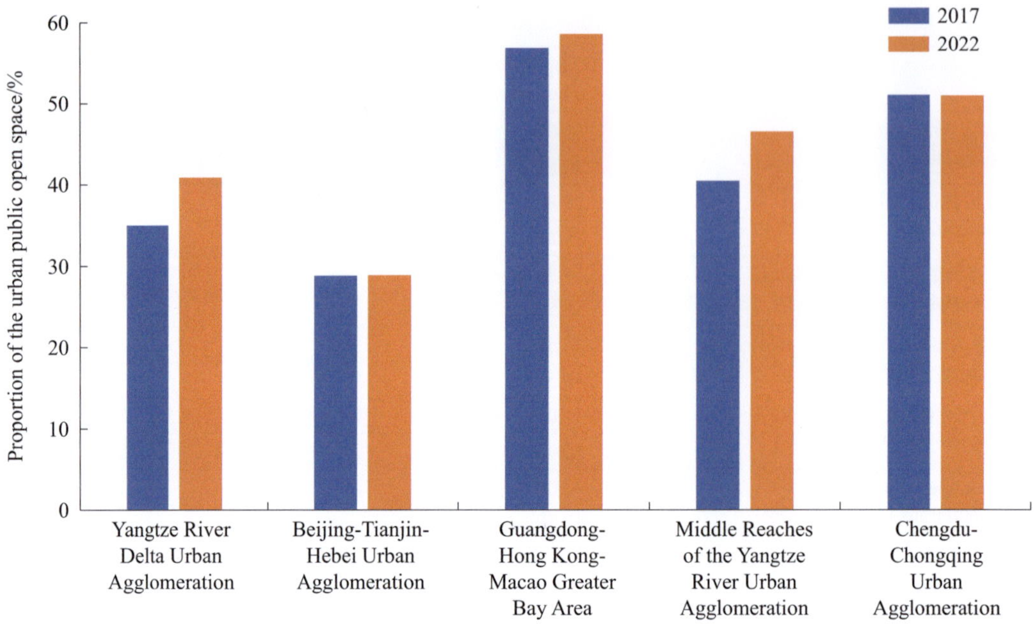

Fig. 5.13 Proportions of urban public open space in China's five major urban agglomerations in 2017 and 2022

sports fields, and green areas) were manually labeled as training samples. Using the deep learning semantic segmentation model and OSM road network data, high-precision open space information extraction was achieved.

– From 2017 to 2022, the proportions of urban public open space and the per capita public open space in China's five major urban agglomerations both increased. Among these, all urban agglomerations except the Chengdu-Chongqing Urban Agglomeration saw an increase in the proportion of urban public open space. The proportion of the per capita public open space increased in all urban agglomerations except the Beijing-Tianjin-Hebei Urban Agglomeration, with the Middle Reaches of the Yangtze River Urban Agglomeration showing the largest increase, rising by 5.99 percentage points.

- **Discussion and Outlook**

Based on the definition of urban public open space in SDG 11.7, this study compares and evaluates the changes in the area of urban public open space in China's five major urban agglomerations between 2017 and 2022 using OSM road network data and high-resolution remote sensing imagery. The method used in this case is simple and easy to implement, and the data are also relatively easy to obtain.

The types of urban public open space covered in the current study are relatively limited, only including five types: public space, water body, sports field, green area, and road. Future research could consider including more types of open space to provide a more comprehensive reflection of the public open space situation in cities. Additionally, the current case mainly focuses on the area changes of urban public open space. Future studies could consider introducing evaluation indicators for the multifunctionality of open spaces, such as ecological value, cultural and historical value, and community service functions. This would help to better understand the comprehensive benefits of urban public open space, and provide valuable insights for more detailed urban planning and management.

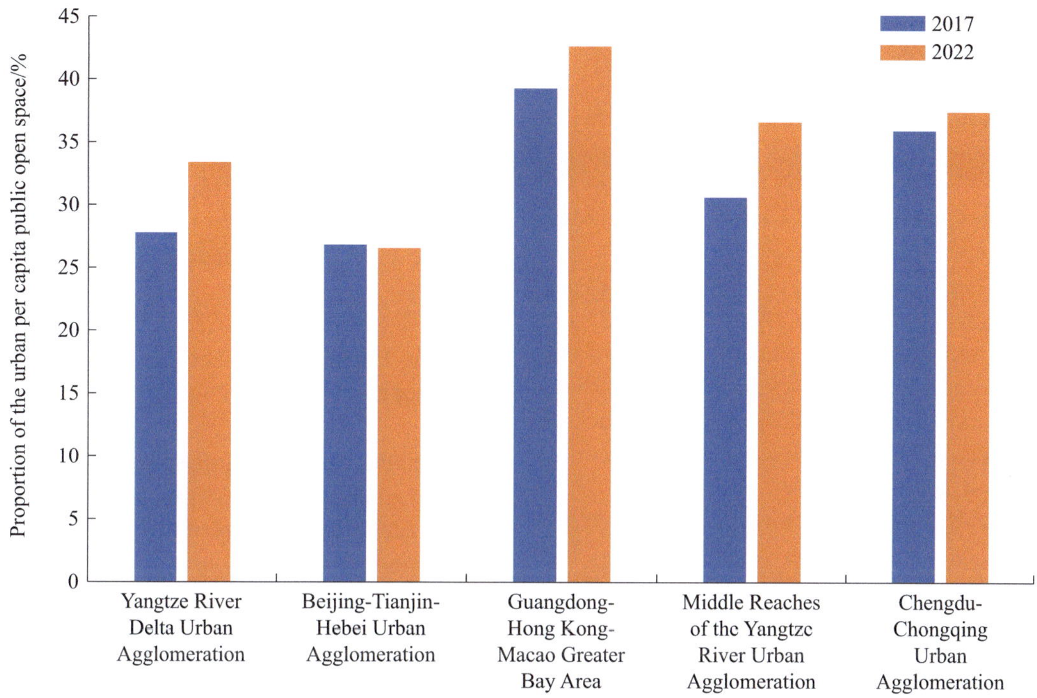

Fig. 5.14 Proportions of urban per capita public open space in china's five major urban agglomerations in 2017 and 2022

5.3.5 Metrics of Urban–Rural Economic Development Disparities in China

Target: SDG 11.a Support positive economic, social, and environmental links between urban, peri-urban, and rural areas by strengthening national and regional development planning.

- **Background**

Promoting urban–rural integration and regional coordinated development is important for China to achieve high-quality development. The urban–rural difference in economic development is one of the important factors affecting urban–rural integration development (Zeng et al. 2018). Effectively measuring the differences between urban and rural economic development can provide important support for China's new urbanization construction and sustainable development strategy. In this field, there are two major issues. One is the difficulty in distinguishing between "urban" and "rural" economies at the town level; the other is that the traditional indicators paid more attention to the income levels, which makes it difficult to comprehensively measure the differences in urban and rural economic development. Therefore, it is imperative to construct newer and more comprehensive indicators of urban–rural economic differences to scientifically promote urban–rural integrated development.

- **Data**

– NPP-VIIRS[1] -like nighttime light data (2000, 2005, 2010, 2015, and 2020).
– The administrative division code provided by the National Bureau of Statistics (2020).

[1] NPP-VIIRS refers to national polar-orbiting partnership-visible infrared imaging radiometer suite.

– Administrative division at the village level (2020).

- **Methods**

1. **The Classification of Urban and Rural town**

This case takes the 11 national urban–rural integration development pilot zones as the basic units, and scientifically classifies each town into urban and rural categories based on the urban–rural classification codes from the National Bureau of Statistics. Since the administrative division code is recorded at the village level, this study divided the town types (Region_type_i) based on the proportion of each town's urban units into four categories: urban area (UR), mixed urban area (MU), mixed rural area (MR), and rural area (RU).

$$P_{\text{urban\_}i} = D_{\text{urban\_}i}/D_i \quad (5.3)$$

$$\text{Region\_type}_i = \begin{cases} P_{\text{urban\_}i} = 1 & \text{UR} \\ 0.5 \le P_{\text{urban\_}i} < 1 & \text{MU} \\ 0 < P_{\text{urban\_}i} < 0.5 & \text{MR} \\ P_{\text{urban\_}i} = 0 & \text{RU} \end{cases} \quad (5.4)$$

where $D_{\text{urban\_}i}$ is the number of village-level divisions belonging to the urban area, while D_i is the number of village-level divisions within the ith town. These two variables can be accessed from the administrative division code directly.

2. **Definition of the Nighttime Light Landscape Metrics**

The 3D landscape metrics have been widely used in digital surface models to measure both horizontal and vertical growth of cities (Huang and Wang 2019). Nighttime light remote sensing data are effective for evaluating socio-economic development, and have a continuous field representation similar to that of digital surface models (Chen et al. 2017). By analogy to the 3D landscape metrics, the nighttime light intensity could be treated as a special landscape of urban development. Specifically, seven traditional 3D landscape metrics are modified into nighttime light landscape metrics, representing seven aspects (Table 5.9), such as economic volume, economic space, and economic homogeneity.

3. **Urban–Rural Difference in Economic Development Analysis**

Based on the seven nighttime light landscape metrics mentioned above, statistical and spatial analysis methods were adopted to analyze differences in urban and rural economic development from seven dimensions: economic volume, economic space, economic abundance, economic dispersion, economic concentration, economic level, and economic homogeneity.

- **Results and Analysis**

1. **Urban–Rural Classification Result**

Different urban–rural integration development pilot zones exhibit distinct distribution characteristics. For example, the eastern area of Fuzhou in Fujian has a fairly balanced distribution of town types (Fig. 5.15). In the Changji junction area of Jilin, the proportion of urban towns (UR and MU) exceeds 40% (Fig. 5.16). In the remaining urban–rural integration development pilot zones, mixed towns (MU or MR) are dominant. This is particularly evident in the western area of Chengdu in Sichuan, where 97.3% of the towns are mixed towns (MU and MR) (Fig. 5.17).

2. **Urban–Rural Nighttime Light Landscape Metrics**

The total nighttime light (TNTL) and total lit area (LArea) in the 11 national urban–rural integration development pilot zones in China have significantly increased in the economic volume and

Table 5.9 The nighttime light landscape metrics adopted in this case

Metrics	Metrics	Economic dimensions	Calculation
Landscape composition metrics	Total nighttime light (TNTL)	Economic volume	$\sum_{i=1}^{N} \text{NTL}_i$
	Total lit area (LArea)	Economic space	$\sum_{i=1}^{N} \text{Area}_i; \text{NTL}_i > 0$
	3D percentage of landscape area (3DPLand)	Economic abundance	$\text{TNTL}/(\max \text{NTL}_i \times N) \times 100\%$
Landscape configuration metrics	3D edge density (3DED)	Economic dispersion	$\sum_{j=1}^{T} \text{LArea}_j / \sum_{j=1}^{T} \text{TNTL}_j$
	3D largest patch index (3DLPI)	Economic concentration	$\max \text{TNTL}_j / \sum_{j=1}^{T} \text{TNTL}_j$
Nighttime light intensity metrics	Average nighttime light (ANTL)	Economic level	TNTL/N
	Deviation of nighttime light (DNTL)	Economic homogeneity	$\sum_{i=1}^{N} (\text{NTL}_i - \text{ANTL})^2/N$

*: NTL_i, Area_i, and N indicate the ith pixel's NTL intensity, area, and the amount of NTL pixels within a specific town, respectively. LArea_j and TNTL_j, refer to the jth town's lit area and total NTL intensity. T is the number of towns in 11 national urban–rural integration development pilot zones

economic space from 2000 to 2020, but the trend of each type is different (Fig. 5.18). Taking the eastern area of Fuzhou in Fujian as an example, TNTL increased by nearly 20 times and LArea also increased by 8 times. In addition, the MR towns have the fastest development speed, their proportions of TNTL and LArea increased almost two times. The expansion of TNTL and LArea in the other three types of towns (MU, UR, and RU) remained relatively stable. Overall, the growth rates of TNTL and LArea in rural-related towns (MR and RU) were much faster than those in urban-related towns (MU and UR), which indicates that the processes of urbanization and rural revitalization have narrowed the economic gap between urban and rural areas.

By analyzing the 3D percentage of landscape area (3DPLand) and average nighttime light (ANTL), the 11 national urban–rural integration development pilot zones in China have significantly increased the economic abundance and average economic level from 2000 to 2020 (Fig. 5.19). First, before 2010, the economic abundance and average economic level of UR towns were much higher than those of other types of towns, and the economic abundance gradually tended to saturation. Second, after 2010, the economic abundance and average economic level of MU, MR, and RU towns showed significant growth, and the growth rate exceeded that of UR towns, which indicates that these towns have developed in terms of both economic level and economic abundance, and the gap is gradually narrowing. Therefore, it can be concluded that the economic development gap between urban and rural areas is gradually shrinking in both economic level and economic abundance dimensions.

By analyzing the 3D edge density (3DED) and 3D largest patch index (3DLPI) of China's national urban–rural integration development pilot zones, it was found that in the Changji junction area of Jilin, the land space required for economic development in rural-related towns has significantly decreased, but the primacy has also declined. In contrast, the land space required for economic development in urban-related towns remains at a low level without significant changes, and the primacy remains stable (Fig. 5.20). It can be concluded that a decoupled relationship was found between urban–rural economic development and land

Fig. 5.15 Distribution map of towns divided into four types in the eastern area of Fuzhou, Fujian, and the number of towns classified into these four types

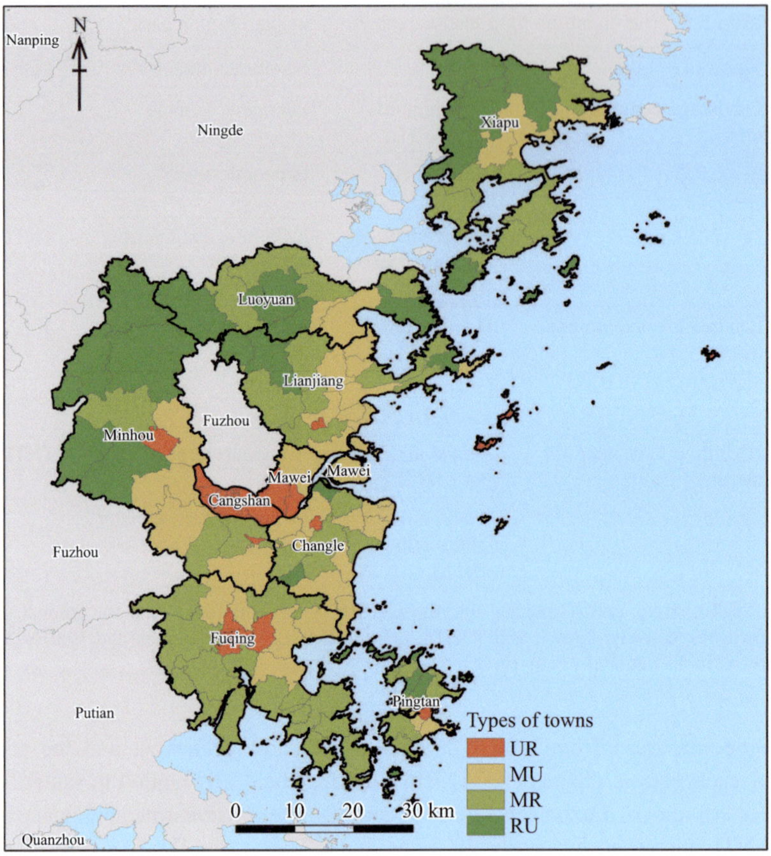

(a) Distribution of urban and rural towns

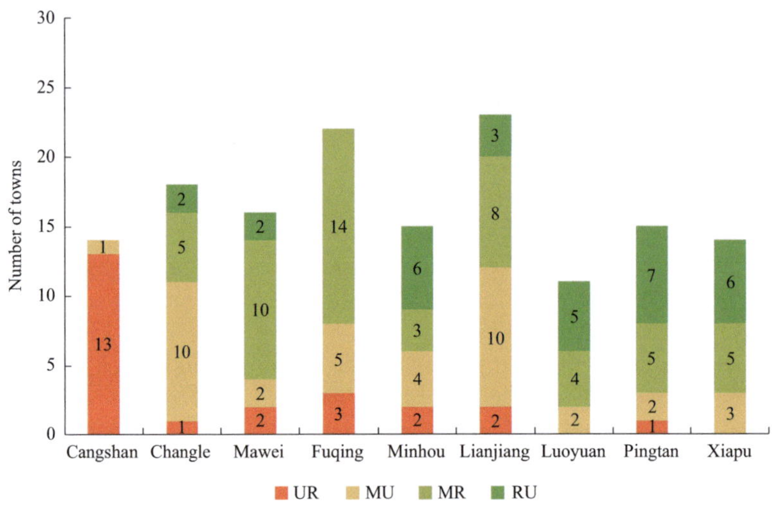

(b) Number of urban and rural towns

Fig. 5.16 Distribution map of towns divided into four types in the Changji junction area of Jilin and the number of towns classified into these four types

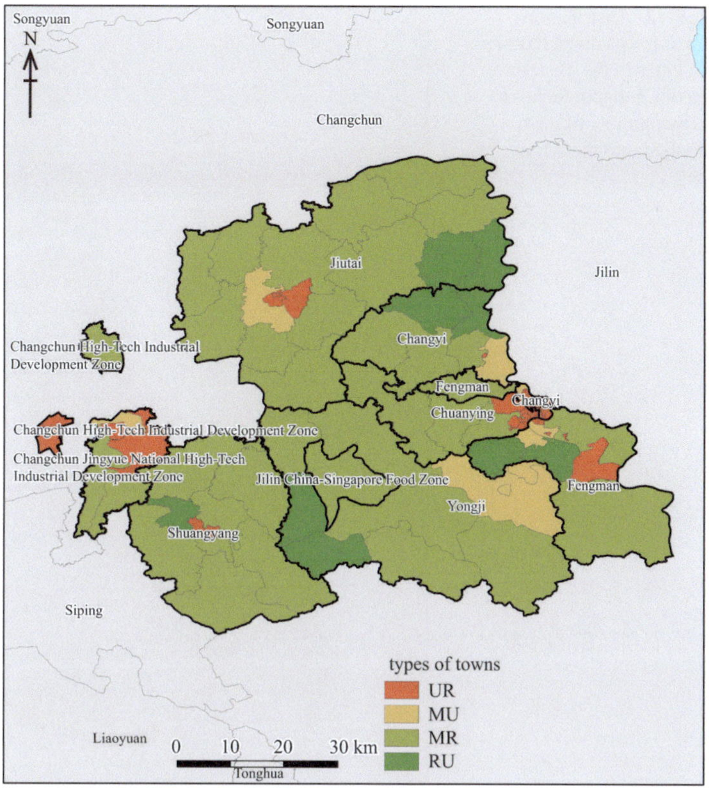

(a) Distribution of urban and rural towns

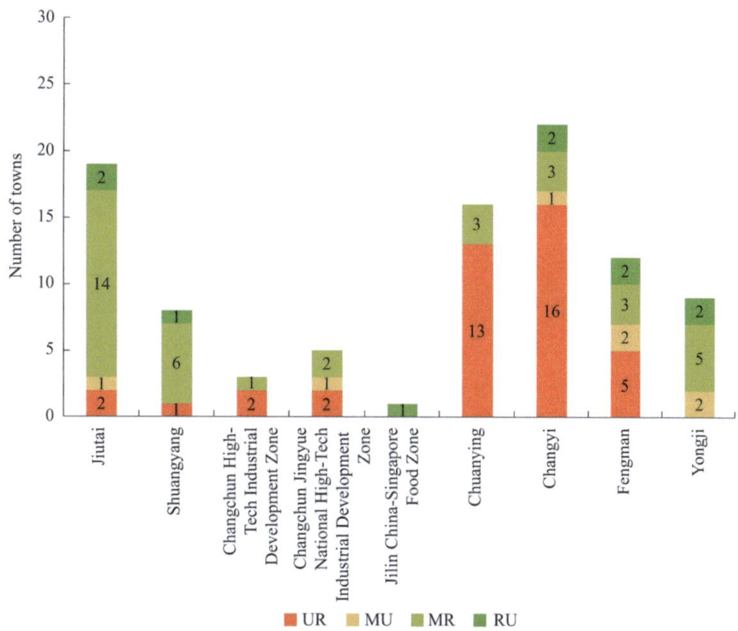

(b) Number of urban and rural towns

Fig. 5.17 Distribution map of towns divided into four types in the western area of Chengdu, Sichuan, and the number of towns classified into these four types

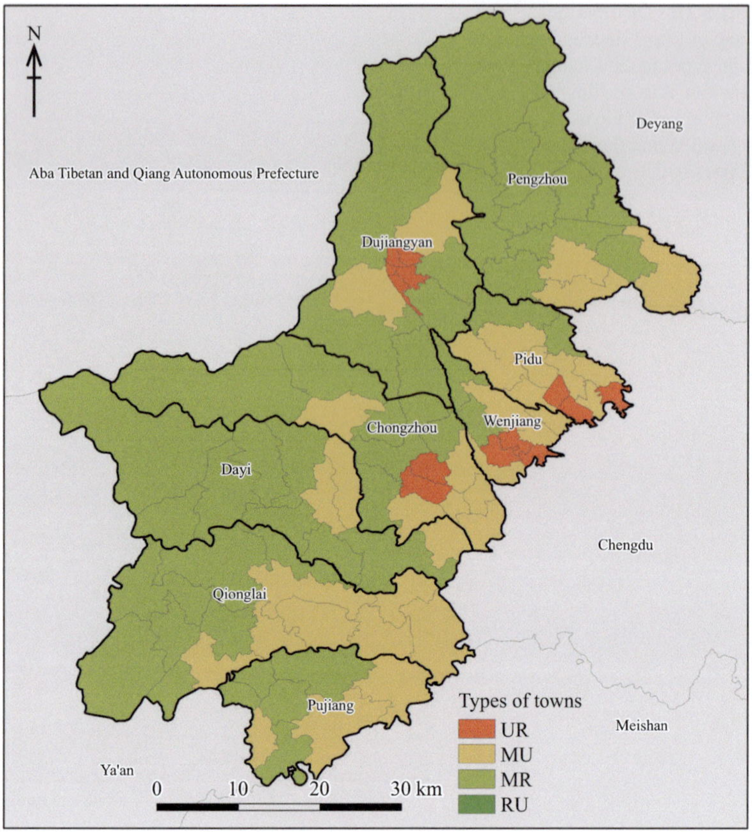

(a) Distribution of urban and rural towns

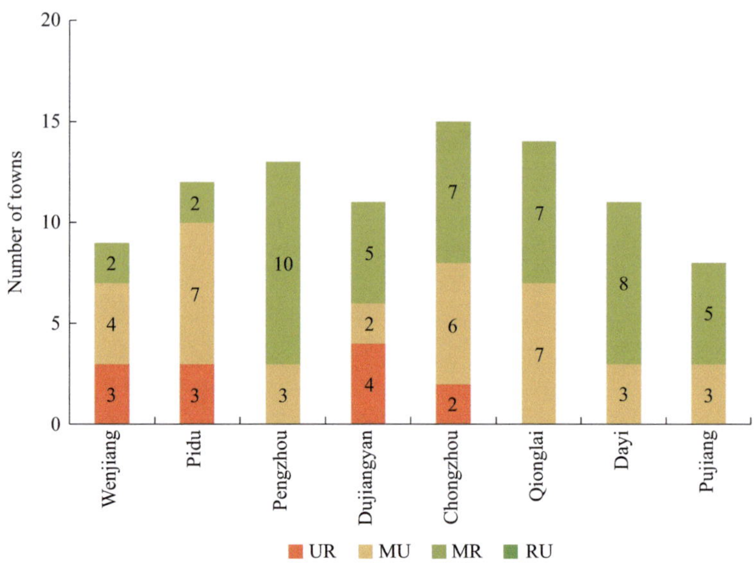

(b) Number of urban and rural towns

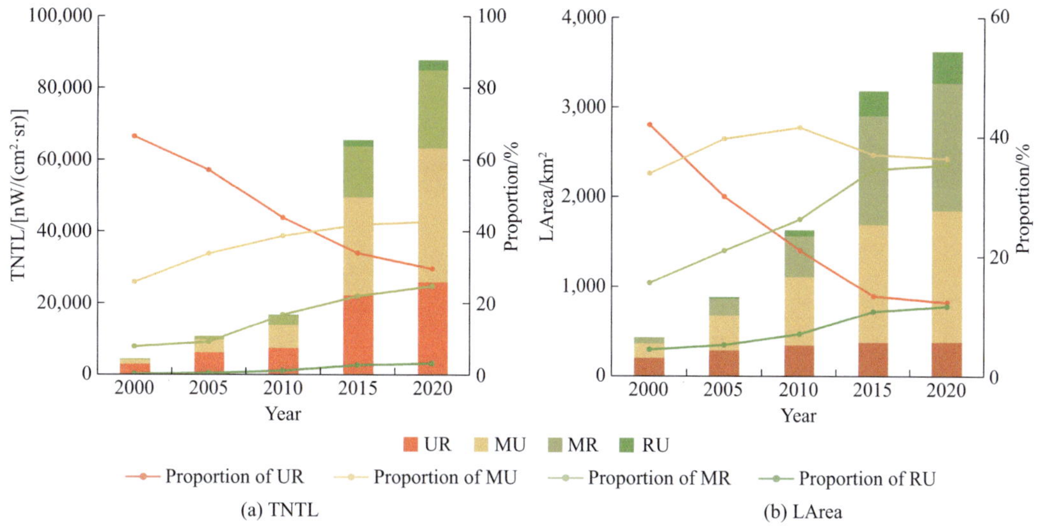

Fig. 5.18 Values of TNTL and LArea and their proportions for each town type in the eastern area of Fuzhou, Fujian, from 2000 to 2020

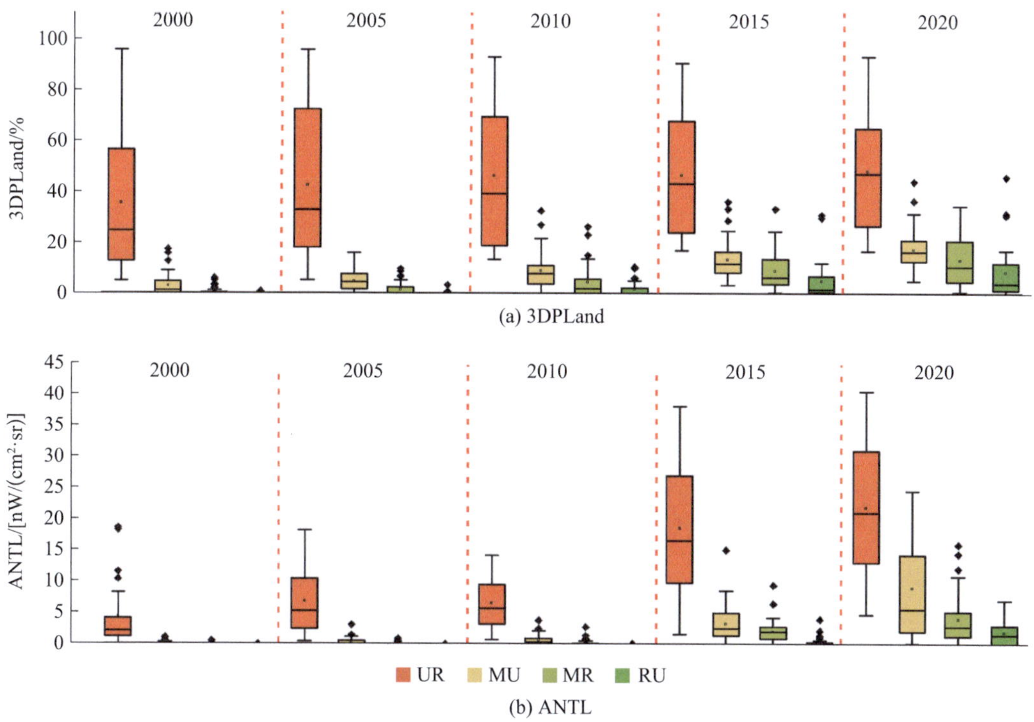

Fig. 5.19 Box plot of the 3DPLand and ANTL intensity for each town type in the eastern area of Fuzhou, Fujian, from 2000 to 2020

resource consumption, but the urban–rural gap still exists. Meanwhile, the gap in the primacy of urban–rural economies is gradually narrowing, indicating that the policies promoting urban–rural economic integration have begun to show preliminary success.

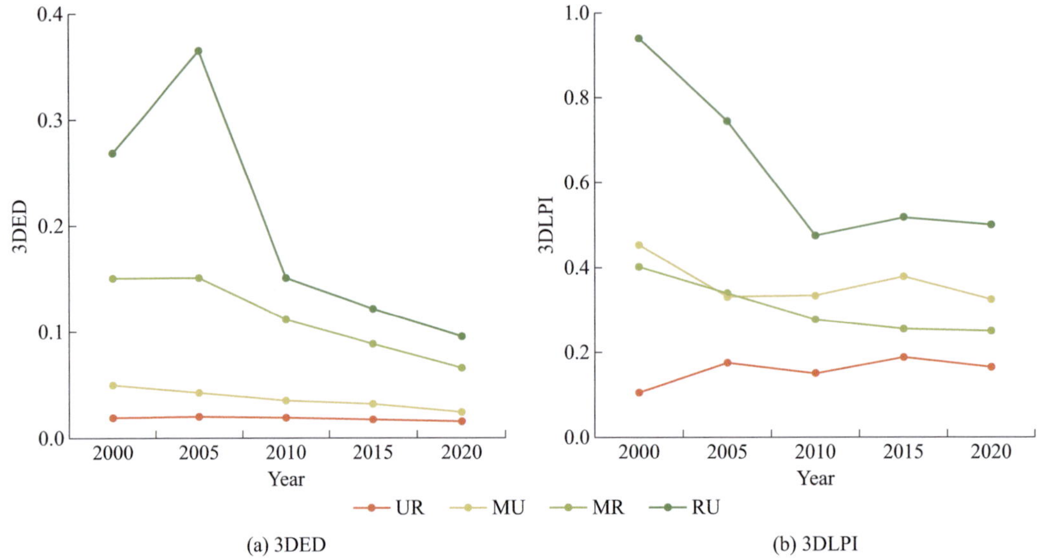

Fig. 5.20 Line chart of the 3DED and 3DLPI for each town type in the Changji junction area of Jilin from 2000 to 2020

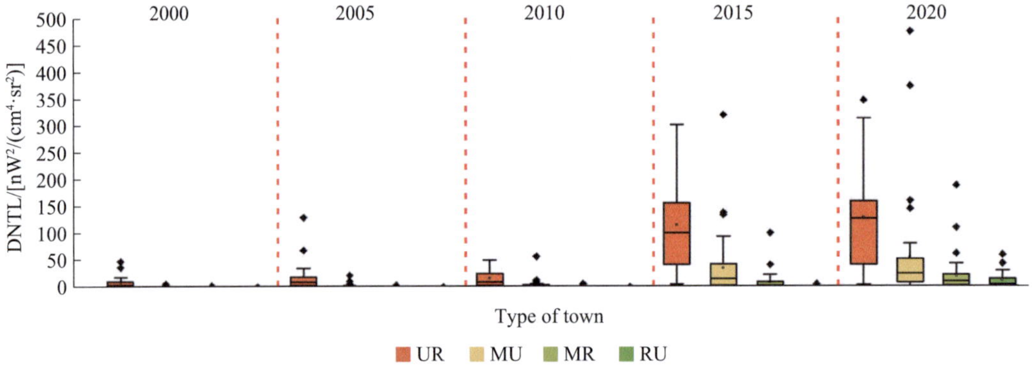

Fig. 5.21 Box plot of the DNTL for each town type in the eastern area of Fuzhou, Fujian, from 2000 to 2020

The deviation of nighttime light (DNTL) value has a decreasing trend at the 11 national urban–rural integration development pilot zones in China significantly from 2000 to 2020 (Fig. 5.21, taking the eastern area of Fuzhou in Fujian as an example), but the urban–rural disparity in economic balance has grown larger. The economic homogeneity of rural-related towns is much higher than that of urban-related towns. In addition to UR towns, the economic homogeneity of the other three types of towns is also related to their average economic levels. For example, the fitted line between DNTL and ANTL intensity for MU towns has the largest slope value across all years (Fig. 5.22). This implies that with the economic development, the economic balance of MU will soon fall far below those of MR and RU, which may be related to the different driving forces of economic development between urban and rural areas.

3. **Spatial Patterns of Urban and Rural Economic Development**

Before 2010, the economic volume and economic space were highly concentrated in urban-related towns. After 2010, the urban-related towns had a rapid and intense urbanization trend, where the economic volume and economic space had

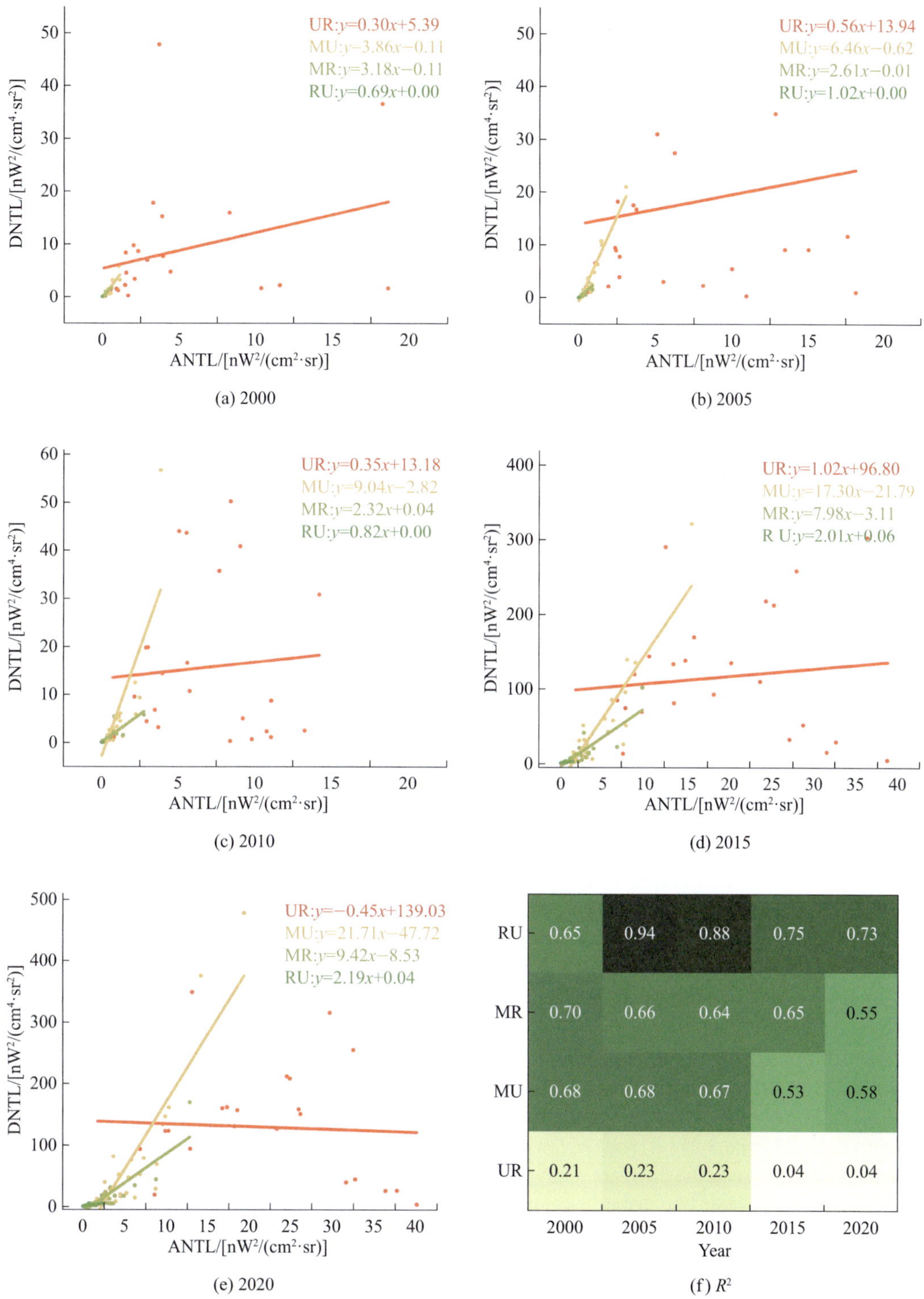

Fig. 5.22 Fitting the relationship between ANTL and DNTL for towns from 2000 to 2020

a faster development, and this also promoted the development of surrounding towns which further narrowed the urban–rural economic development gap (Figs. 5.23, 5.24 and 5.25).

Furthermore, before 2010, the growth of economic volume in all pilot zones mainly relied on the expansion of economic space. After 2010, this reliance showed three different changes. First, in the Jiangsu Ning-xi-Chang junction area (Fig. 5.23), the Shandong Jiqing partial area, the Zhejiang Jiahu area, the Guangdong Guangqing junction area, Henan Xuchang, and Jiangxi Yingtan, the growth of economic volume still relied on the expansion of economic space, with the growth rate of economic space far exceeding the growth rate of economic volume. This indicates that these areas require a better improvement in land-use efficiency. Second, in the eastern area of Fuzhou in Fujian (Fig. 5.24), the Zhejiang Jiahu area, the Changji junction area of Jilin, and the western area of Chengdu in Sichuan, although the growth of economic volume still relied on the expansion of economic space, the growth rates of both were more consistent. Third, in the Xi'an-Xianyang junction area of Shaanxi and the western area of Chongqing, the economic volume and economic space began to decouple, with reduced reliance on economic development on economic space (Fig. 5.25). This indicates that the economic development in these areas is gradually shifting toward improving labor productivity and innovation performance, which aligns closely with China's transition from quantity-driven to quality-driven economic development.

The 3DPLand and DNTL values for most towns in the 11 pilot zones are small, indicating low economic abundance and economic heterogeneity [Figs. 5.24 and 5.25 (iii) and (iv)], without significant changes over the 20 years. Only a few urban towns are in a state of economic saturation, suggesting that the economy in most towns has not yet been fully developed, and the development model is relatively homogeneous.

Highlights

- By analogy with traditional landscape metrics, a set of nighttime light landscape metrics has been constructed in the 3D space based on nighttime light intensity, enabling the examination of urban–rural economic development disparities from multiple aspects.
- By analyzing the changes in the nighttime light landscape indices from 2000 to 2020, it was found that urban–rural differentiation in economic development still exists in the 11 national urban–rural integration development pilot zones in China but showing a decreasing trend.
- The reliance of economic development on the expansion of economic space in the pilot zones has declined, indicating that economic growth is gradually freeing itself from heavy dependence on land resources, shifting toward a high-quality development path focused on improving labor productivity and innovation performance.

- **Discussion and Outlook**

This case study uses new nighttime light landscape metrics to reveal that urban–rural differences in economic development in the 11 national urban–rural integration development pilot zones of China exist across all seven dimensions, and shows a decreasing trend. Furthermore, given that traditional landscape indices have already been full-developed that provides a solid theoretical foundation for further improving the nighttime light landscape metrics. Therefore, in the future, other landscape metrics [such as the landscape shape index (LSI), proximity index, skewness, etc.] can be introduced to enrich the nighttime light landscape metrics and further expand the evaluation dimensions of urban–rural economic development disparities.

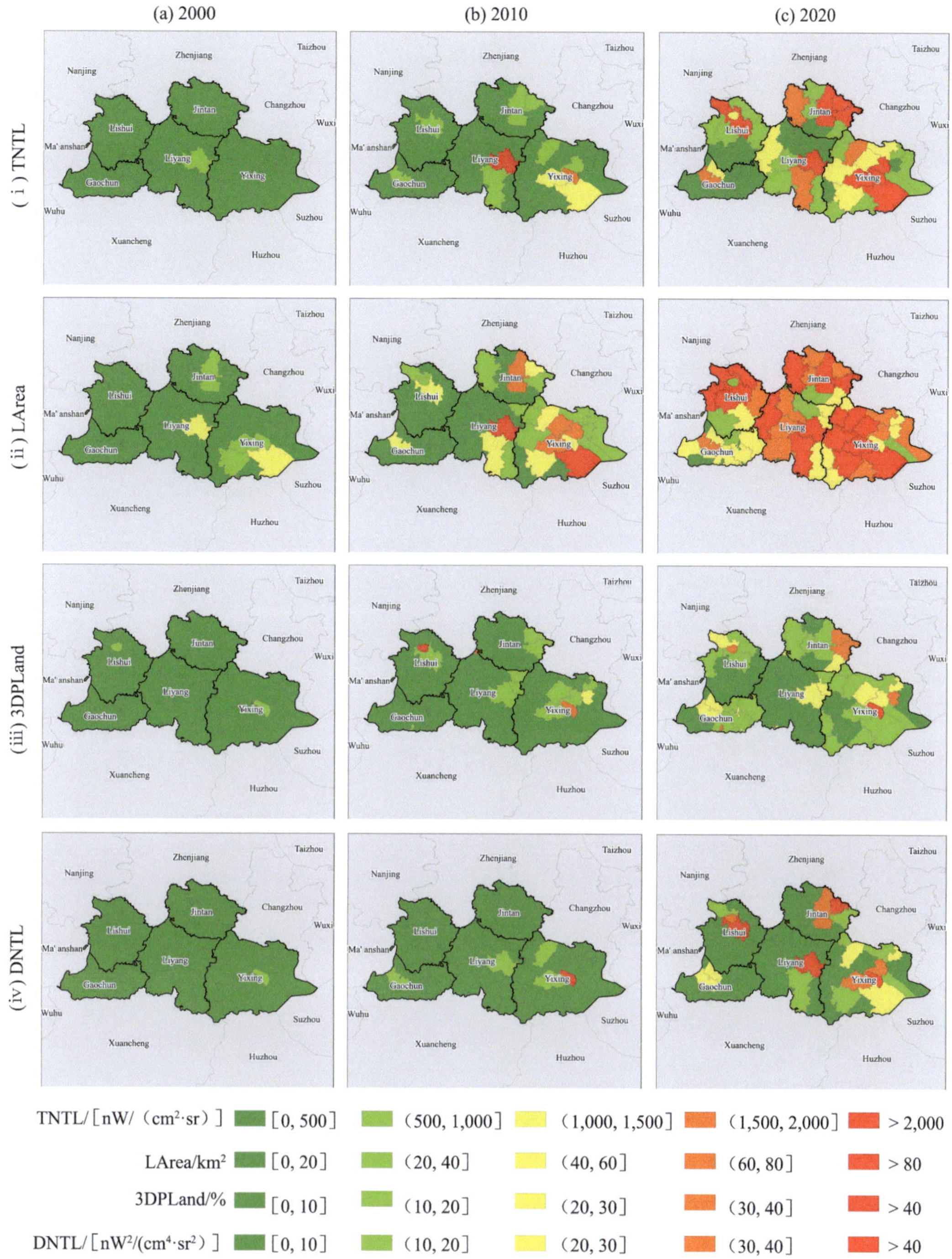

Fig. 5.23 Spatial distribution map of landscape indices in the Ning-xi-Chang junction area of Jiangsu

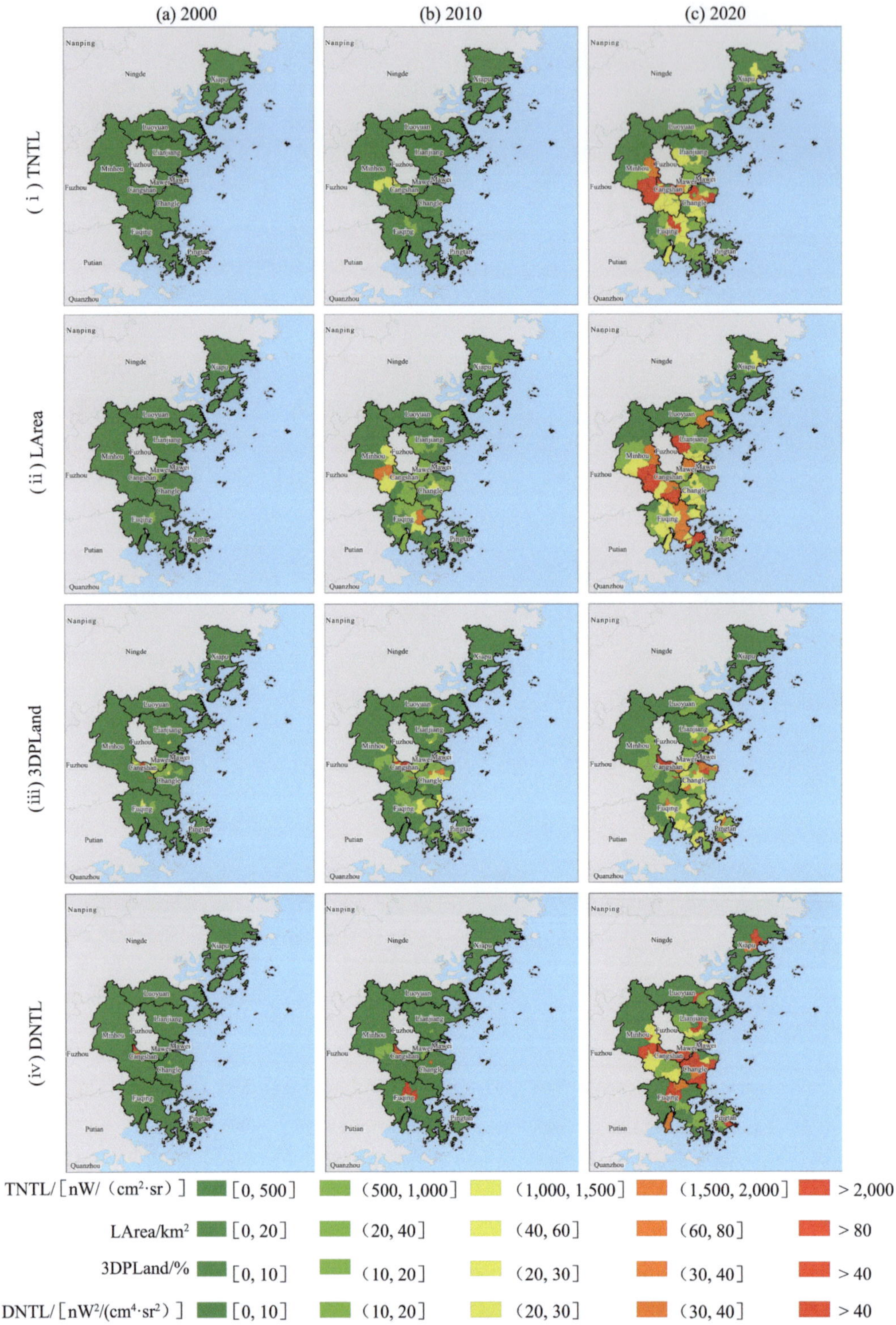

Fig. 5.24 Spatial distribution map of landscape indices in the eastern area of Fuzhou, Fujian

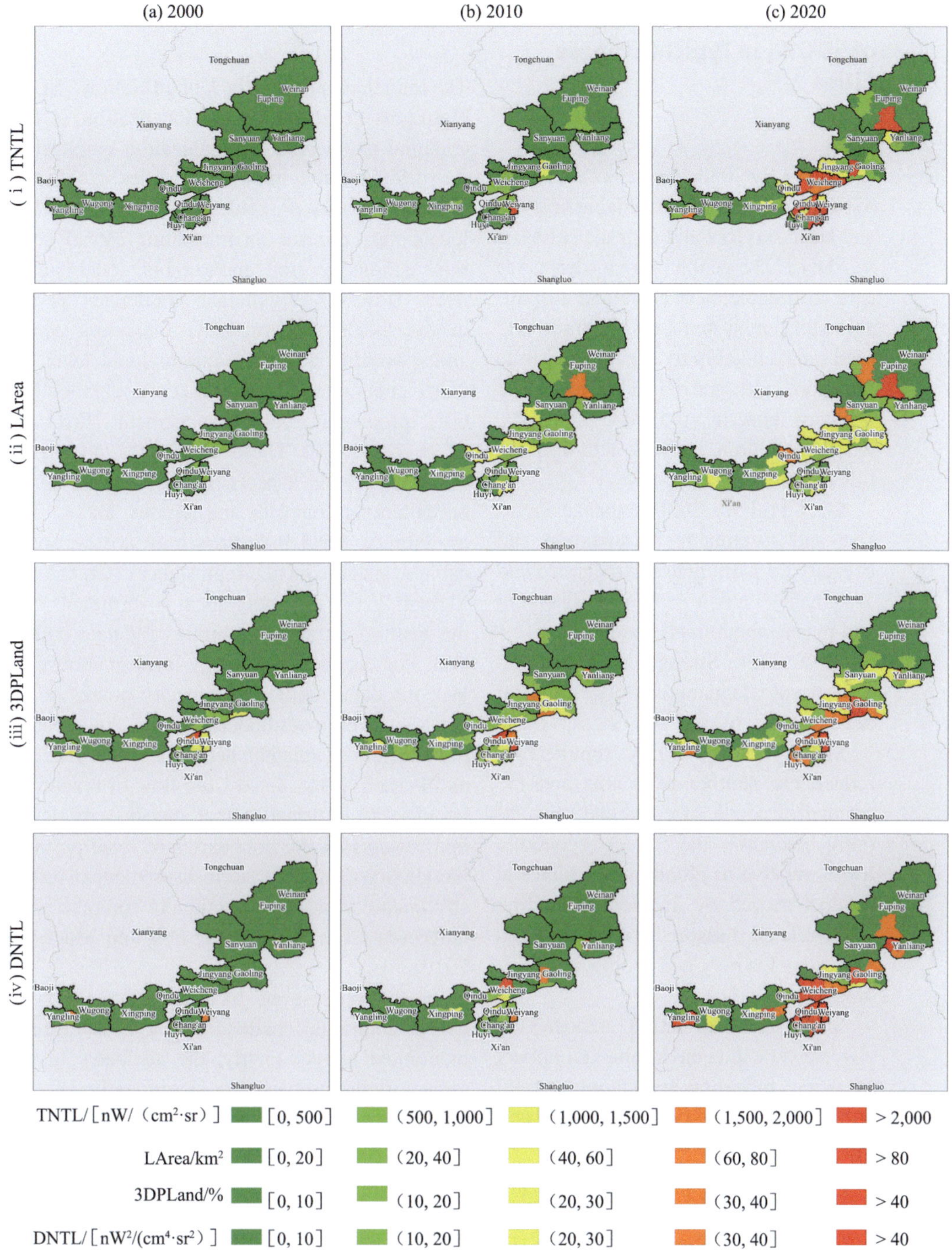

Fig. 5.25 Spatial distribution map of landscape indices in the Xi'an-Xianyang junction area of Shaanxi

5.3.6 Comprehensive Assessment of SDG 11 in Typical Chinese Cities

Target: SDG 11.1 By 2030, ensure access for all to adequate, safe and affordable housing and basic services and upgrade slums.

SDG 11.2 By 2030, provide access to safe, affordable, accessible and sustainable transport systems for all, improving road safety, notably by expanding public transport, with special attention to the needs of those in vulnerable situations, women, children, persons with disabilities and older persons.

SDG 11.3 By 2030, enhance inclusive and sustainable urbanization and capacity for participatory, integrated and sustainable human settlement planning and management in all countries.

SDG 11.4 Strengthen efforts to protect and safeguard the world's cultural and natural heritage.

SDG 11.5 By 2030, significantly reduce the number of deaths and the number of people affected and substantially decrease the direct economic losses relative to global gross domestic product caused by disasters, including water-related disasters, with a focus on protecting the poor and people in vulnerable situations.

SDG 11.6 By 2030, reduce the adverse per capita environmental impact of cities, including by paying special attention to air quality and municipal and other waste management.

SDG 11.7 By 2030, provide universal access to safe, inclusive, accessible, and green public spaces, in particular for women and children, older persons, and persons with disabilities.

- **Background**

The implementation of the United Nations' SDGs highlights urbanization and localization as two essential perspectives. Urbanization emphasizes the differentiated impacts of cities of various scales and types on achieving the SDGs, while localization focuses on translating global goals into actionable and measurable local initiatives. However, significant challenges persist in the localization of SDG indicators across cities worldwide. At the international level, all cities encounter issues related to data collection, processing, and management (Fox and MacLeod 2023). These challenges are manifested in several aspects. Firstly, due to differences among countries and regions, and in data availability, SDG indicators may not be applicable to specific local circumstances (Greene and Meixell 2017). Secondly, local governments lack the institutions and capabilities for data collection. For instance, many city-level governments lack the data monitoring capacity possessed by national-level governments, further exacerbating the problem of mismatched data scales (Barnett and Parnell 2016). Lastly, the flow of resources, people, and information that sustains city development crosses the boundaries of local political jurisdictions, making the measurement of urban sustainability progress a uniquely complex challenge (da Cruz et al. 2019; Fox and MacLeod 2019).

In the process of localizing the SDGs in China, additional challenges arise beyond those mentioned above. Firstly, the attributes of the indicators themselves vary significantly, encompassing both quantitative and qualitative evaluation criteria, as well as guiding and aspirational indicators that are difficult to assess. Secondly, certain SDG indicators differ in expression from China's statistical metrics or have never been statistically collected, making it challenging to obtain the necessary data and rendering these indicators inapplicable for evaluating China's SDG progress (Shao et al. 2021). Thirdly, while the United Nations Sustainable Development Solutions Network (UNSDSN) has conducted index-based fitting for all SDG indicators at the national level, there is a lack of corresponding efforts at the city level, leaving urban managers without integrated decision-support tools related to sustainable development.

Against the backdrop of advocating for global urban sustainability and promoting high-quality urban development alongside the modernization of governance capabilities, urban assessment serves as an innovative governance approach. It is also an effective tool to support high-quality urban development and refined management in the future. With the advancing localization of the SDGs in cities worldwide, the significance of urban assessment for achieving the localization of SDGs in China has become increasingly prominent. On the one hand, the eight thematic components of the urban assessment indicator system are classified as corresponding to related SDGs, aligning qualitatively with the SDGs. On the other hand, quantitatively, urban assessment indicators not only facilitate the computation of SDG indicators but also contribute to the refinement of SDG indicators through further decomposition. Moreover, advancements in Earth observation technologies and spatio-temporal big data science provide increasingly diverse data sources for the statistics of urban assessment indicators, expanding and enhancing the coverage and reliability of indicator statistics.

To address the deficiencies in indicator data and comprehensive index results for the localization of SDG 11, and the lack of clarity in assessment methods at the city level, this case aims to supplement the relevant data from two perspectives: (1) Based on the "check-ups" for pilot cities, the SDG 11 indicator system is localized, and data from pilot cities are collected, using the latest data collection methods, as shown in Fig. 5.26. (2) By referring to existing United Nations' research, a comprehensive index is developed by fitting multiple indices, providing Chinese urban managers with quantitative analysis results for comprehensive decision-making. It also serves as a tool for monitoring the process of achieving the SDGs by 2030. Data, collected from 59 pilot cities that had "check-ups" in 2022, went through spatial analysis and generated results of the localized comprehensive index for SDG 11.

- **Data**

  - The 2022 China urban assessment pilot cities dataset;
  - Analysis dataset of the 2022 Tsinghua Urban Health Index;
  - Data of the 2022 proportion of population with convenient access to public transport in China by the Institute of Geographic Sciences and Natural Resources Research, Chinese Academy of Sciences.

- **Methods**

The localization of SDG indicators is the process of generating SDG indicator sets tailored to specific regions based on the United Nations' Global Indicator Framework. Its purpose is to align and adapt the SDGs with local development realities and needs, transforming them into locally actionable indicator groups or sets. Starting from the target level of the SDG indicator system and the urban assessment system, the process delves into the thematic and indicator levels. Through keyword matching, content alignment, and measurability analysis, the content of each level is systematically reviewed. Following further screening or refinement, a localized set of SDG indicators is constructed.

At the thematic level, the thematic layers of SDG 11 and the urban assessment system are matched based on keyword relevance. Through an analysis of the connotations of each theme and considering the region's development stage, the specific meanings of each theme are determined. By mapping, the seven sub-goals of SDG 11 are aligned with the eight thematic components of urban assessment, ensuring that the indicator set comprehensively covers all major SDG 11 sub-goals, and facilitating quantitative evaluation. During the process of content matching, "many-to-many" relationships may exist between themes. In such cases, the focus should shift from the thematic level to the indicator level, where the content of individual indicators is matched. Combining the measurability of indicators, urban

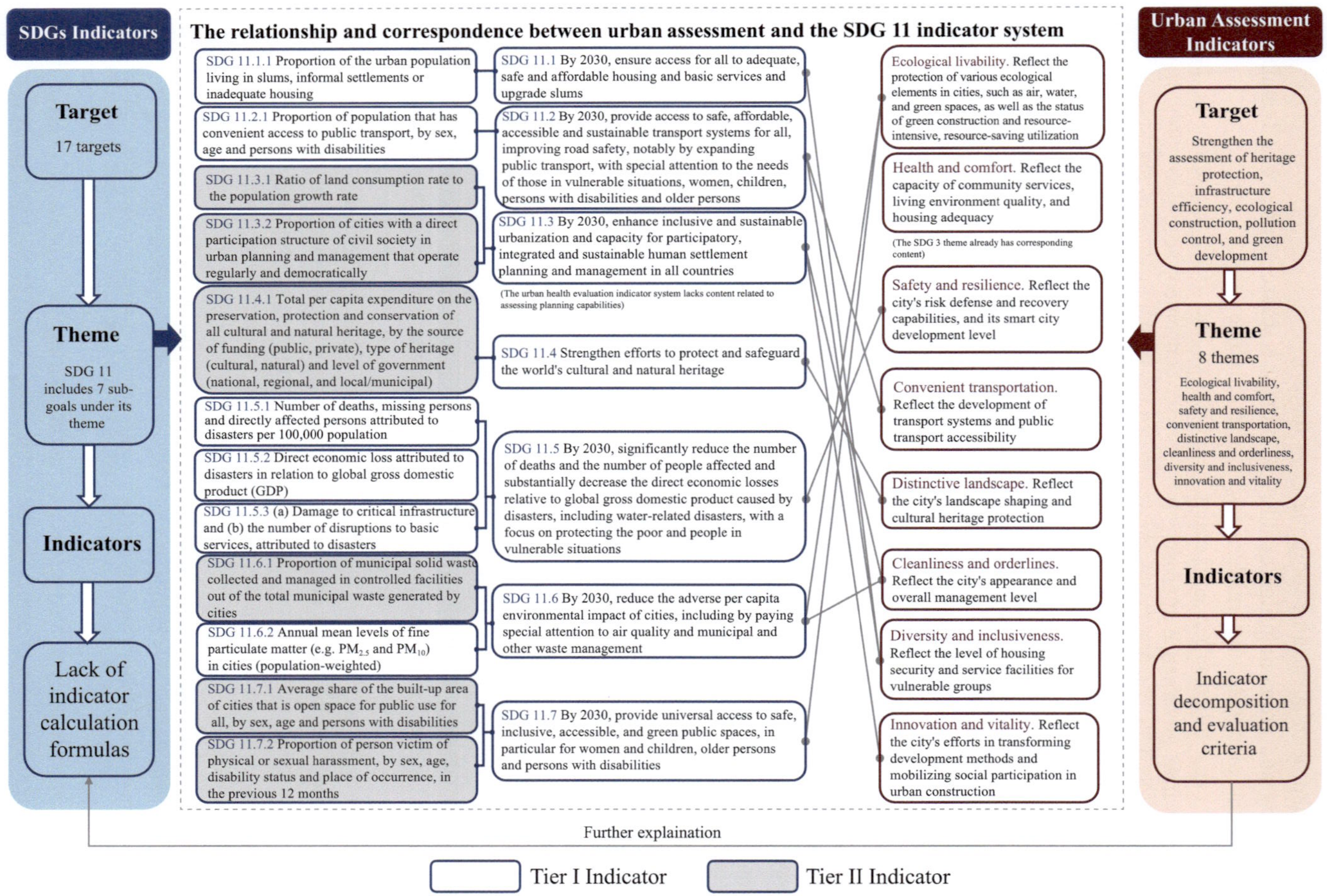

Fig. 5.26 An indication of the SDG 11 localization approach supported by urban assessment

assessment metrics are aligned with SDG 11 indicators. At the indicator level, indicators corresponding to the thematic layer are associated, and the measurability of SDG 11 indicators is evaluated. For indicators with data that are difficult to obtain, urban assessment metrics can be considered as substitutes, ensuring that the calculated indicators comprehensively cover SDG 11 indicators in terms of structure, connotation, and operability.

Finally, by integrating the indicator frameworks of urban assessment and the SDGs, and guided by principles such as scientific rigor, systematicity, adaptability, data reliability and accessibility, and scalability (Shao et al. 2021), a localized SDG 11 indicator system supported by urban assessment has been developed, tailored to China's national conditions and informed by global SDG localization practices (Table 5.10). Based on this framework, comprehensive index calculations were performed using 2022 data from pilot cities.[2]

Each calculated indicator is supported by corresponding collection and analysis methods. This case study, following the methodology of the UNSDSN, focuses on introducing the fitting approach for developing a localized comprehensive index for SDG 11, designed for use by comprehensive decision-makers. The specific steps are outlined as follows.

Firstly, the urban assessment indicator system is compared item by item with the sub-goals of SDG 11. Indicators with similar connotations and expressions are linked and matched, and are then incorporated into the sub-goals of SDG 11 to construct a computable indicator system.

Secondly, statistical tests on the indicator data,[3] including outlier reviews are performed, and indicator thresholds for standardization processing are determined. Specifically, the indicator data are transformed into an index ranging from 0 to 1 by setting the minimum and maximum values for each indicator as the "natural zero" and "best performance", respectively, while excluding extreme values (outliers) from each indicator's distribution. Additionally, to ensure comparability among indicators, positive and negative indicators are identified. The formulas are as follows.

$$x_{\text{positive}} = \frac{x - \min(x)}{\max(x) - \min(x)} \times 100 \tag{5.5}$$

$$x_{\text{negative}} = \frac{\max(x) - x}{\max(x) - \min(x)} \times 100 \tag{5.6}$$

Thirdly, the comprehensive index is calculated using the arithmetic mean method. After comparing common weighting approaches such as the entropy method and expert scoring, the methodology of the UNSDSN (2016) is adopted, assigning equal weights to each SDG. This approach allows for the inclusion of new variables for specific SDGs without affecting the relative weights of each goal in the total score. Accordingly, the localized comprehensive index for SDG 11 is derived by aggregating the sustainable development indicators using arithmetic averages. The specific formula for calculating the comprehensive index is as follows.

$$P_i = \sum_{j=1}^{N} x_{i,j} \cdot q_j \tag{5.7}$$

where P_i represents the localized comprehensive index result for SDG 11 of city i, $x_{i,j}$ is the standardized index of the j th SDG 11 indicator for city i, and q_j denotes the weight of the j th SDG 11 indicator.

[2] In terms of data sources for the indicator system, the proportion of the population with convenient access to public transport was obtained from an internal project dataset provided by the Institute of Geographic Sciences and Natural Resources Research, Chinese Academy of Sciences. Data on the number of emergency shelters per capita, the air quality index (AQI), and days of severe air pollution were obtained from government statistical yearbooks. Other indicator data were derived from the internal dataset of the 2022 urban assessment annual evaluation.

[3] Conduct statistical tests to calculate the mean, maximum, minimum, standard deviation, kurtosis, and skewness of the indicators.

Table 5.10 SDG 11 localization indicator system supported by urban assessment

Goals of SDG 11	Localization of SDGs corresponding to urban assessment indicators	Positive /negative indicators
SDG 11.1 By 2030, ensure access for all to adequate, safe and affordable housing and basic services and upgrade slums	The proportion of the population living in slums and urban villages/%	Negative
	The proportion of urban households spending more than 50% of their income on housing/%	Negative
	Coverage rate of housing security services for the permanent population/%	Positive
	Compliance rate of renovated old residential communities/%	Positive
SDG 11.2 By 2030, provide access to safe, affordable, accessible and sustainable transport systems for all, improving road safety, notably by expanding public transport, with special attention to the needs of those in vulnerable situations, women, children, persons with disabilities and older persons	Number of public transport stations per person	Positive
	The proportion of the population with convenient access to public transport/%	Positive
	Share of green transportation modes/%	Positive
	Accessibility facility installation rate on roads/%	Positive
	Average one-way commuting time for urban permanent residents/min	Negative
	The proportion of the population with a commuting distance of less than 5 km/%	Positive
SDG 11.3 By 2030, enhance inclusive and sustainable urbanization and capacity for participatory, integrated and sustainable human settlement planning and management in all countries	Regional development intensity/%	Positive
	Population density based on mobile signaling/(10,000 people/km^2)	Positive
	Urban construction land area with a population density exceeding 15,000 people per square kilometer/km^2	Positive
SDG 11.4 Strengthen efforts to protect and safeguard the world's cultural and natural heritage	Number of historical and cultural districts	Positive
	Number of historical buildings	Positive
	Restoration rate of urban historical and cultural districts/%	Positive
SDG 11.5 By 2030, significantly reduce the number of deaths and the number of people affected and substantially decrease the direct economic losses relative to global gross domestic product caused by disasters, including water-related disasters, with a focus on protecting the poor and people in vulnerable situations	Urban road traffic accident mortality rate /(persons per 10,000 vehicles)	Negative
	Annual urban accident mortality rate /(persons per 10,000 people)	Negative
	Shelter area per person/m^2	Positive
	Number of emergency shelter facilities per person	Positive
	Proportion of permeable surface area in urban areas/%	Positive

(continued)

Table 5.10 (continued)

Goals of SDG 11	Localization of SDGs corresponding to urban assessment indicators	Positive /negative indicators
SDG 11.6 By 2030, reduce the adverse per capita environmental impact of cities, including by paying special attention to air quality and municipal and other waste management	Resource utilization rate of urban household waste/%	Positive
	Reduction in carbon dioxide emissions per capita of GDP/%	Positive
	The proportion of days with good air quality/%	Positive
	Air quality index (AQI)	Negative
SDG 11.7 By 2030, provide universal access to safe, inclusive, accessible, and green public spaces, in particular for women and children, older persons, and persons with disabilities	Coverage rate of complete residential communities/%	Positive
	Coverage rate of community convenience commercial service facilities/%	Positive
	Coverage rate of community elderly service centers/%	Positive
	Coverage rate of inclusive kindergartens/%	Positive
	Coverage rate of urban greenway service radius/%	Positive
	Coverage rate of park green space service radius/%	Positive

- **Results and Analysis**

Based on the data from 59 pilot cities in 2022, the localized comprehensive index for SDG 11 was finalized. As shown in Fig. 5.27, the spatial pattern of the urban comprehensive index exhibits a high similarity to the "diamond structure"[4] of urban economic and social development. Core cities in well-developed urban clusters, such as the Pearl River Delta, Yangtze River Delta, Beijing-Tianjin-Hebei, Chengdu-Chongqing, the Middle Reaches of the Yangtze River, and the western coast of the Taiwan Strait, have higher levels of development. These regions are economically advanced with relatively high urbanization levels, indicating that the economic and social development of cities is a critical factor influencing the achievement of SDG 11. At the city level, Fuzhou, Shanghai, Guangzhou, Shenzhen, and Nanjing have the highest scores, reflecting their efforts and achievements in building inclusive, safe, resilient, and sustainable cities and human settlements. At the regional level, Central China and East China have the highest average scores, followed by South China and Southwest China, while North China, Northwest China, and Northeast China have the lowest scores, indicating distinct differences between the eastern and western regions, as well as northern and southern regions. In summary, the differences in comprehensive index results are driven by a combination of factors, including the economic and social development levels of cities, specific implementation measures, regional characteristics, and policy support.

From the perspective of the seven geographical regions, rated by the average scores of the comprehensive index, from high to low, are as follows: Central China, East China (tied for first), South China, Southwest China, North China, Northwest China, and Northeast China. Overall, cities in the central and eastern regions perform

[4] The "diamond structure" refers to a wedge-shaped pattern in China's urban studies, formed by the four national-level economic and social development growth poles: the Pearl River Delta, the Yangtze River Delta, the Beijing-Tianjin-Hebei region, and Chengdu-Chongqing. Its geometric shape resembles a diamond and is often used to describe the spatial relationships formed by the highly urbanized regions in China.

Fig. 5.27 Spatial distribution of SDG 11 localization comprehensive index results in 2022

better than those in the western and northeastern regions (Fig. 5.28). Specifically, all regions have achieved relatively high levels in ensuring housing and basic services (SDG 11.1), building transport systems (SDG 11.2), reducing negative environmental impacts (SDG 11.6), and providing public spaces (SDG 11.7) (Fig. 5.29). Recent assessments have consistently demonstrated significant progress in the construction of hardware infrastructure, such as housing, transportation, and public space infrastructure, while also achieving new milestones in green development and ecological civilization building.

From the perspective of city size, the average comprehensive index scores, ranked from highest to lowest, are megacities, supercities, medium-sized cities, Type I big cities, Type II big cities, and small cities[5] (Fig. 5.30). As cities with enormous scale, megacities and supercities have a significant advantage in resource aggregation, including population, capital, land, and technology, which contributes to their high comprehensive index scores. Notably, medium-sized cities have surpassed Type I and Type II big cities in average comprehensive index scores,

[5] According to the Notice on Adjusting the Standards for City Size Classification issued by the State Council in November 2014, cities are classified into five categories and seven levels based on the permanent resident population in urban areas. Small cities are defined as those with an urban permanent population of less than 500,000, including Type I small cities with 200,000 to less than 500,000 residents and Type II small cities with fewer than 200,000 residents. Medium-sized cities are those with an urban permanent population of 500,000 to less than 1 million residents. Big cities are defined as having an urban permanent population of 1 million to less than 5 million residents, further categorized into Type I big cities with 3 million to less than 5 million residents and Type II big cities with 1 million to less than 3 million residents. Super cities have an urban permanent population of 5 million to less than 10 million residents, while megacities are those with an urban permanent population of 10 million or more residents (Note: "including" refers to the upper limit, while "excluding" refers to the lower limit in each range).

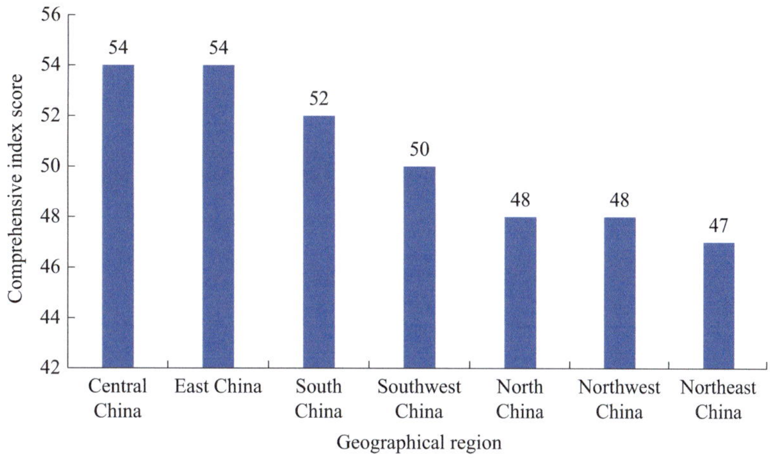

Fig. 5.28 Bar chart of the average scores of SDG 11 localization comprehensive index by geographical regions in 2022

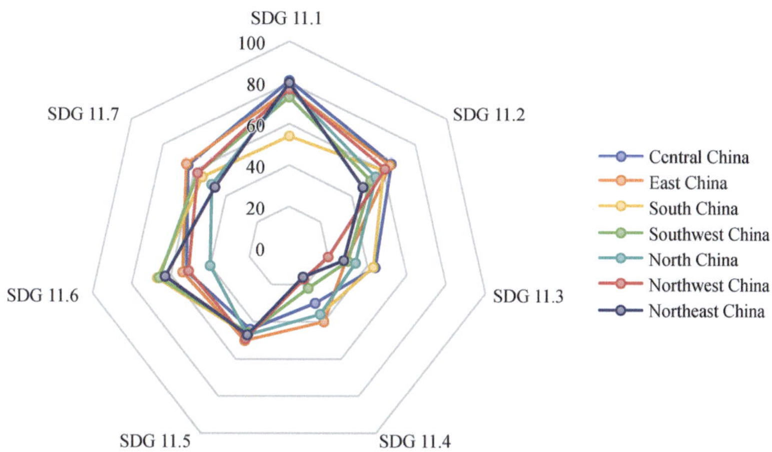

Fig. 5.29 Radar chart of average sub-index scores for SDG 11 localization by geographical regions in 2022

demonstrating clear advantages in specific sub-goals such as housing security (SDG 11.1), transportation systems (SDG 11.2), environmental quality (SDG 11.6), and public spaces (SDG 11.7) (Figs. 5.30 and 5.31). This reflects the superior development level and prospects of medium-sized cities in terms of sustainable development, with balanced progress in housing, transportation, and environmental quality. In contrast, larger cities may face challenges related to excessive resource aggregation, leading to "big city problems" that hinder their sustainable development.

Highlights

- Conduct a comparative study between the urban assessment system and the SDG 11 indicator framework to implement the localization and practical application of the SDGs in the context of China.
- Develop a localized indicator system and comprehensive index for SDG 11 to support the formulation of sustainable development-related policies at the city level in China, as well as the monitoring and evaluation of existing public policies, including urban planning and development strategies.
- Analyze and compare the development levels of pilot cities, summarizing the characteristics of different geographical regions and city sizes. In terms of the seven geographical regions, the average comprehensive index scores rank from highest to lowest as follows: Central China and East China (tied for first), South China, Southwest China, North China, Northwest China, and Northeast China, with

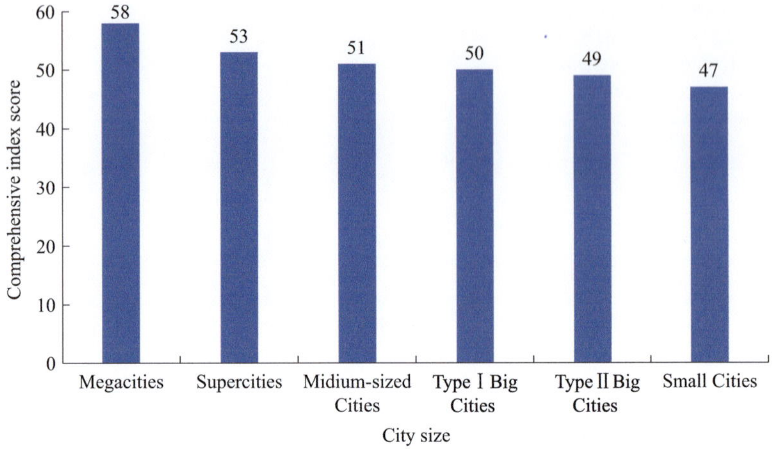

Fig. 5.30 Bar chart of the average scores of SDG 11 localization comprehensive index by city size in 2022

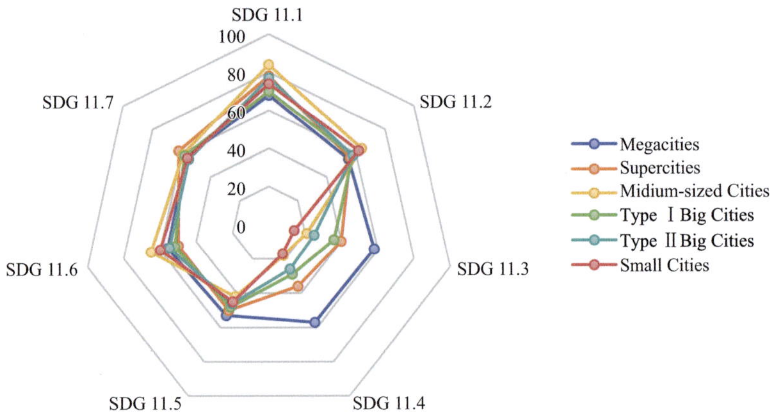

Fig. 5.31 Radar chart of average sub-index scores for SDG 11 localization by city size in 2022

cities in the central and eastern regions generally outperforming those in the western and northeastern regions. Regarding city size, the average comprehensive index scores rank from highest to lowest as follows: megacities, supercities, medium-sized cities, Type I big cities, Type II big cities, and small cities.

- **Discussion and Outlook**

Scientific evaluation of the progress of the SDGs is essential for advancing their achievement and represents a core challenge faced by countries worldwide. Due to differences in development stages, statistical systems, and data availability across nations, the United Nations' SDG indicator framework can't be directly applied to national-level evaluation and monitoring. Therefore, it is necessary to develop localized SDG indicator systems tailored to each country's specific conditions and context, establish unified statistical standards, and promote the collection and integration of multi-source data. This work is crucial for enabling quantitative assessments of development goals, serving as a key mechanism for advancing concrete localized actions, and providing comprehensive decision-making tools urgently needed by local policymakers. From both international and domestic experience, especially in a vast country like China, significant disparities and imbalances in development indicators persist within the same nation. This is due to two main factors: one is the regional development differences, where the eastern coastal regions are more economically developed, and have better infrastructure compared to the relatively underdeveloped central and western regions; the other is the urban–rural development disparities, with urban areas having significantly better infrastructure, education, and healthcare compared to rural areas. By conducting comprehensive assessments of indicator systems to identify gaps in specific dimensions of SDGs across regions, local governments can be better supported in formulating relevant public policies. Moreover, cities, as the primary battlegrounds of urbanization, are the core administrative units for implementing such policies. Therefore, the localization of tools for monitoring and evaluating urban planning, construction, and management is critical. These tools must be adaptable to local conditions at both the overall index and individual indicator levels.

Currently, exemplified by urban assessment, an increasing number of departments and institutions are utilizing indicator systems to study urban issues and strengthen their capacity for quantitative monitoring and evaluation. These efforts aim to establish and improve evaluation systems for urban construction management and the quality of human settlements, thereby promoting urban sustainability. This aligns closely and is highly consistent with the objectives of SDG 11. Among numerous studies and publications, future efforts should focus on key issues in the localization of SDGs, dedicating more resources to addressing critical and challenging aspects of localization. This will facilitate more scientific and objective quantitative assessments of SDG progress, and allow reliable indicators and data derived from China's practical experience to inform and support SDGs, contributing China's wisdom and solutions to the global SDG agenda. It will also provide valuable working examples for the international community. Besides, efforts should be made to strengthen international cooperation and experience exchange, learning from the successful practices of other countries in achieving SDGs. Participation in more international collaborative projects should also be prioritized, leveraging project platforms to share experiences with other nations, playing an active role at the international level, and collaboratively addressing global sustainable development challenges.

5.4 Summary

This chapter explores progress in sustainable cities and communities, focusing on urban land use efficiency, urban air pollution, open public spaces, urban–rural regional development, and a comprehensive assessment of SDG 11. Research findings reveal the following aspects: Inclusive and sustainable urban development is crucial, with an emphasis on enhancing the sustainability of human settlement planning and management; from 2010 to 2022, $PM_{2.5}$ levels nationwide decreased significantly, showcasing the effectiveness of China's comprehensive air pollution control efforts and reducing the risks of $PM_{2.5}$ exposure for residents; between 2017 and 2022, the proportion of public open spaces in China's five major urban clusters increased, along with the availability of public open spaces per capita in each cluster; the urban–rural economic development gap in China has narrowed across various dimensions, significantly improving the level of urban–rural integration; localized SDG 11 composite indices for cities exhibit diverse characteristics, reflecting the varying progress of sustainable development in different regions. These findings provide critical support for detailed analysis of regional indicator progress and evidence-based decision-making for sustainable cities and communities.

Based on the research in this chapter, we propose the following recommendations: (1) Governments should encourage and support the use of renewable energy sources such as solar, wind, and hydropower. Tax incentives, subsidies, and concessions may be carried out to attract investors and businesses to transition to renewable energy. Investments in environmental education and awareness campaigns should be prioritized to increase public understanding of environmental issues. Besides, green consumption habits should be cultivated to promote low-carbon lifestyles. (2) In comprehensive efforts to improve global air quality and conduct attribution studies, attention should be paid not only to areas with existing well-established monitoring networks and complete data but also to regions with inaccurate or incomplete data, such as desert areas like the Gobi Desert. The causes and sources of $PM_{2.5}$ in non-urban areas should be further analyzed to promote comprehensive and effective $PM_{2.5}$ reduction. (3) Efforts should be made to strengthen the assessment of the per capita negative impacts of air pollution, with a focus on integrated exposure to multiple pollutants. The precision and spatio-temporal resolution of indicators should also be improved. (4) Future research should be conducted to expand the types of urban public open spaces and incorporate multifunctionality evaluation metrics. This would allow the diversity and comprehensive benefits of urban public open spaces to be fully reflected, providing more detailed references for urban planning and management. (5) Diversified economic models should be developed in rural areas to further reduce dependence on land for economic development. (6) Localized assessments of urban sustainability in China should be strengthened, and monitoring of changes in urban sustainable development and capacity optimization should be improved. Moreover, greater attention should be given to the sustainable development of cities in western and northeastern regions, particularly in housing, public services, environmental pollution control, and risk resilience. (7) Localized SDG 11 indicator systems that align with national circumstances and languages should be developed to promote the collection and integration of multi-source data and facilitate the quantitative assessment of development goals.

Looking forward, with the continuous development of Big Earth Data technologies, we will be able to monitor and evaluate urban development trends more comprehensively and accurately. This will allow us to make more effective use of information on population flow and resource utilization, provide rich data for sustainable cities and communities, inform policy on new indicator development, and offer a scientific basis for government planning and decision-making.

References

Anselin L (2013) Spatial econometrics: methods and models. Springer Science & Business Media, Dordrecht

Bai KX, Li K, Ma ML et al (2022) LGHAP: the long-term gap-free high-resolution air pollutant concentration dataset, derived via tensor-flow-based multimodal data fusion. Earth Syst Sci Data 14(2):907–927

Barnett C, Parnell S (2016) Ideas, implementation and indicators: epistemologies of the post-2015 urban agenda. Environ Urban 28(1):87–98

Bonsu NO (2020) Towards a circular and low-carbon economy: insights from the transitioning to electric vehicles and net zero economy. J Clean Prod 256:120659

Chen MX, Chen LK, Cheng JF et al (2022a) Identifying interlinkages between urbanization and sustainable development goals. Geogr Sustain 3(4):339–346

Chen S, Huang QX, Muttarak R et al (2022b) Updating global urbanization projections under the shared socioeconomic pathways. Sci Data 9:137

Chen ZQ, Yu BL, Song W et al (2017) A new approach for detecting urban centers and their spatial structure with nighttime light remote sensing. IEEE Trans Geosci Remote Sens 55(11):6305–6319

Creß C, Bing ZS, Knoll AC (2021) Intelligent transportation systems using external infrastructure: a literature survey. IEEE Trans Intell Transp Syst 25(7):6309–6327

da Cruz NF, Rode P, McQuarrie M (2019) New urban governance: a review of current themes and future priorities. J Urban Aff 41(1):1–19

Elhorst JP (2014) Spatial econometrics: from cross-sectional data to spatial panels. https://link.springer.com/book/10.1007/978-3-642-40340-8 [2024-08-01]

Fox S, MacLeod A (2019) Bristol and the SDGs: a voluntary local review of progress 2019. Cabot Institute for the Environment, University of Bristol, Bristol

Fox S, MacLeod A (2023) Localizing the SDGs in cities: reflections from an action research project in Bristol, UK. Urban Geogr 44(3):517–537

Goodchild M, Haining R, Wise S et al (1992) Integrating GIS and spatial data analysis: problems and possibilities. Int J Geogr Inf Syst 6(5):407–423

Greene S, Meixell B (2017) Hacking the sustainable development goals. https://www.urban.org/research/publication/hacking-sustainable-development-goals [2023-04-29]

Grossman GM, Krueger AB (1995) Economic growth and the environment. Q J Econ 110(2):353–377

He QS, He WS, Song Y et al (2018) The impact of urban growth patterns on urban vitality in newly built-up areas based on an association rules analysis using geographical "big data." Land Use Policy 78:726–738

Huang X, Wang Y (2019) Investigating the effects of 3D urban morphology on the surface urban heat island effect in urban functional zones by using high-resolution remote sensing data: a case study of Wuhan, central China. ISPRS J Photogramm Remote Sens 152:119–131

Jiang YX, Zhang YY, Liu Y et al (2023) A review of urban vitality research in the Chinese World. Trans Urban Data Sci Technol 2(2–3):81–99

Jin Y (2007) Evaluation of China's urban economic vitality. Geogr Sci 27:9–16

Kelly G (2012) Sustainability at home: policy measures for energy-efficient appliances. Renew Sustain Energy Rev 16(9):6851–6860

Li FZ, Yao N, Liu DN et al (2021a) Explore the recreational service of large urban parks and its influential factors in city clusters: experiments from 11 cities in the Beijing-Tianjin-Hebei region. J Clean Prod 314:128261

Li ZY, Guo YY, Han ZZ (2021b) Comparative study on statistical measurement of urban vitality. World Surv Res 8:74–80 (In Chinese)

Liang QL, Wang W, Mu JS et al (2020) Artificial intelligence in China: proceedings of the international conference on artificial intelligence in China. Springer, Singapore

Liu BY, Wang LL, Zhang L et al (2024) Evaluating urban and nonurban $PM_{2.5}$ variability under clean air actions in China during 2010–2022 based on a new high-quality dataset. Int J Dig Earth 17(1):2310734

Liu HY, Song YR (2020) Financial development and carbon emissions in China since the recent world financial crisis: evidence from a spatial-temporal analysis and a spatial Durbin model. Sci Total Environ 715:136771

Ozdil NC, Vejre H, Bilsel FC (2020) Emergence and evolution of the urban public open spaces of Ankara within the urban development history: 1923 to present. J Plan Hist 19(1):26–51

Pan A, Zhang WN, Shi XP et al (2022) Climate policy and low-carbon innovation: evidence from low-carbon city pilots in China. Energy Econ 112:106129

Selden TM, Song DQ (1994) Environmental quality and development: Is there a Kuznets curve for air pollution emissions? J Environ Econ Manag 27(2):147–162

Shafik N, Bandyopadhyay S (1992) Economic growth and environmental quality: time-series and cross-country evidence. https://documents.worldbank.org/en/publication/documents-reports/downloadstats?docid=833431468739515725 [2024-08-01]

Shahtahmassebi AR, Li CL, Fan YF et al (2021) Remote sensing of urban green spaces: a review. Urban Forestry & Urban Greening 57:126946

Shan M, Wang Y, Lu YL et al (2023) Uncovering $PM_{2.5}$ transport trajectories and sources at district within city scale. Jo Clean Product 423:138608

Shao CF, Chen SH, Gao JL et al (2021) Design of China's sustainable development evaluation index system based on the SDGs. China Popul Resour Environ 31(4):1–12 (In Chinese)

Sims NC, England JR, Newnham GJ et al (2019) Developing good practice guidance for estimating land degradation in the context of the United Nations sustainable development goals. Environ Sci Policy 92:349–355

Sun LQ, Chen J, Li QL et al (2020) Dramatic uneven urbanization of large cities throughout the world in recent decades. Nat Commun 11(1):5366

Tan S, Yang J, Yan JY et al (2017) A holistic low carbon city indicator framework for sustainable development. Appl Energy 185(2):1919–1930

United Nations Sustainable Development Solutions Network (2016) SDG index and dashboards report 2016. https://sdgtransformationcenter.org/reports/sdg-index-2016 [2023-04-29]

United Nations (2015) Transforming our world: the 2030 agenda for sustainable development. https://sdgs.un.org/2030agenda [2024-07-24]

Wang CJ, Wang F, Zhang XL et al (2017) Examining the driving factors of energy related carbon emissions using the extended STIRPAT model based on IPAT identity in Xinjiang. Renew Sustain Energy Rev 67:51–61

Wang LL, Liu BY, Li R et al (2023) Prediction of daily $PM_{2.5}$ and ozone based on high-density weather stations in China: nonlinear effects of meteorology, human and ecosystem health risks. Atmos Res 293:106889

Wang Q, Li LJ (2021) The effects of population aging, life expectancy, unemployment rate, population density, per capita GDP, urbanization on per capita carbon emissions. Sustain Product Consum 28:760–774

Wang Y, Li L, Kubota J et al (2016) Does urbanization lead to more carbon emission? Evidence from a panel of BRICS countries. Appl Energy 168:375–380

Wei J, Li ZQ, Lyapustin A et al (2021) Reconstructing 1-km-resolution high-quality $PM_{2.5}$ data records from 2000 to 2018 in China: spatiotemporal variations and policy implications. Remote Sens Environ 252:112136

Wei J, Li ZQ, Wang J et al (2023) Ground-level gaseous pollutants (NO_2, SO_2, and CO) in China: daily seamless mapping and spatiotemporal variations. Atmos Chem Phys 23(2):1511–1532

Xia J, Li RYM, Zhan XG et al (2022) A study on the impact of fiscal decentralization on carbon emissions with U-shape and regulatory effect. Front Environ Sci 10:964327

Yang J, Huang X (2021) The 30 m annual land cover dataset and its dynamics in China from 1990 to 2019. http://irsip.whu.edu.cn/resources/CLCD.php [2024-08-07]

Yang J, Huang X (2022) The 30 m annual land cover datasets and its dynamics in China from 1990 to 2021. https://zenodo.org/records/5816591#.Y-4bEsisGvM [2024-08-07]

Yao MF, Liu HF, Feng X (2011) The development of low-carbon vehicles in China. Energy Policy 39(9):5457–5464

Zeng C, Song Y, He QS et al (2018) Urban-rural income change: influences of landscape pattern and administrative spatial spillover effect. Appl Geogr 97:248–262

Zhang YJ (2011) The impact of financial development on carbon emissions: an empirical analysis in China. Energy Policy 39(4):2197–2203

Zhu XQ, Zhang ZQ, Chen XW et al (2022) Nexus of mixed-use vitality, carbon emissions and sustainability of mixed-use rural communities: the case of Zhejiang. J Clean Prod 330:129766

6 SDG 13 Climate Action

6.1 Background

Global greenhouse gas concentrations continue to rise, leading to record-breaking global average temperatures. This trend results in an increase in extreme weather events and an acceleration of glacier and ice cap melting, causing rising sea levels and posing a threat to biodiversity (WMO 2023). The Global Risks Report 2023 indicates that the three greatest risks facing humanity in the next decade are the inability to mitigate climate change, the inability to adapt to climate change, and extreme natural disasters, all of which are closely related to climate change. Climate change is already seriously threatening human survival and development (WEF 2023). The Sustainable Development Goals Report 2022 of the United Nations points out that the world is on the edge of a climate catastrophe, and the window for avoiding it is closing. Global greenhouse gas emissions are still increasing, and all countries need to take immediate action to shift from the tipping point of a climate catastrophe to a turning point toward a sustainable future (United Nations 2022). China released the National Climate Change Adaptation Strategy 2035 in 2022, promoting enhanced adaptive capacity and proactive responses to climate change.

SDG 13 (Climate Action) aims to take urgent action to combat climate change and its impacts. However, it is currently the most data-deficient one among all 17 SDGs, with only about 20% of countries having relevant data (United Nations 2022), and data with spatial and temporal information are even scarcer. Therefore, there is an urgent global need for data that can reflect the overall progress toward SDG indicators while providing spatial details and temporal trends to inform decisions on disaster response and climate change mitigation.

The reports from 2019 to 2022 focused on five indicators: disaster losses (SDG 13.1.1), national disaster risk reduction strategies (SDG 13.1.2), proportion of local governments with disaster risk reduction strategies (SDG 13.1.3), greenhouse gas emissions per year (SDG 13.2.2), and climate change education (SDG 13.3.1). They provided indicator calculation methods, spatio-temporal data products, and scientific decision support at both the Chinese and global scales.

This year's report for the first time includes China's climate change response strategies (SDG 13.2.1/SDG 13.b.1), and provides new developments in flood disaster impacts and public awareness of climate change, providing scientific data support for formulating disaster prevention and reduction, and greenhouse gas emission reduction strategies.

H. Guo, *Big Earth Data in Support of the Sustainable Development Goals (2023)—China*, Sustainable Development Goals Series, https://doi.org/10.1007/978-981-95-2636-9_6

6.2 Main Contributions

This chapter focuses on the three specific goals of SDG 13, based on multi-source data such as satellite remote sensing, surveys, and statistics. Through three cases, it provides model methods for indicator monitoring and analysis of SDG 13, as well as data products with rich spatio-temporal information. Based on these, decision recommendations are proposed as follows (Table 6.1).

6.3 Case Studies

6.3.1 Spatial Distribution and Impact Analysis of Flood Disasters in China

Target: SDG 13.1 Strengthen resilience and adaptive capacity to climate-related hazards and natural disasters in all countries

- **Background**

SDG 13.1 is one of the critical targets of the SDGs (United Nations 2015). Against the backdrop of global warming, the frequency of extreme weather disasters worldwide and the economic losses caused by them have multiplied, threatening the security of lives and property of the people, and restricting the sustainable development of society and economy. In terms of disaster frequency, economic losses, and affected populations, flood disasters are among the most severe natural disasters globally. Among all economic losses caused by natural disasters worldwide, flood disasters account for the highest proportion, reaching 40%. China is one of the countries most prone to frequent and severe flood disasters, with approximately two-thirds of its territory potentially affected by various types and degrees of flooding. Remote sensing technology provides an effective technical means for flood disaster monitoring (DeVries et al. 2020; Zhu et al. 2022).

The years 2020 and 2021 represent the most severe flood disaster years in southern and northern China over the past decade, respectively. During the main flood season of 2020, southern China experienced the most severe flooding since 1998. According to the China Flood and Drought Disaster Prevention Bulletin in recent years, half of the top ten natural disasters in the country were flood disasters, including the catastrophic heavy rainfall and flooding in the Yangtze River and Huai River basins in July, severe rainstorms and flooding in Sichuan-Chongqing and Shaanxi-Gansu-Yunnan regions in mid-August, heavy rainfall and flooding in the South China regions in early-to-mid-June, and heavy rainfall and flooding in Southwest China in late June. These events caused the total affected area of crops to account for 76.9% of the total affected area of the top ten natural disasters nationwide, and the total economic losses accounted for 86.0% of the total economic losses from the top ten natural disasters. Compared to 2020, northern China experienced above-average precipitation in 2021, ranking second in history, only behind 1964. Flood disasters accounted for half of the top ten disasters nationwide in 2021, including the catastrophic heavy rainfall in Henan in mid-to-late July, severe autumn flooding in the Middle and Lower Reaches of the Yellow River, heavy rainfall and flooding in Shanxi in mid-to-late July, heavy rainfall and flooding in Hubei in early-to-mid-August, and heavy rainfall and flooding in Shaanxi in mid-to-late August. The total affected area of crops accounted for 80.2% of the total affected area of the top ten natural disasters nationwide, and the total economic losses accounted for 84.9% of the total economic losses from the top ten natural disasters. Multi-source and multi-temporal remote sensing data can be used to extract the pre-disaster and post-disaster water body ranges, obtaining the spatial distribution of flood inundation areas.

- **Data**

- Satellite data from Sentinel-1 and Sentinel-2 for 2020 and 2021.

Table 6.1 Case names and their main contributions

Targets	Case studies	Contributions
SDG 13.1 Strengthen resilience and adaptive capacity to climate-related hazards and natural disasters in all countries	Spatial distribution and impact analysis of flood disasters in China	Data product: losses due to flooding from 2010 to 2021, and flood inundation distribution in China in typical years Decision support: the overall change of major flood disaster indicators in China from 2010 to 2021 shows a downward trend, and meteorological, hydrological, and climate information helps disaster prevention and reduction
SDG 13.2 Integrate climate change measures into national policies, strategies, and planning	China's climate change strategies and actions	Data product: a dataset of China's climate change policy actions from 2007 to 2023 has been constructed Decision support: China has integrated measures to address climate change into national policies, strategies, and plans, and has earnestly fulfilled its international obligations concerning climate change
SDG 13.3 Improve education, awareness-raising, and human and institutional capacity on climate change mitigation, adaptation, impact reduction, and early warning	Survey on climate change education and communication in China	Decision support: presenting a map of public awareness regarding climate change issues across the seven geographic regions of China

Sentinel-1A C-band data were utilized with a spatial resolution of 5 m × 20 m and a revisit cycle of 12 days. Sentinel-2A visible light and near-infrared data were utilized with a spatial resolution of 10 m and a revisit cycle of 10 days.
Data source: Copernicus Open Access Hub (https://scihub.copernicus.eu/).
– Satellite data from Gaofen-3 for 2020 and 2021.
Gaofen-3 satellite data consist of domestically produced C-band synthetic aperture radar (SAR) data with a spatial resolution of 5 m.
Data source: China Center for Resources Satellite Data and Application (https://data.cresda.cn/#/home).
– GlobeLand30 land cover map in China for 2020.
GlobeLand30 is the world's first 30 m resolution global land cover dataset developed by China. The dataset includes 10 major land cover types, and this case study utilized the China regional land cover map from GlobeLand30.
Data source: National Catalogue Service for Geographic Information (https://www.webmap.cn/commres.do?method=globeIndex).
– China Flood and Drought Disaster Prevention Bulletin from 2010 to 2021.
Source: Ministry of Water Resources of the People's Republic of China (http://www.mwr.gov.cn/sj/tjgb/zgshzhgb/).
– China Statistical Yearbook from 2010 to 2021.
Source: National Bureau of Statistics (http://www.stats.gov.cn/).

- **Methods**

Multi-source data, including Gaofen-3 and Sentinel-1, underwent preprocessing steps such as terrain correction and filtering. The region-growing method was employed to extract the maximum monthly water body coverage area in the region, which was then subtracted from the normal water body range to determine the monthly flood coverage range, area, and spatio-temporal variations across China. Subsequently, the flood disaster impacts were assessed by integrating population distribution and socioeconomic data, and conducting spatial distribution surveys of flood disasters in China for 2020 and 2021. Furthermore, statistical data from the China Flood and Drought Disaster Prevention Bulletin and the China Statistical Yearbook (2010–2021), including data on affected population, deaths and missing persons, direct economic losses, year-end total population, and GDP, were utilized to analyze trends in key flood disaster indicators such as affected population, deaths and missing persons, affected crop area, and direct economic losses during 2010–2021. The effectiveness of China's flood prevention and mitigation measures was also evaluated.

- **Results and Analysis**

1. **Spatial distribution analysis of flood disasters in 2020 and 2021**

Using multi-source remote sensing data, the flood inundation distribution ranges in China for 2020 and 2021 were extracted (Fig. 6.1). The results indicate that the flood disasters in 2020 mainly occurred in the Pearl River Basin and Taihu Basin in June, the Huai River Basin in July, the Yangtze River Basin and Yellow River Basin in August, and the Songliao River Basin in September. In 2021, the flood disasters were primarily distributed in the Songliao River Basin, Henan Province, and Hebei Province in July, the Taihu Basin in August, and the Yangtze River Basin, Yellow River Basin, and Haihe River Basin in September.

2. **Analysis of changes in major flood disaster losses from 2010 to 2021**

The spatial distribution of flood disaster losses in China in 2020 and 2021 is shown in Fig. 6.2. In 2020, Anhui, Sichuan, Jiangxi, and Hubei provinces were severely affected by floods, with direct economic losses reaching 163.93 billion yuan, accounting for 61.4% of the national total. The affected crop area was 3,986.1 khm^2, accounting for 55.4% of the national total, and the completely destroyed crop area was 904.9 khm^2, accounting for 68.5% of the national total. A total of 41,000 houses collapsed, accounting for 45.6% of the national total, and 42.47 million people were affected, accounting for 54.4% of the national total, with 103 deaths and missing persons, representing 36.9% of the total deaths and missing persons caused by floods nationwide. In 2021, Henan, Shaanxi, Sichuan, and Shanxi provinces were most severely affected, with direct economic losses from floods reaching 190.78 billion yuan, accounting for 77.6% of the national total. The affected crop area was 2,112.1 khm^2, accounting for 44.4% of the national total, and the completely destroyed crop area was 451.5 khm^2, accounting for 51.8%. A total of 126,000 houses collapsed, accounting for 82.7% of the national total, and 34.572 million people were affected, accounting for 58.6% of the national total, with 508 deaths and missing persons, representing 86.1% of the total deaths and missing persons caused by floods nationwide.

From the changes in major flood disaster indicators in China from 2010 to 2021 (Fig. 6.3), an overall downward trend can be observed. The affected population decreased from 210.8468 million in 2010 to 59.0101 million in 2021. The number of deaths decreased from 3,222 in 2010 to 512 in 2021, while the number of missing persons dropped from 1,003 in 2010 to 78 in 2021. In 2010, the affected crop area was 17,866.69 khm^2, and the completely destroyed crop area was 2,469.77 khm^2. By 2021, the affected crop area had decreased to 4,760.43 khm^2, and the completely destroyed crop area had decreased to 872.35 khm^2. Direct economic losses declined from 374.543 billion yuan in 2010 to 245.892 billion yuan in 2021, and the percentage of direct economic losses to GDP decreased from 0.93% in 2010 to 0.22% in 2021. Compared to the average values in 2010–2020, the flood-related affected population, deaths, and missing persons, affected

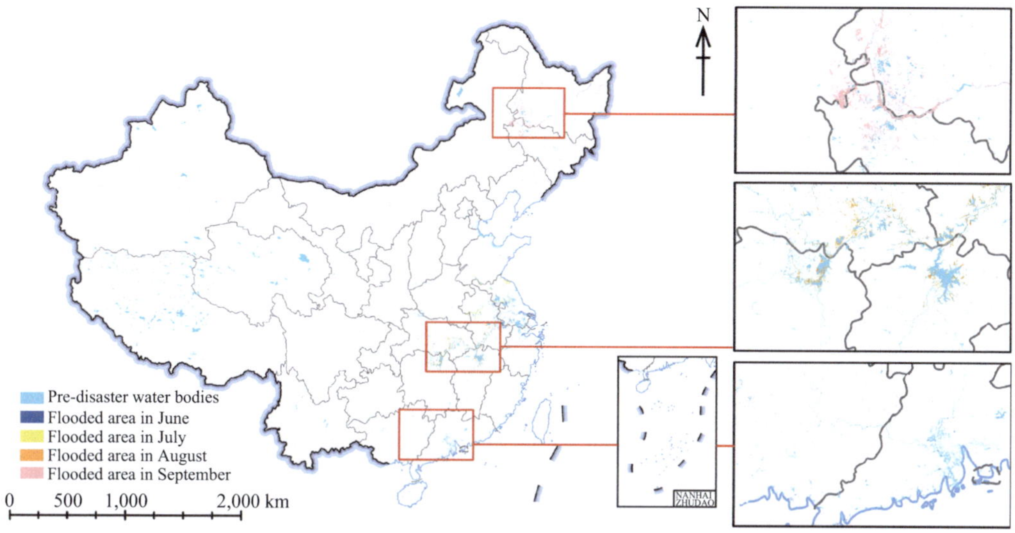

(a) 2020

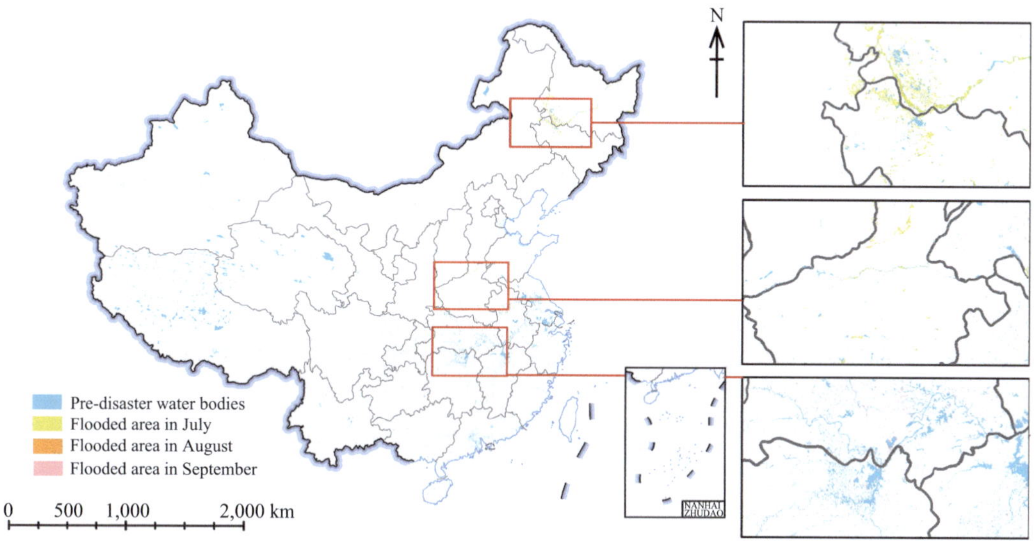

(b) 2021

Fig. 6.1 Spatial distribution of flood inundation areas in China in 2020 and 2021

crop area, completely destroyed crop area, and percentage of direct economic losses to GDP in 2021 decreased by 37.1%, 36.5%, 44.9%, 31.3%, and 40.8%, respectively. Overall, the total losses from flood disasters showed a declining trend from 2010 to 2021.

3. **Flood prevention and mitigation benefits of national flood disaster defense measures**

To ensure flood control safety and reduce disaster losses, the nation adheres to the principles of being "systematic, coordinated, scien-

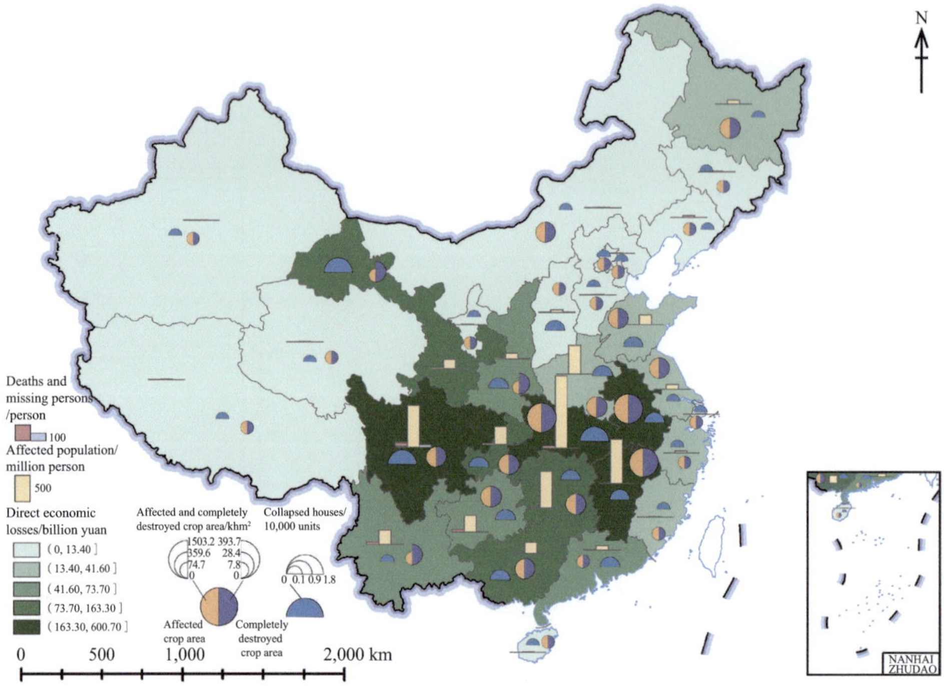

(a) 2020

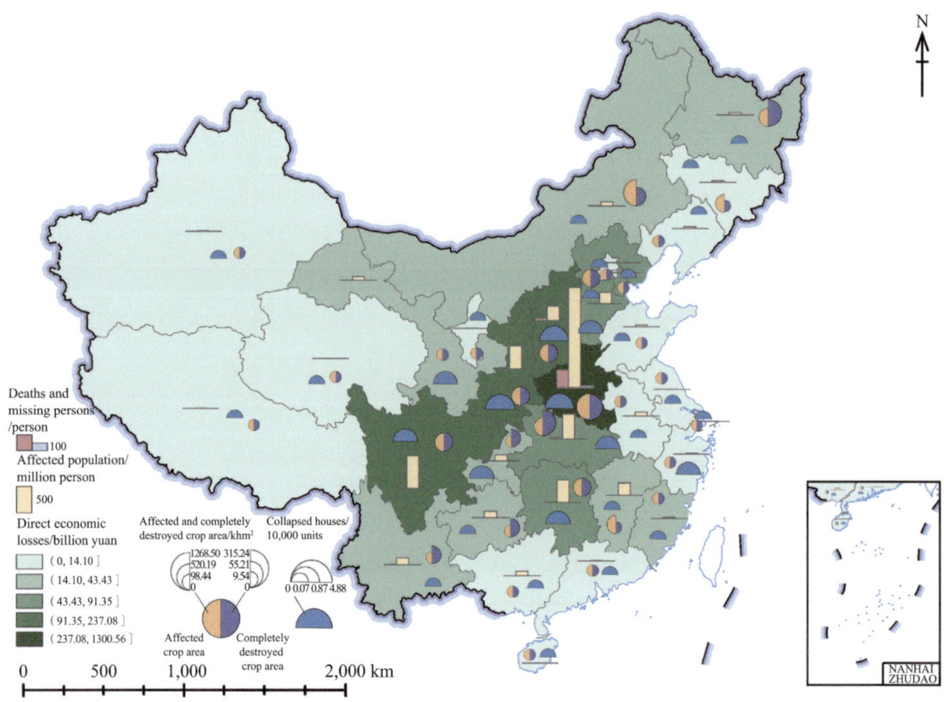

(b) 2021

Fig. 6.2 Spatial distribution of flood disaster losses in China in 2020 and 2021. *Note* No data for Taiwan Province

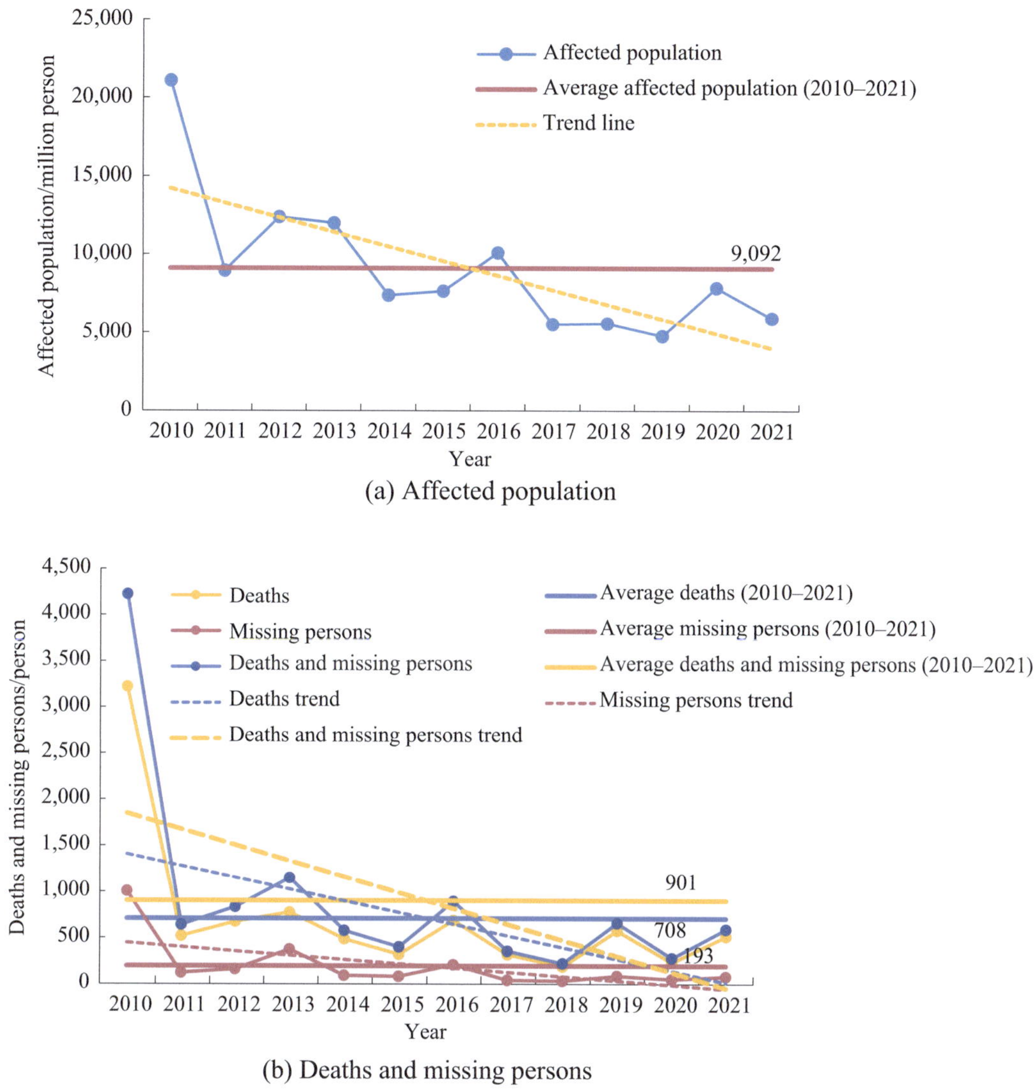

Fig. 6.3 Changes in major flood disaster indicators in China from 2010 to 2021

tific, and safe" in flood monitoring and early warning, consultation and response, flood control scheduling, and emergency rescue support. During the 2021 flood season, 4,347 large and medium-sized reservoirs were utilized for flood retention, storing 139 billion m^3 of floodwater. Additionally, 11 national flood storage and detention areas were activated, storing 1.328 billion m^3 of floodwater. The measures prevented flooding in 1,494 urban areas, protected 1,689.4 khm^2 of farmland, and reduced the affected population by 15.246 million people. The flood prevention and mitigation benefits of China's flood disaster defense measures from 2010 to 2021 are shown in Fig. 6.4. Statistical analysis indicates that from 2010 to 2021, the average reduction in the affected population was 25.955 million people, and the average reduction in the flooded farmland area was 2,887.3 khm^2.

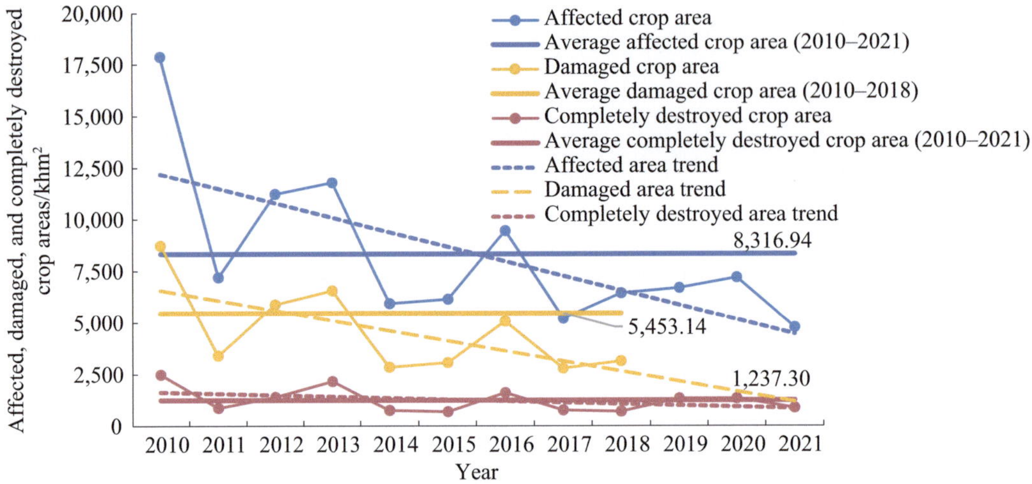

(c) Affected, damaged, and completely destroyed crop areas

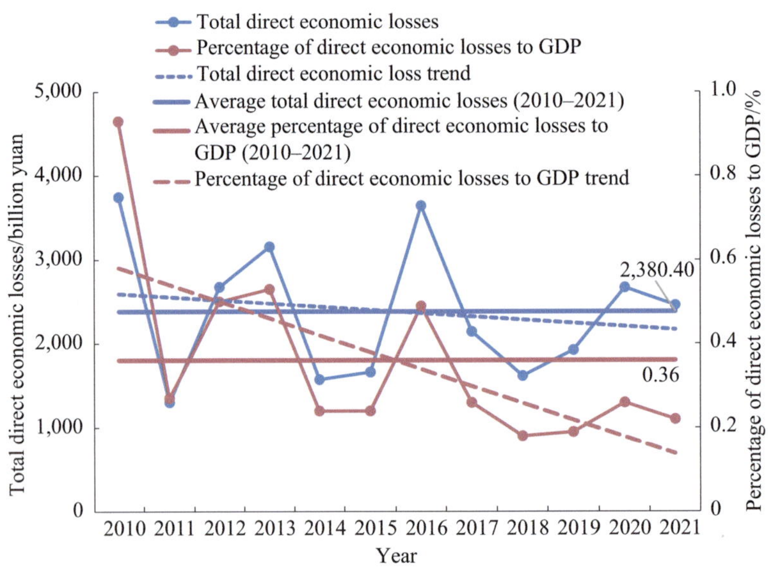

(d) Direct economic losses

Fig. 6.3 (continued)

Highlights

- Based on the Sendai Framework for Disaster Risk Reduction 2015–2030 and Big Earth Data, the first evaluation of the SDG 13.1.1 indicator was conducted in China to assess the number of disaster-related deaths, missing persons, and directly affected individuals per 100,000 persons due to floods.
- From 2010 to 2021, flood-related losses in China exhibited a declining trend. Compared to the average values for 2010–2020, in 2021, the affected population, deaths, and missing persons, affected crop area, completely destroyed crop area, and percentage of direct economic losses to GDP decreased by 37.1%, 36.5%, 44.9%, 31.3%, and 40.8%, respectively.

Fig. 6.4 Flood prevention and mitigation benefits of China's flood disaster prevention measures (2010–2021)

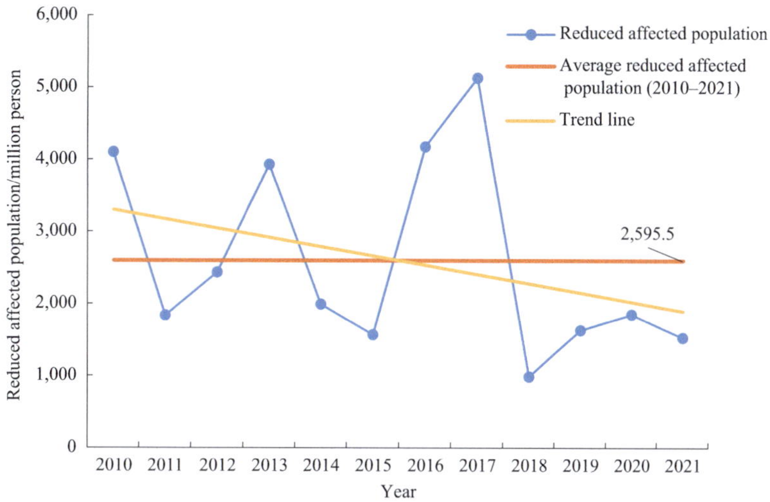

(a) Reduced affected population

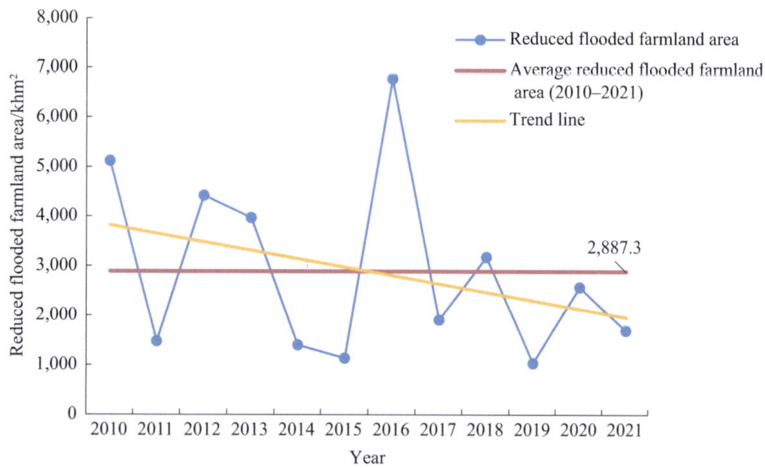

(b) Reduced flooded farmland area

– From 2010 to 2021, China's flood disaster prevention measures improved year by year, and the impact of flood disasters gradually weakened. The average reduction in the affected population during this period was 25.955 million people, and the average reduction in flooded farmland area was 2,887.3 khm^2.

- **Discussion and Outlook**

To quantitatively analyze the casualties and property losses caused by flood disasters, and to promote practical actions and effective measures for flood disaster prevention and mitigation, this case study responds to the SDG 13.1.1 indicator, which evaluates the number of disaster-related deaths, missing persons, and directly affected individuals per 100,000 people. It aims to assess the capacity of countries to resist and adapt to climate-related and natural disasters. Based on Sentinel-1, Gaofen-3, optical data, and statistical survey data, this case study extracts the spatial distribution of flood inundation in China for the typical years 2020 and 2021. It also analyzes changes in key flood disaster indicators, including affected population, deaths and missing persons, affected crop area, and direct economic losses from 2010 to 2021. Furthermore, it evaluates the flood prevention and mitigation benefits of national flood disaster prevention measures.

Using multi-source remote sensing data, this case study analyzes the spatial distribution of flood inundation in China for the typical years 2020 and 2021. In terms of rainfall distribution, China's annual precipitation exhibits a distinct pattern of "more in the south, less in the north; more in the east, less in the west". Most areas in Northwest China receive less than 400 mm of total annual precipitation, whereas annual precipitation in southern regions generally exceeds 800 mm. Similarly, flood disasters in China also show a pattern of "more in the east, less in the west", necessitating further analysis of the correlation between floods and meteorological factors such as precipitation.

From 2010 to 2021, the nationwide flood situation showed a downward trend, with meteorological, hydrological, and climate information effectively supporting disaster prevention and mitigation efforts. Although natural disasters cannot be completely eliminated, disaster defenses can minimize the impacts and losses caused by them. Timely, accurate, and rapid meteorological disaster warnings form the first line of defense against floods. Meteorological and hydrological departments collaborate to share information and jointly issue warnings, providing effective safeguards against flood disasters. Human interventions, such as flood retention by large and medium-sized reservoirs and the use of national flood storage and detention areas, have significantly reduced disaster losses (Aerts et al. 2018).

It is important to note that, due to climate change, extreme weather and flood events are becoming increasingly frequent and intense worldwide. China is also more vulnerable than ever to multiple associated hazards, further exacerbated by population growth, urbanization, and environmental degradation. According to the World Meteorological Organization, by 2030, an estimated 50% of the global population will live in coastal areas prone to floods, storms, and tsunamis. As extreme heavy rainfall events are likely to continue in the future, the impacts of severe flood disasters remain inevitable. While China's flood disaster prevention measures have made positive progress, further analysis based on historical flood disaster data is necessary. Evaluating the effectiveness of various flood prevention and rescue measures will provide robust support for improving flood disaster defense efforts.

6.3.2 China's Climate Change Strategies and Actions

Target: SDG 13.2 Integrate climate change measures into national policies, strategies, and planning

- **Backgrounds**

Climate change represents a significant challenge to global sustainable development. Since the 1980s, an increasing number of countries have gradually recognized the severity of climate change and have taken proactive measures. Policies serve as a potent instrument for governments worldwide in addressing climate issues, with the formulation and issuance of relevant policies providing robust assurance for the implementation of climate action. SDG 13.2.1, which tracks the number of countries that have communicated nationally determined contributions, long-term strategies, national adaptation plans, and adaptation communications to the secretariat of the United Nations Framework Convention on Climate Change (UNFCCC), is designed to monitor and guide policy actions taken by countries in response to climate change. China has consistently placed great emphasis on tackling climate change. In China, climate action has evolved into a government-led endeavor with multi-stakeholder participation. However, there has not yet been an evaluation of the completion status of SDG 13.2.1, including assessments of the effectiveness of China's actions and the compilation of relevant policy datasets. This case study primarily analyzes and evaluates the current state of China's climate action from the perspective of policy data.

- **Data**

Since 2007, government documents related to climate action have been issued by China's national and provincial governments, the Ministry of Ecology and Environment, the China Meteorological Administration, and other relevant departments covering areas such as green low-carbon development, clean energy, ecological conservation, and water resource management, all of which fall under the scope of addressing climate change. In 2007, China promulgated and implemented its first comprehensive policy document on climate change, the National Program for Addressing Climate Change, which laid out a framework for responding to climate change within the context of sustainable development. Therefore, this case study selects 2007 as the starting point for the policy data, marking the inception of China's systematic approach to climate policy.

- **Methods**

Design method of evaluation indicators: In alignment with SDG 13.2.1, evaluation indicators were designed to correspond. These include the reporting of national long-term strategies to the secretariat of UNFCCC, the work conducted by countries on adapting to climate change, adaptation communications, and the possession of nationally determined contributions.

Policy dataset construction and analysis method: Based on over 50 search terms collected during preliminary work on tracking the trends of climate change, relevant government documents were retrieved and gathered from official government websites, the PKU Law database, and other sources. A total of 2,144 government documents related to China's climate change actions from 2007 to 2023 (as of May 15, 2023) were selected, including plans, schemes, regulations, decisions, programs, and reports. This selection formed the first dataset of China's climate change policy actions. A statistical analysis was conducted based on various attribute information contained within these government documents.

- **Results and Analysis**

1. **China submitted its nationally determined contributions (NDCs) and long-term strategies to the secretariat of the UNFCCC**

China actively implements the UNFCCC and has carried out a series of strategies, measures, and actions to address climate change. In 2007, China issued its first policy document on climate change, the National Program for Addressing Climate Change, emphasizing a response to climate change within the framework of sustainable development and giving equal emphasis to mitigation and adaptation. Following this, the Chinese government successively released action plans addressing climate change across various sectors including forestry, industry, and scientific innovation. Specific theme policies were also introduced, targeting the control of greenhouse gas emissions, promoting green and low-carbon circular development, implementing energy conservation and carbon reduction, and strengthening meteorological disaster monitoring, early warning, and prevention. In 2015, China submitted the Enhanced Actions on Climate Change: China's Intended Nationally Determined Contributions to the UNFCCC secretariat. In 2021, it further submitted China's Achievements, New Goals and New Measures for Nationally Determined Contributions and China's Mid-Century Long-Term Low Greenhouse Gas Emission Development Strategy, demonstrating China's determination and efforts in actively addressing global climate change.

2. **China has also released the National Climate Change Adaptation Strategy, coordinating efforts for climate change adaptation nationwide**

In 2013, China issued its first strategic plan specifically focusing on climate change adaptation, the National Climate Change Adaptation Strategy, signifying the elevation of climate

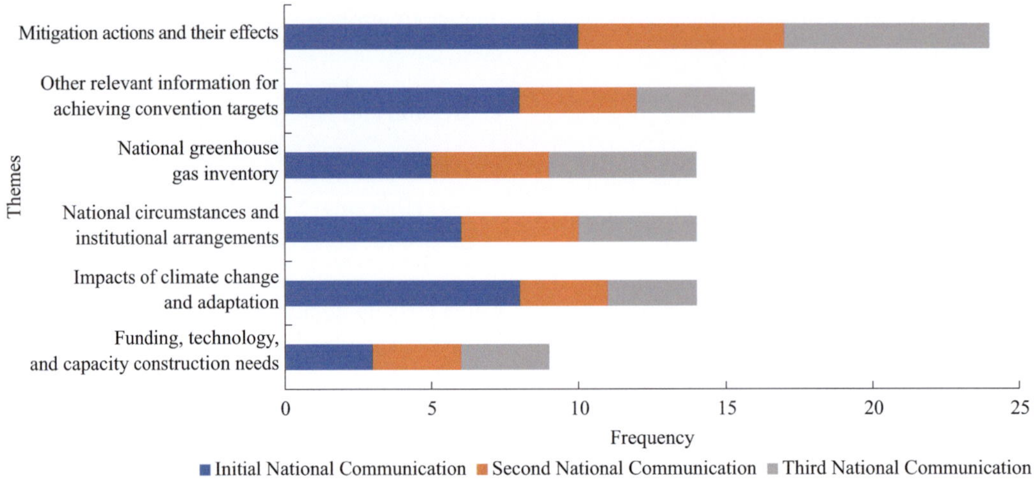

Fig. 6.5 Theme distribution of China's three national communications chapters

change adaptation to the level of national strategy. In 2022, the National Climate Change Adaptation Strategy 2035, jointly issued by 17 government departments, was released, reflecting China's prioritization of proactive climate change adaptation as one of the key focuses of its climate action. During the 13th Five-Year Plan period, local governments formulated climate change-related plans to guide the implementation of climate change mitigation and adaptation efforts. A series of policies and measures to enhance climate change adaptation capabilities were issued in various sectors, including agriculture, water resources, forestry, human health, disaster prevention, and reduction, with the aim of enhancing resilience to climate change in natural ecosystems and economic and social sectors. Currently, provinces are formulating provincial climate change adaptation action plans based on the Guidelines for the Preparation of Provincial Climate Change Adaptation Action Programs.

3. **China actively fulfills its international obligations and communicates its actions and achievements in addressing climate change to the international community**

As a non-Annex I party to the UNFCCC, China has fully disclosed its actions and achievements in addressing climate change to the international community. As of August 2023, China has submitted several documents to the UNFCCC secretariat, including the Third National Communication on Climate Change of the People's Republic of China and the Second Biennial Update Report on Climate Change of the People's Republic of China, and completed the Fourth National Communication on Climate Change of the People's Republic of China and the Third Biennial Update Report on Climate Change of the People's Republic of China, demonstrating significant progress in mitigation actions (Fig. 6.5).

Highlights

- A dataset of China's climate change policy actions from 2007 to 2023 has been constructed.
- China has integrated measures to address climate change into national policies, strategies, and plans, and has earnestly fulfilled its international obligations concerning climate change. The proportion of provincial governments that have developed and implemented local climate change strategies reaches 100%.

- **Discussion and Outlook**

By constructing a dataset of China's climate change policy actions from 2007 to 2023 and employing a policy text quantification method, the progress in implementing indicators of China's climate change policy actions has been evaluated. To mitigate and adapt to global warming, and promote the sustainable development of the economy and society, the Chinese government has implemented an active national strategy to address climate change, contributing its due share to global climate action through concrete measures. The establishment of the national park system serves as an exemplary practice of China's efforts in tackling climate change.

6.3.3 Survey on Climate Change Education and Communication in China

Target: SDG 13.3 Improve education, awareness-raising, and human and institutional capacity on climate change mitigation, adaptation, impact reduction, and early warning

- **Background**

Mitigating climate change requires broad public participation. Social media, with its decentralized and participatory nature, is becoming an important platform for information exchange, opinion formation, and public engagement. On social media, information about climate change is spreading widely, especially in recent years as people frequently encounter extreme weather events. The public has spontaneously engaged in discussions, debates, and scientific popularization about climate change. Therefore, understanding how the public discusses climate change on social media is crucial. As China increasingly becomes a key leader in global climate change efforts, presenting a case study that showcases the focus and attitudes of the Chinese public in their discussions about climate change is necessary.

- **Data**

Sina Weibo is one of China's largest social media platforms with 450 million monthly active users. This case study uses Weibo post data, with the data collection period set from January 1, 2022, to December 31, 2022.

- **Methods**

1. **Data collection**

This case study adopted a random sampling approach, with the sampling rule set to randomly select one day per week. Python software was used to collect data from Sina Weibo for the specified 2022 time period. The selected keywords included "climate change", "global warming", "extreme weather", "Dual Carbon", etc. The main data columns included user ID, post time, post content, user IP location (provincial/municipal level), number of shares, number of likes, number of comments, etc. After data cleaning to remove duplicates and irrelevant data, a total of 53,788 valid data entries were obtained for the period from January 1, 2022, to December 31, 2022.

2. **Data preprocessing**

In this case, the Python-Jieba library was used to perform Chinese word segmentation on the data. A common stop-word list was applied to filtering out meaningless pronouns, conjunctions, and adverbs, creating a corpus for the experiment.

3. **Quantitative analysis**

Based on the identification of the IP addresses of the poster, this case study analyzes the overall status and regional characteristics of public discussions on various climate change topics.

4. **Topic modeling**

The Python-Gensim library was used in this case, employing the term frequency-inverse document frequency (TF-IDF) method to extract features from Weibo texts. The latent Dirichlet allocation (LDA) topic model was then applied to performing topic modeling on the Weibo texts, aiming to uncover the thematic structure related to climate change hidden within the document set.

5. **Sentiment analysis**

This case uses the Baidu Smart Cloud (https://cloud.baidu.com) text sentiment analysis application programming interface (API) to analyze the sentiment tendencies of Weibo posts. The main focus is on classifying the sentiment polarity of the post content into categories (positive, negative, or neutral).

- **Results and Analysis**

1. **The climate change communication capacity across China's seven geographical regions is quite different, and the three major economic belts of China demonstrate stronger climate change communication capacity, while regions with more severe natural environments lack the capacity to actively set climate change topics**

In economically developed regions such as South China, North China, and East China, the communication capacity on climate change issues is stronger. The public in the Pearl River Delta, Yangtze River Delta, and Beijing-Tianjin-Hebei economic belts is more active in engaging with climate change topics on social media, showing more interest, shares, and comments, with communication capacity scores all above 80 points. Among them, South China exhibits the highest level of activity, with the public taking the initiative to set climate change topics and engage in discussions, achieving a communication capacity score of 85.03 points (Table 6.2).

In contrast, regions like Northwest China, Southwest China, Central China, and Northeast China, which frequently experience natural disasters or extreme weather events due to climatic factors, show lower levels of proactive engagement. In these areas, the public tends to passively receive information rather than actively set topics. The number of active users commenting and sharing climate change-related posts is also lower. Specifically, in Central China, where high temperatures and heat waves are common, climate change topics are less prevalent on social media. The number of comments and shares is only higher than that of Northeast China, with a communication capacity score of only 73.96 points.

2. **Economic development and the celebrity effect are the two major drivers of viral spread of climate change issues, and the introduction of the "Dual Carbon" goals has effectively increased public attention to climate change in China**

This case study used the LDA method to extract topics from documents processed by TF-IDF (Table 6.3). After conducting a topic analysis of the data, it was found that public discussions on climate change on Sina Weibo in China often revolve around economic issues. The public generally perceives climate change as an economic problem and emphasizes the importance of addressing it through economic and technological solutions. Furthermore, the introduction of China's "Dual Carbon" goals has led to the emergence of new job categories, such as "carbon traders", sparking discussions in the job market about the transformation of traditional industries and the demand for new types of talent. This indirectly reflects the impact of policies on the economy and market (Fig. 6.6). Through Topic 5 and Topic 6, we can better understand the complex views of the Chinese public on climate change—seeing it both as a threat and as a significant opportunity for economic transformation. Topic 1 and Topic 4, to some extent, reflect the celebrity effect in climate change discussions. For example, the name of celebrities such as "Junkai Wang", "Yibo Wang", and "Lei Wu" appear as keywords. Junkai Wang, Yibo Wang, and Lei Wu are well-known male young celebrities in China, each with more than 40 million Weibo followers. Their names frequently appear alongside words such as "charity" and "ambassador" because they actively participate in climate action

Table 6.2 Regional communication capacity of climate change on Chinese social media

No	Regions	Resident population/ billion	Number of posts	Number of likes	Number of comments	Number of shares	Active users	Communication capacity score/ points	Communication capacity rank
1	North China	0.28	4,783	11,370	2,931	14,610	3,901	82.45	2
2	Northeast China	0.11	1,498	574	327	369	1,318	77.33	5
3	East China	0.48	12,749	19,117	3,817	11,151	10,782	81.57	3
4	Central China	0.38	3,888	2,417	451	1,718	3,359	73.96	7
5	South China	0.23	6,063	18,895	3,431	6,430	5,144	85.03	1
6	Southwest China	0.26	3,439	1,746	650	5,281	2,978	76.39	6
7	Northwest China	0.21	1,458	3,062	719	2,615	1,216	78.31	4

Table 6.3 Topic classification of climate change communication on Chinese social media

Topic	Topic names	Keywords
1	Public awareness Campaign	Ocean, sea level, environment agency, goodwill ambassador, United Nations, Earth, glacier, climate, Junkai Wang, renewable energy
2	Biodiversity protection	Protection, rescue, wildlife, public welfare, diversity, ambassador, climate change, forest protection
3	Safety guidelines	Green, Sichuan, climbing, summer, lightning, safety precautions, emergency, guidelines, weather, extreme, storm, Sun-Earth system
4	Celebrity communication	Reporter, weather, Yibo Wang, Bingbing Wang, Lei Wu, Earth, scientists, Arctic, refugees, witch hunts
5	Dual carbon and energy conservation	Green, dual carbon, carbon peak, energy, goals, carbon neutrality, climate change, energy conservation, ecological environment, economy, high temperature
6	Industrial economy	Artificial intelligence, internet, talent gap, career change, electricity, carbon dioxide

charity events, which triggers a large number of related discussions. This indicates that young celebrities play a significant role in influencing public opinion in China, especially regarding the views of younger generations on climate change.

3. **Unlike the polarized discussions about climate change on social media in Western countries, where there is a clear division between the "truth" and "falsehood" camps, the Chinese public on social media exhibits a positive attitude toward the country's "Dual Carbon" goals and efforts to mitigate climate change**

This case study utilizes Baidu's text sentiment analysis API to analyze the sentiment of climate change-related posts on Weibo. Overall, the sentiment analysis reveals that 70.08% of the posts express positive emotions, 27.48% express negative emotions, and 2.45% are neutral. This indicates that discussions on Chinese social media are characterized by positivity, optimism, and confidence (Fig. 6.7).

Highlights

– This case study proposes that the dissemination of sustainable development education and climate change science knowledge in China's

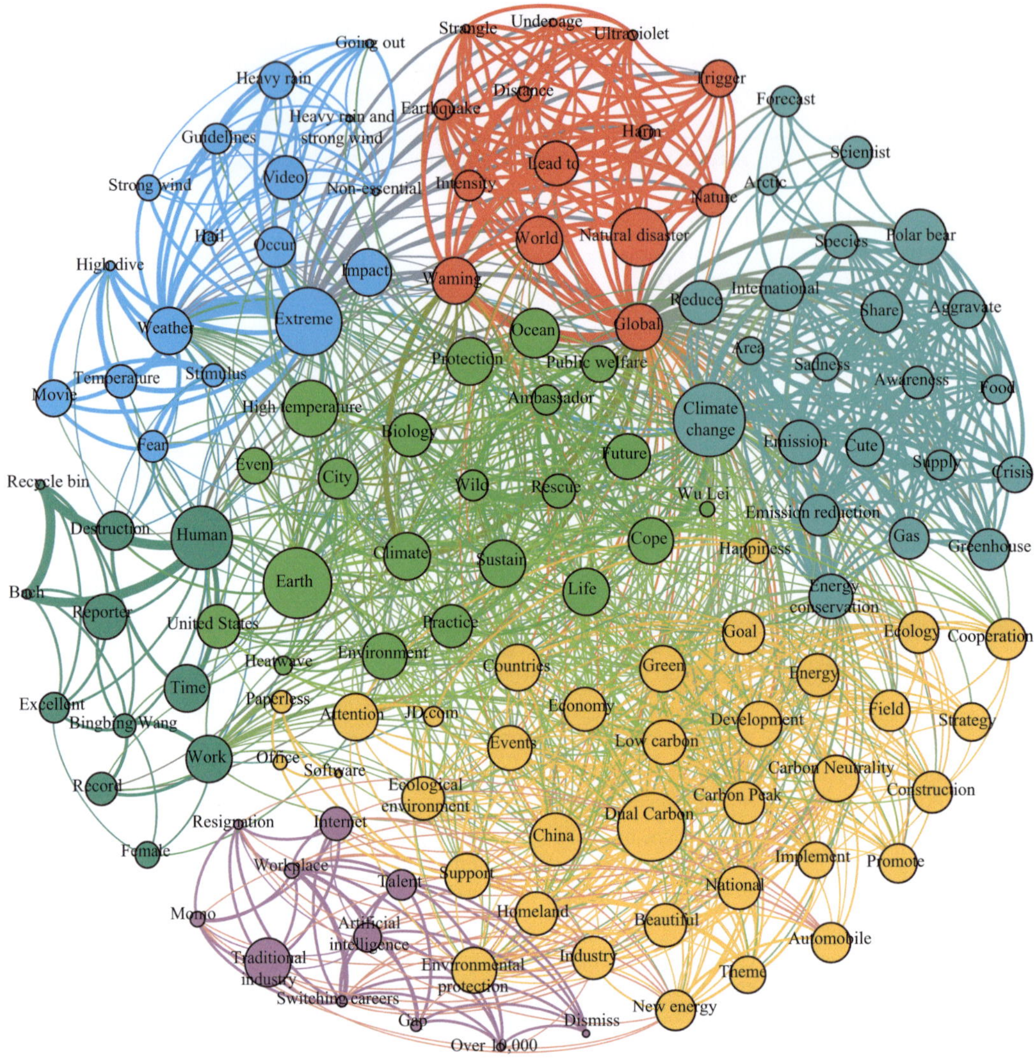

Fig. 6.6 Co-occurrence network of post topics on Weibo

primary and secondary education stages relies on media communication and the diffusion of information through social networks. As a key platform for the daily information acquisition of the Chinese public, social media serves not only as an important source of climate change information, but also as a space to gauge public attention to climate change issues and emotional preferences. To some extent, it reflects the public's intention to engage in climate action.

– This case study employed big data mining techniques and methods to collect the 2022 annual data on climate change topics from Sina Weibo, China's most influential social media platform. A quantitative analysis was used to assess the public's attention to climate change, while the keyword and topic analysis was employed to present the themes of public discussions on climate change. A sentiment analysis was conducted to assess the public's confidence in China's "Dual Carbon" climate policy and action. This case study presents a comprehensive overview of the public's understanding of climate change issues across China's seven geographic regions.

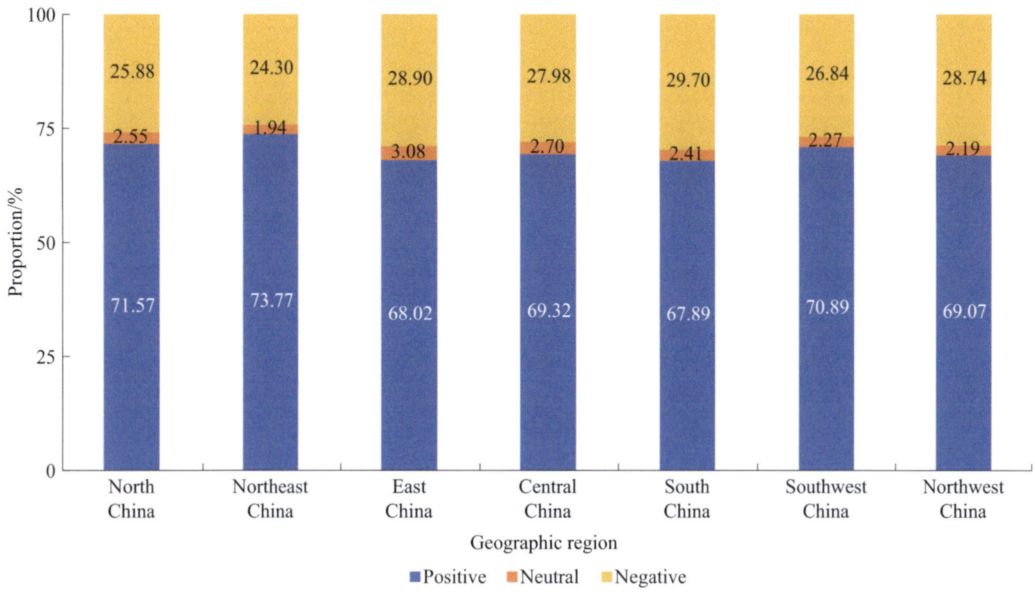

Fig. 6.7 Sentiment preferences of climate change-related posts across seven geographic regions

- **Appendix and Notes**

1. **Provinces and municipalities included in the seven geographic regions**

   North China: Beijing, Tianjin, Hebei, Shanxi, Nei Mongol Autonomous Region

   Northeast China: Heilongjiang, Jilin, Liaoning

   East China: Shanghai, Jiangsu, Zhejiang, Anhui, Jiangxi, Shandong, Fujian, Taiwan

   Central China: Henan, Hubei, Hunan

   South China: Guangdong, Guangxi Zhuang Autonomous Region, Hainan, Hong Kong Special Administrative Region, Macao Special Administrative Region

   Southwest China: Chongqing, Sichuan, Guizhou, Yunnan, Xizang Autonomous Region

   Northwest China: Shaanxi, Gansu, Qinghai, Ningxia Hui Autonomous Region, Xinjiang Uygur Autonomous Region

2. **Standardized algorithm for climate change communication influence across different geographic regions**

   Standardization or normalization of data: First, each column of data is divided by the population of the corresponding region to control the interference of the population size. Then, each column of data is either standardized or normalized to a common scale. Standardization can be achieved by calculating the difference between each data point and the mean of that column, divided by the standard deviation. Normalization is achieved by dividing each data point by the range of that column (i.e., the maximum value minus the minimum value).

   Summing and aggregating: After standardizing or normalizing the data, the values are summed column by column to obtain a total value. This results in a single column representing the aggregated influence of climate change communication across the region.

   Ranking: Based on the total values in the aggregated column, the original data are ranked to identify the regions with the highest and lowest communication influence on climate change.

3. **Baidu text sentiment analysis API**

   The Baidu text sentiment analysis API (https://cloud.baidu.com) was used to analyze the sentiment of Weibo posts. It is useful for determining overall sentiment trends in social media discussions, offering insights into public

Table 6.4 Examples of content and sentiment bias

No	Content	Data source	Sentiment bias
1	Skiing is one of the most popular winter sports. However, under the reality of global warming, most ski resorts need to rely on artificial snow-making to compensate for the lack of natural snow	https://finance.sina.com.cn/tech/roll/2022-12-30/doc-imxykzct2524483.shtml [2022-13-30]	Negative
2	June 24, 2022, marked a historic milestone in implementing the 2030 Sustainable Development Agenda. During the 14th BRICS Leaders' Meeting, a global development high-level dialog was held, achieving several agreements. The "Bamboo for Plastics" initiative proposed by the International Bamboo and Rattan Organization was included in the results of the dialog, aiming to reduce plastic pollution and address climate change, opening new paths for global development	https://baijiahao.baidu.com/s?id=1736953393007127753&wfr=spider&for=pc [2024-09-30]	Positive
3	On June 29, in Huai'an, Jiangsu, during a strong wind, a woman went to close a door that was blown open and retrieve items from outside. When she tried to shut the door, it was shattered by the wind, and she was blown away. Fortunately, she was unharmed. *Reminder: In extreme weather, please protect your personal safety*	https://weibo.com/1389537561/4786107105676594 [2024-09-30]	Neutral

opinion and emotional engagement on various topics, including climate change. The API assesses the emotional bias of posts, categorizing them as negative, positive, or neutral, and provides the corresponding confidence level for each classification (Table 6.4).

- **Discussion and Outlook**

In recent years, most of the Chinese public has directly or indirectly been affected by the increased frequency of extreme weather events (such as storms, floods, and droughts) caused by climate change. However, unlike the polarized climate narratives often found on social media in Western societies, understanding the common issues and regional differences in climate change communication on Chinese social media can more effectively spread climate change information and offer new perspectives for enhancing public awareness and action. Based on this case, we propose the following recommendations.

(1) Build a multi-dimensional climate change communication narrative using media platforms. In addition to the positive coverage of China's climate action in mainstream national media, it is recommended that local media in various provinces and cities strengthen the dissemination of local climate change issues. This will help promote awareness, scientific understanding, and communication about local climate problems across different geographic regions.

(2) Leverage opinion leaders to create a more diverse climate change communication network. As key communicators of information, media platforms should go beyond relying on scientists and actively involve influential youth leaders and celebrities from various sectors in climate change discourse. By harnessing their influence, these opinion leaders can better encourage the public, especially young people, to engage in climate change issues and participate in climate actions, such as the transition to a low-carbon economy.

(3) Focus on geographical differences and accumulate multi-region climate change communication data. There is a need to further capture variations between different geographic regions, gaining insights into the daily behaviors and lifestyles of local populations, as well as their motivations for engaging with climate change issues and participating in climate actions. A deeper understanding of these factors will help identify the barriers and drivers for broader climate action across regions, enabling more tailored

strategies to promote climate change engagement. These strategies can help enhance the public's climate change awareness, facilitate the involvement of various stakeholders, and encourage active participation in climate actions across China.

6.4 Summary

This chapter mainly focuses on three themes: reducing climate-related disasters, actively responding to climate change, and strengthening climate change education. Through the use of Big Earth Data, it calculates three SDG indicators, namely disaster losses (SDG 13.1.1), climate change response strategies (SDG 13.2.1), and climate change education (SDG 13.3.1), and provides corresponding spatio-temporal data products. On the basis of summarizing the continuous monitoring results of climate action-related indicators, a mid-term evaluation was conducted, and it was found that China's disaster prevention and reduction strategy has been relatively complete and has achieved good results. The "Dual Carbon" strategy and climate change response strategy are gradually improving, but implementing the tasks is still quite difficult.

Based on the data and conclusions in this chapter, we make the following recommendations.

(1) The flood monitoring, forecasting, and early warning system in China, as well as the flood control dispatch system for large and medium-sized reservoirs, storage, and detention areas, have greatly reduced the impact of flood disasters. In the future, when dealing with extreme climate change conditions such as floods, it is necessary to further strengthen the three-dimensional monitoring capabilities of the space, sky, and ground, and improve the accuracy of forecasting and early warning.
(2) Strengthen the education of young people's climate change scientific literacy and media literacy, focus on the media dissemination of climate change scientific knowledge education, actively establish enthusiastic young public welfare leaders for climate action, and achieve the positive guidance and mainstreaming goals of SDG 13.3.1 under the dual drive of education and communication as soon as possible.

China and the world have achieved certain results in disaster prevention and reduction under SDG 13, and in the future, we need to focus on the implementation progress of greenhouse gas emissions reduction. On the one hand, strengthen the data monitoring capability for greenhouse gas emissions; on the other hand, effective policies should be formulated to accelerate the progress of emission reduction, while increasing forest land protection and farmland management, and enhancing the ecosystem's ability to collect greenhouse gases.

References

Aerts JCJH, Botzen WJ, Clarke KC et al (2018) Integrating human behaviour dynamics into flood disaster risk assessment. Nat Clim Chang 8:193–199

DeVries B, Huang CQ, Armston J et al (2020) Rapid and robust monitoring of flood events using Sentinel-1 and Landsat data on the Google Earth Engine. Remote Sens Environ 240:111664

United Nations (2015) Transforming our world: the 2030 agenda for sustainable development. https://sustainabledevelopment.un.org/post2015/transformingourworld [2024-10-09]

United Nations (2022) The sustainable development goals report 2022. https://desapublications.un.org/publications/sustainable-development-goals-report-2022 [2023-10-09]

WEF (2023) Global risks report 2023. https://www.weforum.org/publications/global-risks-report-2023/ [2024-07-24]

WMO (2023) State of the global climate 2022. https://library.wmo.int/doc_num.php?explnum_id=11593 [2024-07-24]

Zhu H, He QS, Li JY et al (2022) Remote sensing monitoring and analysis of flood in flood storage area based on multi-source remote sensing data. J Hohai Univ (Natl Sci) 50(4):50–57 (In Chinese)

7 SDG 14 Life Below Water

7.1 Background

The ocean plays a crucial role in regulating the global water cycle, controlling the climate, protecting biodiversity, and providing habitats for many important species. In addition, marine products supply at least 20% of animal protein to 3.1 billion people worldwide (FAO 2016), which is particularly important for the livelihoods of coastal regions and small island developing countries with poorer economic conditions.

The 2030 Agenda includes SDG 14 as one of its 17 SDGs, which focuses on the conservation and sustainable use of oceans and marine resources. However, on the global scale, the implementation of most specific targets under SDG 14 has not been very successful. According to the results of The Second World Ocean Assessment Report released by the United Nations on April 21, 2021, many pressures from human activities have continued to degrade the ocean since the First World Ocean Assessment in 2015. The Lisbon Declaration, adopted at the United Nations Ocean Conference in 2022, agreed to intensify ocean actions based on science and innovation to address the current ocean emergencies. The declaration states that, in order to tackle the severe state of the ocean, all parties must be more ambitious, take stronger and more timely actions to improve the health of ocean and coastal ecosystems, and enhance their sustainability and resilience. The declaration calls on all parties to take further steps in areas such as strengthening data collection, reducing greenhouse gas emissions, and innovating financing mechanisms to achieve a sustainable ocean economy.

In the earlier stages, we utilized technologies and methods such as Big Earth Data to conduct in-depth research on specific targets and related indicators under SDG 14.1, SDG 14.2, etc., which lacked data despite having established methods. We explored approaches to acquire relevant data and computational models for typical regions in China and globally to fill the gaps in the United Nations' data. At the same time, we carried out long-term dynamic monitoring and analysis of specific targets with certain regional characteristics in China, such as SDG 14.5 and SDG 14.7. In this chapter, through the analysis of dynamic changes in nearshore microplastics, seagrass beds, bay ecosystem health, and nearshore raft aquaculture, we further improve monitoring methods for SDG 14-related indicators. We also build corresponding technologies and modeling systems to provide refined monitoring and assessment products, thus enabling the implementation of measures to promote the achievement of the SDGs.

H. Guo, *Big Earth Data in Support of the Sustainable Development Goals (2023)—China*, Sustainable Development Goals Series, https://doi.org/10.1007/978-981-95-2636-9_7

7.2 Main Contributions

This chapter conducts monitoring and assessment of indicators related to SDG 14.1 (reducing marine pollution), SDG 14.2 (protecting marine ecosystems), and SDG 14.7 (sustainable management of aquaculture) in China and surrounding regions through four case studies. The main contributions are shown in Table 7.1.

7.3 Case Study

7.3.1 Distribution Characteristics and Time-Varying Changes of Microplastics in China's Offshore

Target: SDG 14.1 By 2025, marine pollution of all kinds from land-based activities is prevented and significantly reduced, including marine debris and nutrient pollution.

- **Background**

As a new type of pollutant, marine microplastics pose a serious threat to the marine environment and ecological health. The scholars and the public at home and abroad are concerned about the problem, and it has become one of the current international research hotspots (Ding et al. 2022; Gao et al. 2023; Sun et al. 2022). The Fifth Session of the United Nations Environment Assembly, held from February 28 to March 2, 2022, listed plastic pollution as the primary environmental issue. At the "One Ocean Summit" held in February 2022, leaders of many countries stressed the pollution of plastics and microplastics. On May 4, 2022, the General Office of the State Council of the People's Republic of China issued the Action Plan for the Treatment of New Pollutants (Document [2022] No. 15), which listed microplastics as one of the key new pollutants under priority control, and emphasized the strengthening of research on new theories and new technologies related to microplastics (General Office of the State Council of the People's Republic of China 2022). China is one of the countries with a relatively large amount of plastic use, and China has a long coastline. Therefore, it is of great significance to carry out the analysis of the distribution and change trend of microplastics in our offshore areas for pollution tracking, prevention, and control. This case focuses on SDG 14.1 and carries out relevant research on the distribution and timely change of microplastics in China's offshore, providing important theoretical basis and technical support for accurate and objective assessment of the impact of microplastics on China's offshore marine environment, pollution ecological effects, and human health.

- **Data**

- Monitoring data of China's offshore microplastics from 2015 to 2022 (Sun et al. 2018; Zhu et al. 2018; Zhang et al. 2020a; Gu et al. 2022; Liu et al. 2022; Yu et al. 2022).
- The data of the Bulletin of Marine Ecology and Environment Status of China from 2015 to 2022 (https://www.mee.gov.cn/hjzl/sthjzk/jagb/).
- The literature data related to microplastics in China's offshore waters published in international journals, which mainly include the *Journal of Hazardous Materials*, *Chemosphere*, *Environmental Pollution*, *Marine Pollution Bulletin* and *Science of the Total Environment*, etc.

- **Methods**

The occurrence status of microplastics in China's offshore environment was analyzed by using the main indicators such as the abundance, particle size, and type of microplastics, and the distribution characteristics and pollution status of microplastics in China's offshore waters were studied based on the comprehensive monitoring data of China's offshore and the data

Table 7.1 Case studies and their main contributions

Targets	Tiers	Case studies	Contributions
SDG 14.1 By 2025, marine pollution of all kinds from land-based activities is prevented and significantly reduced, including marine debris and nutrient pollution	Tier II	Distribution characteristics and time-varying changes of microplastics in China's offshore	Data product: the dataset on the current status and interannual variations of microplastics in China's offshore waters Method and model: construct evaluation indicators for microplastic pollution and analyze their correlation with plastic waste Decision support: help to evaluate the regional distribution differences and pollution prevention and control of microplastics in China's coastal areas
SDG 14.2 By 2020, sustainably manage and protect marine and coastal ecosystems to avoid significant adverse impacts by strengthening their resilience, etc., and take actions to restore them to achieve healthy and productive oceans	Tier II	Spatio-temporal distribution dynamics monitoring of seagrass beds in China	Data product: distribution map of seagrass beds in China from 1985 to 2020 Method and model: support vector machine (SVM) supervised classification based on Google Earth Engine (GEE) Decision support: provide data support for scientific protection and management of seagrass beds
		Comprehensive assessment of typical coastal bay ecosystem health in China	Data product: long-term monitoring dataset of ecosystem elements in typical bays near the coast of China (Jiaozhou Bay) Method and model: eutrophication assessment and nearshore ecosystem health assessment model Decision support: propose measures and suggestions for long-term maintenance of healthy and sustainable development in coastal areas from a scientific perspective, reveal possible ways to achieve high-quality development of the marine economy under effective land-sea coordinated coastal prevention and control measures, and provide typical cases for the sustainable development research of typical bays in China's coastal areas
SDG 14.7 By 2030, increase the economic benefits derived from the sustainable use of marine resources for small island developing states and least developed countries, including the sustainable management of fisheries, aquaculture, and tourism	Tier I	Remote sensing monitoring and analysis of raft aquaculture in China's coastal waters	Data product: we have produced for the first time a 10 m spatial resolution dataset of Chinese offshore raft aquaculture that can distinguish different types of aquaculture in 2017 and 2021 Method and model: extraction method of raft aquaculture area from multi-phase optical remote sensing images Decision support: provide decision-making support for national and local governments in marine aquaculture planning, management, etc.; provide a reference and basis for local governments' sustainable development of aquaculture planning

published in international journals. By analyzing the annual abundance distribution characteristics of microplastics in China's offshore, we can evaluate the annual change of microplastics and objectively evaluate the changing trend of microplastics in China's offshore. In the study, microplastic samples in seawater were mainly collected by trawl. The polymer types of microplastics were identified by using a micro-Fourier Transform Infrared Spectroscope.

- **Results and Analysis**

1. **Distribution of microplastics in seawater**

Based on the comprehensive research results of microplastics in China's offshore in recent years (Fig. 7.1; Ministry of Ecology and Environment of the People's Republic of China 2019, 2020, 2021, 2022; Gu et al. 2022; Liu et al. 2022; Zhang et al. 2020a; Sun et al. 2018), the main types of microplastics in China's offshore were rayon, polyethylene terephthalate (PET), polyethylene (PE), polypropylene (PP), polystyrene (PS), nylon and cellophane. Among them, PS and rayon accounted for a relatively high proportion in the Bohai and the Yellow Sea, PET was dominated in the East China Sea, and PE and PET accounted for a relatively high proportion in the South China Sea. The proportion of microplastic size of 1.0–2.0 mm detected in all seawaters was relatively high, accounting for 28–36%. The proportion

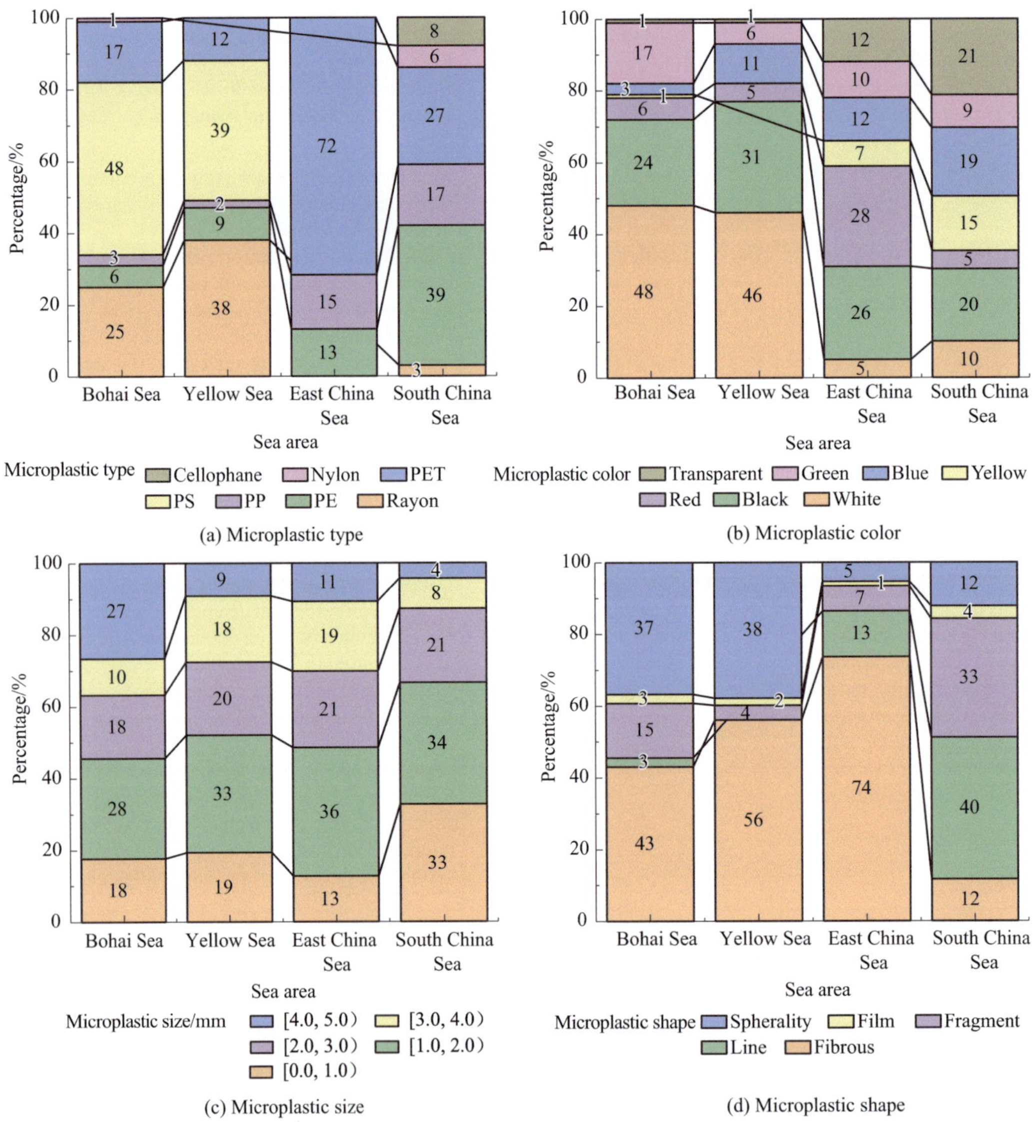

Fig. 7.1 Characteristics of microplastics in China's offshore

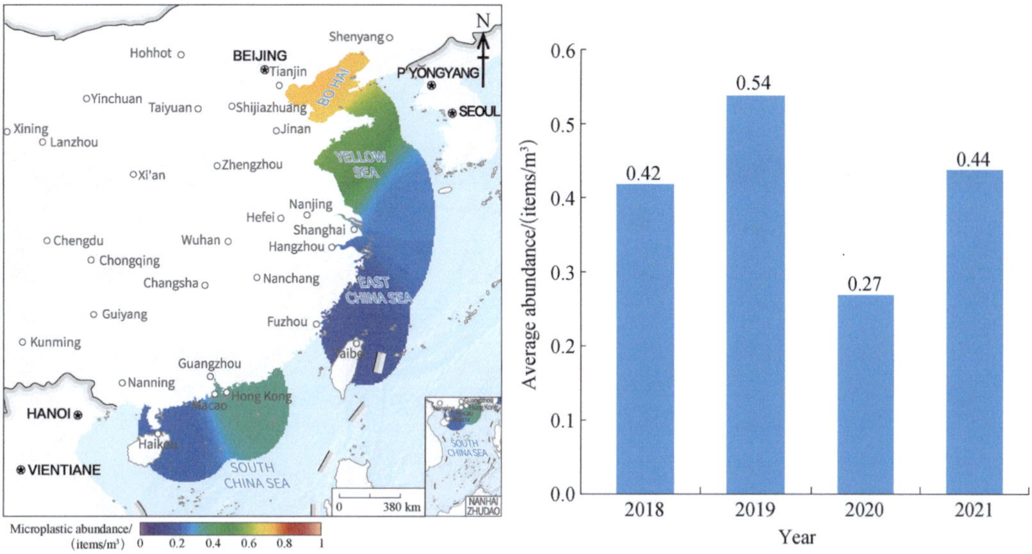

(a) Distribution of microplastic abundance in China's offshore waters in 2021

(b) Annual variation of average microplastic abundance in China's offshore waters

Fig. 7.2 Spatial distribution and interannual average abundance of microplastics in China's offshore waters

of particle size distribution of microplastics in different sea waters was slightly different. The detected microplastic colors were mainly divided into white, black, red, yellow, blue, green, and transparent. The microplastics in the Bohai and the Yellow Sea waters were mainly white and black, the microplastics in the East China Sea waters were mainly red and black, and the microplastics in the South China Sea waters were mainly transparent, black, and blue. The shape of microplastics in the Bohai, the Yellow Sea, and the East China Sea was mainly fibrous, accounting for 43%, 56%, and 74%, respectively, and there was also a high proportion of spherical microplastics in the Bohai and the Yellow Sea. The shape of microplastics in the South China Sea waters was mainly line (40%) and fragment (33%).

The research results indicated that there were differences in the abundance of floating microplastics in different sea areas of China's offshore (Fig. 7.2a). The results showed that the average abundance of microplastics in the seawaters of the Bohai, the Yellow Sea, the East China Sea, and the South China Sea was 0.74 items/m^3, 0.54 items/m^3, 0.22 items/m^3, and 0.29 items/m^3, respectively. There were regional differences in the spatial distribution of microplastics in China's offshore, and the average abundance of microplastics in the Bohai waters was relatively high. On the whole, the abundance of microplastics in China's offshore was at a medium–low level. From 2018 to 2021, the average abundance of floating microplastics in all monitoring sections of China's coastal waters was 0.42 items/m^3 (2018), 0.54 items/m^3 (2019), 0.27 items/m^3 (2020), and 0.44 items/m^3 (2021), respectively (Fig. 7.2b). Compared with similar international surveys in recent years, the average abundance of microplastics in China's coastal waters was at a medium-to-low level.

2. **Correlation analysis of microplastic and plastic garbage**

The sources of plastic garbage in the ocean are complex, including both land-based plastic waste and sea-based plastic waste. The plastic waste entering the ocean eventually forms secondary microplastics after a series of physical and chemical actions. Therefore, plastic waste in the

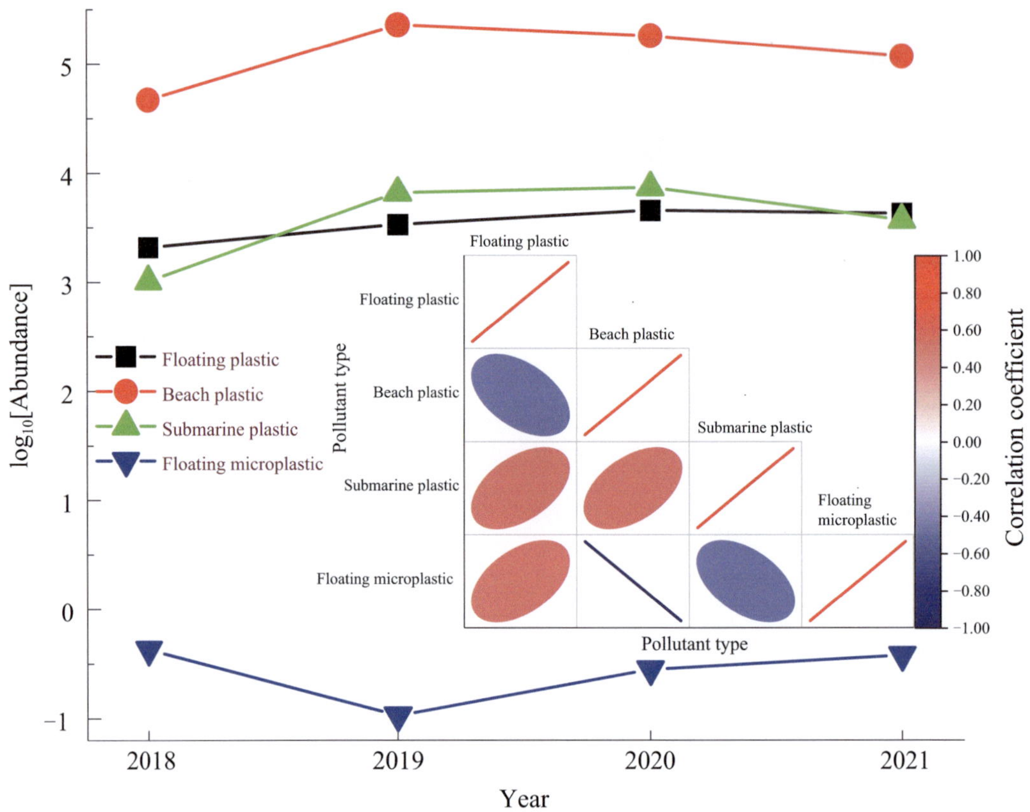

Fig. 7.3 Annual distribution and correlation of plastic waste and microplastics in China's offshore waters

marine environment is one of the main sources of marine microplastics (Zhu et al. 2018; Yu et al. 2022). This case study analyzed the interannual changes of floating plastic waste, plastic waste on the shore, and plastic waste on the seabed in China's offshore from 2018 to 2021. The results showed that there were differences in the annual abundance of floating plastic garbage, beach plastic garbage, and submarine plastic garbage in China's offshore. The annual average abundance was 3,587 pieces/km^2, 144,177 pieces/km^2, and 4,684 pieces/km^2, respectively. The results showed that the current trend of change in plastic waste tended to be flat, and the development trend was good. The annual change trend of microplastics in China's offshore had a positive correlation with the floating plastic garbage, indicating that the floating plastic garbage had a certain "abundance contribution" to the distribution of microplastics in China's offshore seawater (Fig. 7.3).

Highlights

- The main types of microplastics in China's offshore were rayon, PET, PE, PP, PS, nylon, and cellophane. The types and distribution of microplastics in the Bohai and the Yellow Sea were similar, but different from those in the East China Sea and the South China Sea. The average abundance of microplastics in China's offshore waters was at a medium-to-low level, with a small annual variation and a good development trend.
- The annual abundance distribution of floating plastic garbage, beach plastic garbage, and submarine plastic garbage in China's offshore environment was relatively stable. The abundance of microplastics in China's offshore waters had different degrees of correlation with the abundance of floating plastic garbage, beach plastic garbage, and submarine plastic garbage. Among them,

the floating plastic garbage has a certain "abundance contribution" to the distribution of microplastics in China's offshore waters.

- **Discussion and Outlook**

The problem of marine microplastic pollution has become one of the hot topics of the international public. At present, governments and scientific researchers at home and abroad have paid much attention to it. China is a developing country with a large population, and its offshore microplastic pollution has attracted universal attention worldwide. In addition to being affected by the degree of population density, the pollution status of microplastics offshore is also partly related to a country's pollution control, and the pollution status of microplastics in the offshore of different regions is also different. Most of the microplastic and plastic pollution in the sea comes from the discharge of plastic waste in coastal and inland rivers, so the treatment of production and domestic plastic waste in coastal and river areas is to solve the problem of microplastic and plastic pollution at the root. In order to further strengthen the management of plastic pollution, the National Development and Reform Commission and the Ministry of Ecology and Environment jointly issued the Opinions on Further Strengthening the Management of Plastic Pollution (NDRC Huanzi [2020] No. 80), promoting China's plastic pollution management to move toward full chain management. In September 2021, the National Development and Reform Commission and the Ministry of Ecology and Environment released the 14th Five-Year Plan for Plastic Pollution Control Action Plan and other documents.

In order to better serve the pollution prevention and control work of marine microplastics, it is an important prerequisite to clarify the source and trend of marine microplastics for governance and response. However, if you want to obtain a correct understanding of its source and trend, it is very important to pay attention to the distribution status and time-varying rule of China's offshore microplastics for a long time. In addition, an important source of marine microplastics is the miniaturization of plastic waste, so it is also very important to analyze the spatial distribution and correlation between marine microplastics and plastic waste. This study focused on the environmental occurrence characteristics and annual change rules of microplastics in China's offshore, explored the correlation between marine microplastics and plastic waste in different spaces of the offshore environment, and could provide data and theoretical support for better monitoring and predicting the pollution of microplastics in China's offshore. However, there is a long way to go to solve the environmental problems of marine microplastics. In the future, it is necessary to develop and improve the monitoring and sampling methods in line with international standards, and study the migration and settlement behavior of microplastics in the ocean in combination with ocean hydrodynamics. At the same time, in the process of dealing with the pollution of marine microplastics, we should pay attention to the power of the masses, actively carry out popular science propaganda, let the masses understand that the reduction of microplastic and plastic pollution is closely related to the improvement of life, truly reduce plastic pollution from the source, and promote the high-quality and sustainable development of the marine environment.

7.3.2 Spatio-Temporal Distribution Dynamics Monitoring of Seagrass Beds in China

Target: SDG 14.2 By 2020, sustainably manage and protect marine and coastal ecosystems to avoid significant adverse impacts by strengthening their resilience, etc., and take actions to restore them to achieve healthy and productive oceans.

- **Background**

Seagrass beds, along with mangroves and coral reefs, are considered the three typical marine ecosystems. They are among the most productive ecosystems on the Earth and play an indispensable role in China's "Blue Carbon" plan. Currently, there is limited research on China's seagrass beds. From 2015 to 2021, the Institute of Oceanology, Chinese Academy of Sciences, and other research institutions conducted the first nationwide survey of seagrass beds. However, the historical changes and main driving factors of seagrass beds are still unclear. Therefore, monitoring the spatio-temporal distribution dynamics of China's seagrass beds through remote sensing and big data is of great significance for scientifically analyzing the changes in seagrass beds and for their standardized management and protection.

- **Data**

– Sentinel-2 multispectral satellite data in 2016–2021.
– Landsat terrestrial satellite data in 1985–2015, including Landsat-5 multispectral satellite data in 1985–2012, and Landsat-8 multispectral satellite data in 2013–2015.
– SPOT satellite data in 1986–2014, including SPOT-1 multispectral satellite data in 1986–1990, SPOT-2 multispectral satellite data in 1990–2008, SPOT-3 multispectral satellite data in 1994–1997, SPOT-4 multispectral satellite data in 1998–2013, and SPOT-5 multispectral satellite data in 2002–2014.
– Key-hole spy satellite data in 1980–1984.

- **Methods**

The remote sensing image interpretation work in this case is mainly implemented on the GEE. The aforementioned data are provided by the GEE and have completed all preprocessing such as atmospheric correction. The method used in this case is a pixel-based supervised classification, specifically using the SVM model, which is widely applied and has good classification results. This model uses randomly selected pixels from the images as samples. To more effectively identify different land cover types, this case calculates multiple predictors for sample training in SVM classification, including blue band, green band, red band, near-infrared spectrum (NIR), short wave infrared region (SWIR), normalized difference water index (NDWI), modified normalized difference water index (MNDWI), automated water extraction index (AWEI), NDVI, and enhanced vegetation index (EVI). After classification, post-processing is conducted on the SVM classification results, and obvious misclassifications and omissions are corrected through visual interpretation. Post-processing and the drawing of land cover type distribution information are all implemented in ArcGIS 10.7 software.

- **Results and Analysis**

Based on conventional surveys and integrating multi-source satellite remote sensing data, this case uses a high-performance computing platform for Big Earth Data and a machine learning model to extract and analyze the spatio-temporal distribution dynamics of seagrass beds in China.

Combining the survey results of the National Seagrass Resource Survey (2015–2021) and remote sensing monitoring results, the total area of seagrass beds in China's offshore areas is approximately 27,010 hm^2. Among them, Liaoning has about 3,205 hm^2, Hebei about 9,171 hm^2, Tianjin about 466 hm^2, Shandong about 4,193 hm^2, Fujian about 470 hm^2, Guangdong about 1,538 hm^2, Guangxi about 665 hm^2, Hainan about 6,728 hm^2, and Taiwan about 574 hm^2 (Zhou et al. 2023; Li et al. 2022).

Based on the Big Earth Data computing platform, using a series of satellite data from Landsat, SPOT, and Sentinel from 1985 to 2020, and applying the SVM algorithm for supervised classification of China's offshore areas, we obtained the distribution and area dynamic change information of seagrass beds from 1985 to 2020 (Fig. 7.4). The results show that from 1985 to 2020, the area of seagrass beds in China (excluding Sansha

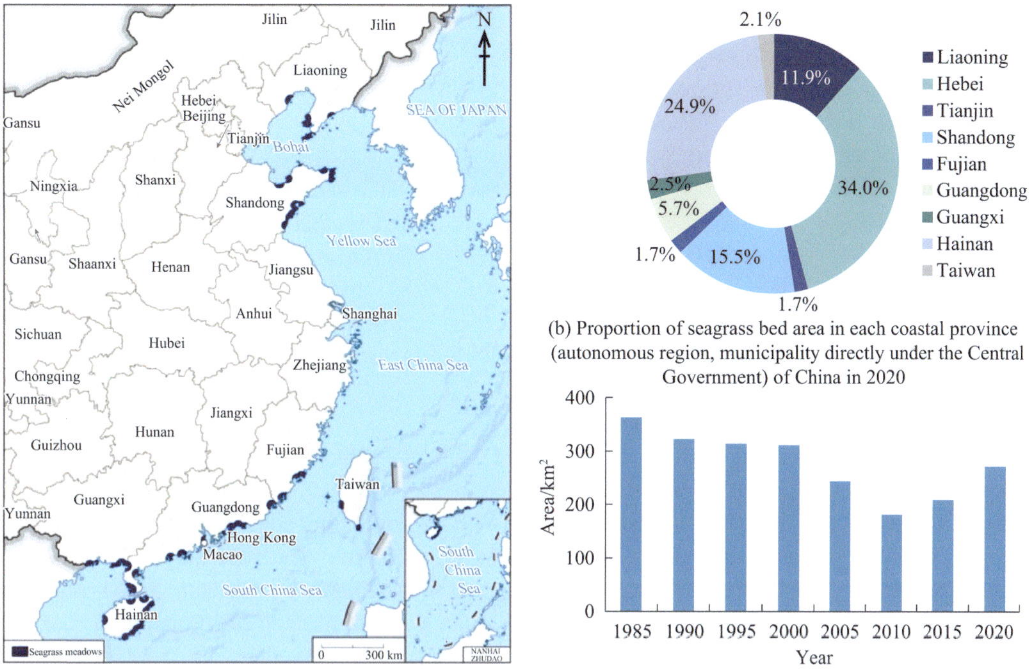

(a) Spatial distribution of seagrass beds in 2020

(b) Proportion of seagrass bed area in each coastal province (autonomous region, municipality directly under the Central Government) of China in 2020

(c) Changes in seagrass bed area from 1985 to 2020

Fig. 7.4 Distribution and area changes of seagrass beds in China based on remote sensing monitoring

City) generally showed a trend of decline (1985–2010) followed by a slight recovery (2010–2020). Land reclamation, human activity pollution, and biological invasions are the main reasons for the shrinking of seagrass beds (Zhou et al. 2023). Since 2010, relevant Chinese departments have paid attention to and protected seagrass bed ecosystems, leading to the recovery of many key seagrass bed areas. However, due to the interference of seagrass species and different environmental factors, there are significant differences in the changes in seagrass beds in the north and south of China. For example, the largest seagrass bed in northern China, dominated by eelgrass in the Caofeidian area, has gradually decreased in area since the 1980s due to human activities. In 2015, scientific research teams and local governments jointly carried out protection and restoration efforts (Liu et al. 2016), and the Caofeidian seagrass bed has recovered well and has been restored to the level of the 1980s. In contrast, the small individual species of seagrass such as *Halophila ovalis* and *Halophila beccarii* have obvious seasonal changes, and their degree of degradation and restoration difficulty is much higher than that of eelgrass. For example, the Hepu Dugong National Nature Reserve in the Guangxi Zhuang Autonomous Region established a management station as early as 1996, but the seagrass beds in the area are still in a state of fluctuating decline, and the marine life associated with seagrass beds, such as dugongs, have not been seen for many years (Wu and Zhang 2018).

Highlights

- Based on the Big Earth Data computing platform, for the first time, we obtained the distribution and area dynamic change information of seagrass beds in China from 1985 to 2020 using a series of satellite data from Landsat, SPOT, and Sentinel. The results show that the area of seagrass beds in China (excluding Sansha City) generally showed a trend of decline (1985–2010) followed by a slight recovery (2010–2020).

– Due to the differences in seagrass species and growth environments, the seagrass beds in the Yellow Sea and the Bohai showed a significant recovery from 2010 to 2020, while the seagrass beds in the South China Sea are still in a stage of fluctuating degradation.

- **Discussion and Outlook**

This case has developed a remote sensing classification algorithm based on geographic cloud computing and machine learning, achieving the distribution of data of seagrass beds in China over the past 35 years, providing important data support for the country's "Blue Carbon" plan. Overall, the trend of seagrass beds in China is gradually declining and then slightly recovering, but there are significant differences in different regions. The seagrass beds in the Yellow Sea and the Bohai, dominated by eelgrass, have larger individuals and higher ground coverage, making them easier to monitor from satellites. In recent years, the protection and restoration efforts in this area have been progressing smoothly. In contrast, the seagrass beds in the South China Sea are dominated by the smaller and creeping species of seagrass, such as *Halophila ovalis* and *Halophila beccarii*, which have large seasonal variations, lower coverage, and higher environmental requirements, thus making the uncertainty of extraction from remote sensing greater. This case successfully reveals the overall trend changes of China's seagrass beds. It is recommended to conduct detailed analyses in different regions in the future and scientifically evaluate the impact of human intervention on seagrass ecosystems. At the same time, higher spatiotemporal resolution satellite data, such as Planet or domestic high-resolution satellite data, can be used for quarterly monitoring of key areas of seagrass beds, in order to plan related protection and restoration work more scientifically and thus promote the national "Blue Carbon" plan.

7.3.3 Comprehensive Assessment of Typical Coastal Bay Ecosystem Health in China

Target: SDG 14.2 By 2020, sustainably manage and protect marine and coastal ecosystems to avoid significant adverse impacts by strengthening their resilience, etc., and take actions to restore them to achieve healthy and productive oceans.

- **Background**

With the continuing advance of China's construction of marine eco-civilization, the marine economy and industrial structure have been experiencing a state of deep adjustment and changes. The sustainable development of China's coasts faces challenges related to increasing conflicts between the coastal economic, social development, intensified population, resource depletion, and the marine environment and ecosystems. SDG 14.2 of the 2030 Agenda aims to "sustainably manage and protect marine and coastal ecosystems" to "achieve healthy and productive oceans", and put forward specific requirements for rational development and utilization of marine resources while ensuring the sustained health of marine environments. The Chinese government attaches great importance to this matter as well, and makes substantive process of the 2030 Agenda in all aspects. A series of strategic measures such as "Land-Sea Coordination", "Ecological Civilization", "Beautiful China", and "Healthy Ocean", as well as the concepts of the "maritime community with a shared future" and "global ecological and environmental governance", have been performed to ensure the sustainable development of the Chinese coastal ecosystem. Therefore, under the background of SDGs, studying ecosystem evolution and sustainable development in typical bays in China's coasts could put forward initiatives for the long-term coastal ecosystem health protection from a scientific point of view, and play an important role in understanding the sustainable development of China and global coasts.

Jiaozhou Bay is located on the south side of the Shandong Peninsula, within the administrative boundary of Qingdao, Shandong. The bay is densely populated, with developed industry and agriculture. The ecological environment of the bay is influenced by intense human activities such as urbanization, coastal reclamation, land-based material discharge, international shipping, large-scale engineering, and marine aquaculture. It represents a typical temperate coastal bay in China and effectively captures the developmental history of the eastern coastal region. We take Jiaozhou Bay as a case study to analyze the evolution and sustained development of the coastal ecosystem, to propose measures for ecosystem protection from the perspective of long-term changes in the ecological environment and ecosystem health of Jiaozhou Bay. This case study could provide realistic and ecologically sound management strategies for the sustainable use of coastal seas.

- **Data**

- Observation data of phytoplankton, zooplankton, benthic animals, chlorophyll *a*, chemical elements of seawater, and other environmental/ecological factors in the Jiaozhou Bay in 2003–2021.
- Water quality, aquaculture production, area, and other environmental data for the period of 2003–2021 from environmental situation bulletins by the Ministry of Ecology and Environment.
- Population and GDP data for the period of 2005–2021 from the National Bureau of Statistics.

- **Methods**

Linked to the targets of "sustainably manage and protect marine and coastal ecosystems" and "achieve healthy and productive oceans" of SDG 14.2, this case study analyzed the current state and long-term change trend of various ecological factors based on statistical methods and machine learning technology. Based on the pressure-state-response (PSR) framework, we selected indicators and built a eutrophication assessment model considering exogenous nutrient inputs, the structure of the ecosystem, nutrients, and other relevant aspects.

Aiming at the achievement and maintenance of ocean health, we selected and classified marine ecosystem elements based on the structure, function, and dynamic evolution of the targeted area. We set up indicators such as water quality, biotic environment, and social-economic status, and built a marine ecosystem health report card system to evaluate the ecosystem health condition and long-term trend, with the eventual aim to identify suitable measures (Logan et al. 2020; SDG Working Group of CASEarth 2019).

- **Results and Analysis**

Since 2010, the health of the Jiaozhou Bay ecosystem has shown an overall improvement and reached its best level in nearly 20 years in 2021. The health grade of the Jiaozhou Bay ecosystem was "B" in 2021, indicating a good status. There was no significant difference in the health status among different regions of the bay, as all regions were in the "B" grade of good status (Fig. 7.5). From the perspective of long-term changes, the overall health condition of the Jiaozhou Bay ecosystem has been trending positively since 2010. However, there was a fluctuation in 2019 due to outbreaks of jellyfish and Noctiluca. The health assessment score of the bay in 2021 reached its highest level in the period of 2003–2021 (Fig. 7.6).

The Jiaozhou Bay has achieved a healthy ecosystem and high-quality development of the marine economy through effective land-sea coordination and strengthened nearshore management. In 2021, the marine GDP of Qingdao reached 468.484 billion yuan, ranking first among similar coastal cities in China. While maintaining rapid and high-quality economic growth, the health of Jiaozhou Bay's marine ecosystem continues to improve due to the effective environmental governance and protection measures.

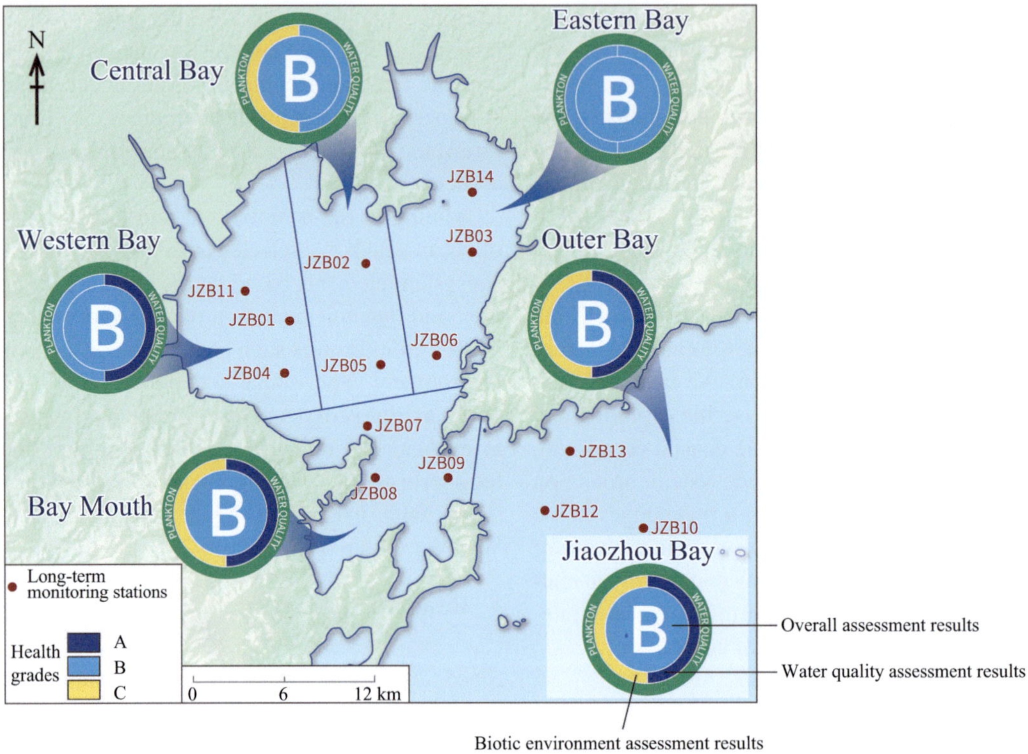

Fig. 7.5 Health assessment results of the Jiaozhou Bay and various marine ecosystems in 2021. *Note* The outer ring shows the overall health condition; the left inner ring represents the plankton condition, and the right indicates the water quality condition

Fig. 7.6 Marine ecosystem health assessment scores of the Jiaozhou Bay in 2003–2021

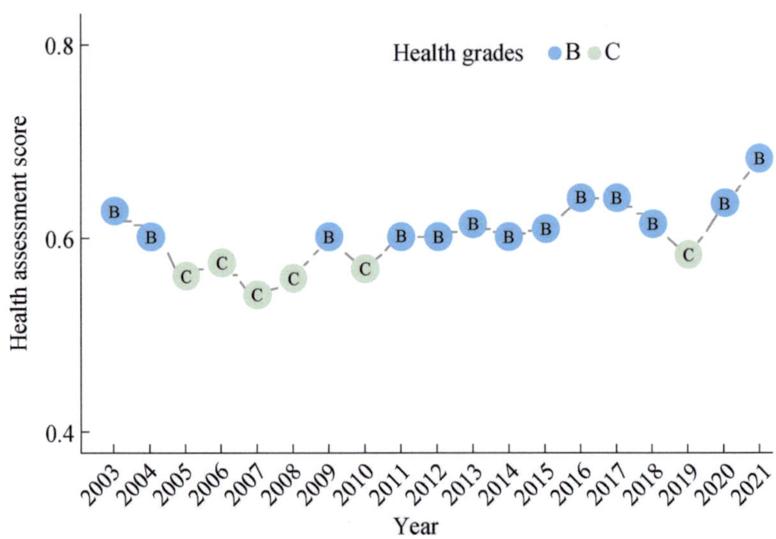

Comprehensive measures, such as the control of nearshore water pollution and the reduction of land-based pollution, have effectively reduced the input of pollutants and lowered the nutrient concentration in the bay. Measures like shoreline management and restoration of bay areas ensure unobstructed flow patterns in the bay's hydrodynamic environment. The stability of benthic shellfish resources and a sound benthic ecosystem guarantee the normal biogeochemical cycles in the bay (Sun et al. 2015). The Jiaozhou Bay has implemented multiple concurrent strategies, providing an important foundation for achieving the goal of sustainable development in the bay.

Highlights

- The Jiaozhou Bay has achieved a healthy ecosystem and high-quality development of the marine economy. The health condition of the Jiaozhou Bay ecosystem has been trending positively since 2010 and reached its highest level in 2021 for the period of 2003–2021.
- Analyze the relationship between urban development and management, bay eutrophication, and changes in ecosystem health. Reveal possible approaches to achieve a healthy ecosystem and high-quality marine economic development through effective land-sea coordination and strengthened nearshore management.

- **Discussion and Outlook**

The current case study aimed at analyzing the relationship between urban development and management and eutrophication and ecosystem health variation in a typical bay in China coasts during different phases of economic development, and revealing the possible approaches to ensure high-quality marine economic development under effective and sea-land coordinated control and preventative measures. With the evolution and elevation of the strategic position of marine economic development, Qingdao has been slated to become a "global marine center city", and the sustainable development of the Jiaozhou Bay ecosystem faces new challenges under the new development situation. Meanwhile, global warming tends to initiate more new environmental issues, affecting various aspects of the structure and function of coastal ecosystems. Therefore, with the increasing compounding pressure from global warming and human activities, scientifically sound regulatory measures and management need to be further dynamically adjusted according to changes in marine ecosystem characteristics, functions, resources, and healthy conditions, and based on current environmental control measures. In order to provide scientifically and ecologically sound support for sustainable development and blue economies, we will develop a more thorough and comprehensive analysis in the future to study possible coastal ecosystem risks and issues, with a focus on understanding ecosystem responses to global warming and the "Dual Carbon" target.

7.3.4 Remote Sensing Monitoring and Analysis of Raft Aquaculture in China's Coastal Waters

Target: SDG 14.7 By 2030, increase the economic benefits derived from the sustainable use of marine resources for small island developing states and least developed countries, including the sustainable management of fisheries, aquaculture, and tourism.

- **Background**

According to The State of World Fisheries and Aquaculture 2018, the contribution of aquaculture to global fishing and aquaculture production has gradually increased, from 25.7% in 2000 to 46.8% in 2016 (FAO 2018). Among them, seawater aquaculture accounts for a large proportion. According to statistics from the Bureau of Fisheries, Ministry of Agriculture and Rural Affairs of the People's Republic of China (2021), in 2020, seawater aquaculture accounted for 59% of the overall market in China, while freshwater aquaculture accounted for 41%. China is a country with developed seawater aquaculture in the world, ranking first in both aquaculture area and the total output. China's mariculture has taken a successful path, which has better enriched the protein sources of the Chinese people. How to ensure the sustainable development of China's mariculture is one of the important issues currently facing.

The main types of raft aquaculture include fish, shellfish, and algae. Among them, the large amount of residual bait, debris, and fish and shrimp feces produced by raft aquaculture of fish leads to eutrophication of the aquaculture water area, providing a suitable ecological environment for red tide organisms and being the root cause of red tide. Algae and shellfish raft farming can consume excess nutrients such as nitrogen, and excessive nitrogen can create hypoxic waters in the ocean where marine organisms find it difficult to survive. By absorbing carbon, they can also help alleviate ocean acidification on a local scale. Therefore, rational planning of marine aquaculture can promote the harmonious development of human aquaculture activities and the marine environment, maintain biodiversity, and promote the sustainable development of marine aquaculture.

This case uses intelligent methods to extract raft aquaculture areas from remote sensing images, in order to understand the spatial distribution, proportion, and changes of raft aquaculture in China. At the same time, it analyzes marine environmental data, marine disaster data, and other statistical and survey data to explore their interrelationships and provide a decision-making basis for the rational layout of raft aquaculture in China.

- **Data**

– Remote sensing data from Sentinel-1 and Sentinel-2 in 2017 and 2021, which are used for extracting the distribution of aquaculture areas and types of aquaculture.
– Remote sensing data from Gaofen-1, Gaofen-2, and Gaofen-3 in 2017 and 2021, which are used to verify the accuracy of extraction in the aquaculture area.
– MODIS data for 2017 and 2021, which are used to obtain information on the marine environment.
– Bulletin of China Marine Disaster (2017–2021, https://www.nmdis.org.cn/hygb/zghyzhgb/), which are used to comprehensively understand the current situation and trends of marine ecological disasters in China, and to analyze the coupling relationship between marine ecological disasters and seawater aquaculture.
– Survey data, fishery statistical data, and others, which are used for correlation analysis.

- **Methods**

This case study focuses on the monitoring needs of raft aquaculture in China's coastal areas. Based on the remote sensing images of Sentinel-1 and Sentinel-2, deep learning methods are used to extract raft aquaculture areas and determine aquaculture types. Higher-resolution remote sensing images of Gaofen-1, Gaofen-2, and Gaofen-3 are utilized, and the accuracy of the extraction method is verified through manual interpretation and field investigations. Then, using the extracted multi-temporal data of raft aquaculture areas, analyze the current situation and changes in raft aquaculture in China's coastal areas. Finally, analyzing the marine environment, marine disasters, and related data in China's coastal areas, combined with the spatial distribution and proportion of aquaculture types in raft aquaculture areas, provides support for the rational layout and sustainable development of raft aquaculture.

The use of deep learning methods for raft aquaculture extraction is an important part of this case study method, and the main key steps include three aspects: first, constructing a sample library containing fish, shellfish, and algae aquaculture types; second, optimizing the deep learning model in a targeted manner (Zhang et al. 2022); third, fusing the raft aquaculture results extracted from different data sources, as shown in Fig. 7.7.

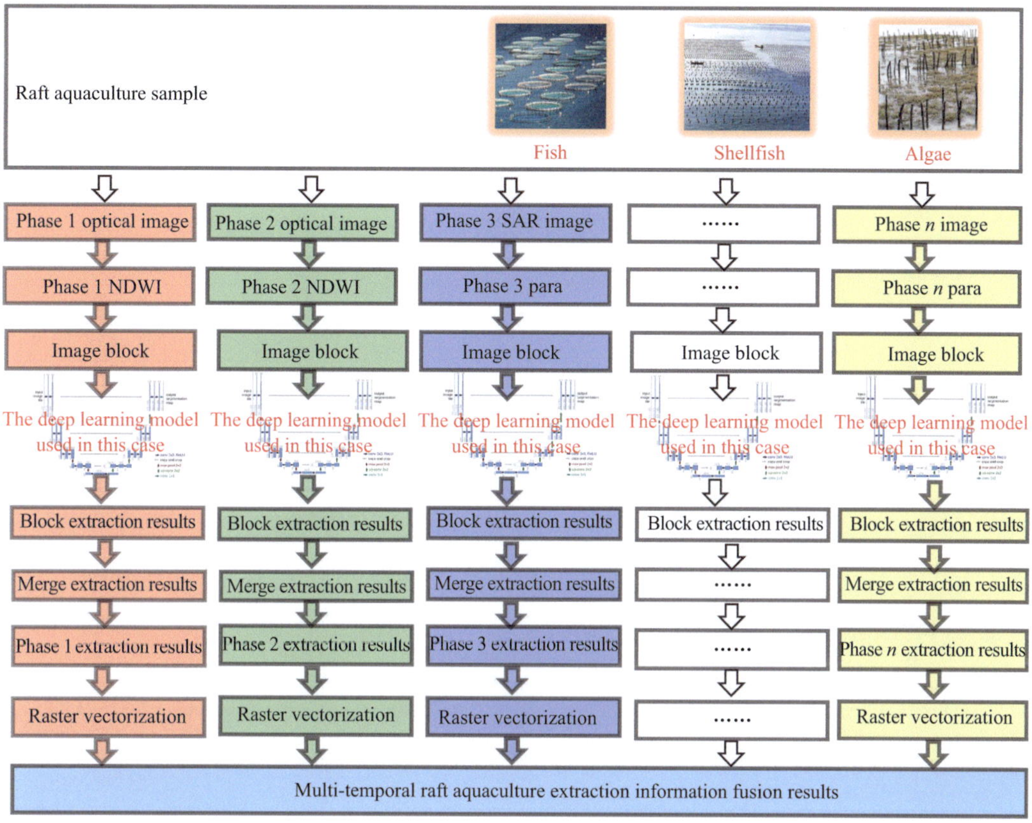

Fig. 7.7 Technical flowchart. *Note* The deep learning model used in this case is detailed in the study of Zhang et al. (2022)

- **Results and Analysis**

From 2017 to 2021, the area of marine raft aquaculture in China was in an orderly growth stage, and the distribution of raft aquaculture areas showed a development trend away from the coastline. In 2017, the remote sensing monitoring area of China's offshore raft aquaculture was about 5,673 km^2. In 2021, the remote sensing monitoring area of China's offshore raft aquaculture was about 5,698 km^2 (Fig. 7.8a), with Shandong, Guangdong, Fujian, Jiangsu, Liaoning and other provinces leading in aquaculture area in 2021 (Fig. 7.8b). By analyzing the distribution of raft aquaculture at distances of 10 and 20 km from the coast, it was found that raft aquaculture tends to move away from the coastline. In 2021, there was a significant increase in raft aquaculture areas 10 km away from the coastline (Fig. 7.8a).

The structure of raft aquaculture in China's coastal areas is increasingly optimized, and the development of ecological and healthy aquaculture models is rapid. The results of remote sensing monitoring of the structure of raft aquaculture in China's offshore waters show that there have been significant changes in the structure of raft aquaculture in China's offshore waters from 2017 to 2021, and the types of aquaculture are more balanced (Fig. 7.8c). By reasonably planning different types of aquaculture, the impact of large-scale single-species aquaculture on ecological degradation and biodiversity reduction has been reduced. For example, the efforts made by Changhai County in Liaoning Province and the shallow waters in northern Jiangsu for sustainable aquaculture, coupled with the continuous introduction of new aquaculture technologies and

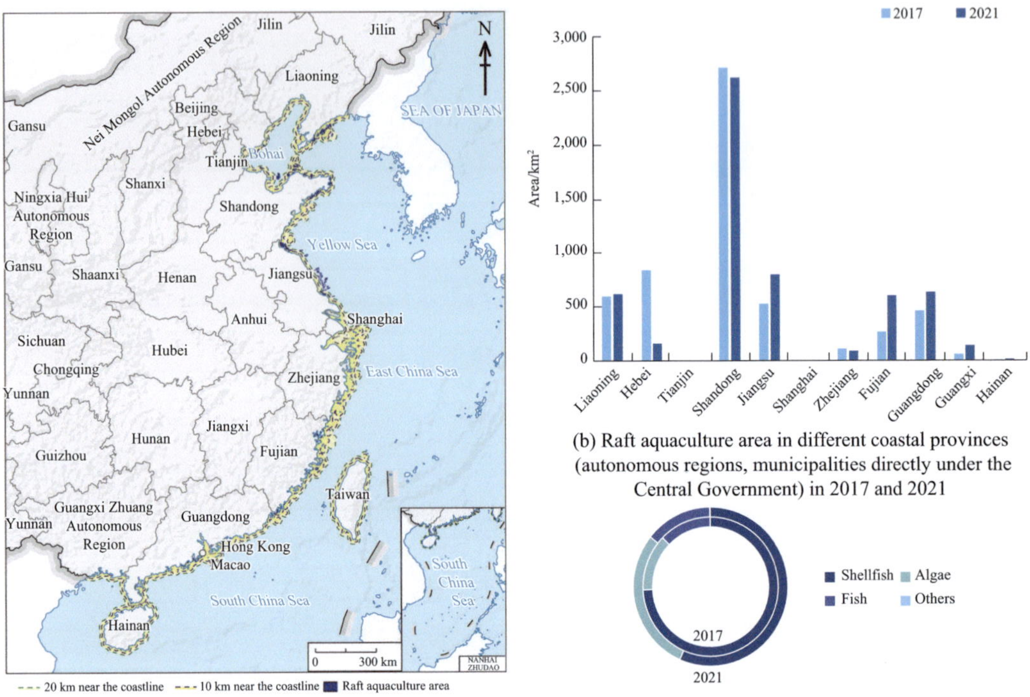

(a) Spatial distribution of raft aquaculture in 2021

(b) Raft aquaculture area in different coastal provinces (autonomous regions, municipalities directly under the Central Government) in 2017 and 2021

(c) Changes in aquaculture types in 2017 and 2021

Fig. 7.8 Spatial distribution and area statistics of raft aquaculture in China's offshore waters

varieties, have further expanded the aquaculture field and rapidly developed ecological and healthy aquaculture models. China's offshore raft aquaculture is currently in a stage of layout optimization and significant improvement in quality.

Highlights

- We have produced for the first time a 10 m spatial resolution dataset of Chinese offshore raft aquaculture that can distinguish different types of aquaculture in 2017 and 2021. This dataset can provide decision-making support for national and local governments in marine aquaculture planning and management, as well as basic data for other scholars to conduct related research.
- The current area of raft aquaculture in China's coastal waters remains relatively stable. From 2017 to 2021, there have been significant structural changes in raft aquaculture in China's coastal areas, with a significant increase in algae cultivation and a decrease in shellfish cultivation. Meanwhile, raft farming tends to move away from the coast. Besides, the layout of raft aquaculture in China's coastal areas tends to be reasonable, and overall it is developing toward ecological aquaculture. However, the marine aquaculture structure dominated by local monoculture aquaculture needs to be optimized.

• Discussion and Outlook

Offshore aquaculture is currently an important component of China's marine aquaculture, and the methods and scales of offshore aquaculture are constantly changing. This case fully utilizes multi-source satellite remote sensing images from Gaofen series and Sentinel series, and adopts a high-precision extraction method for offshore raft aquaculture based on deep learning (Zhang et al. 2020b, 2022) to extract raft aquaculture areas and determine aquaculture types. Through field verification and expert interpretation, algorithm validation is achieved, and multi-temporal raft aquaculture area data are finally obtained to analyze the current situation and changes in raft aquaculture in China's offshore areas.

This case study demonstrates the spatial distribution and changes of raft aquaculture along the coast of China from 2017 to 2021. The results show that the aquaculture area has remained stable, but there have been significant changes in the aquaculture structure. Algae and fish farming have increased significantly, while shellfish farming has decreased accordingly. Meanwhile, raft farming tends to move away from the coast. Besides, China's offshore raft aquaculture is in the stage of layout optimization and quality improvement, and is generally developing toward ecological aquaculture. However, the sustainable development of marine aquaculture, which is mainly based on local monoculture farming, deserves attention. The main reasons for these changes are as follows: Raft aquaculture is mostly far away from the coast because the coastal waters have been designated as ecological reserves, while also considering recreational and leisure purposes. The main reason why China's raft farming has begun to develop toward ecological farming is that farmers realize that pursuing only economic benefits while neglecting ecological balance is not only unsustainable, but may also trigger ecological disasters and lead to a decrease in production.

7.4 Summary

This chapter focuses on themes such as reducing marine pollution, protecting marine ecosystems, and sustainably managing aquaculture. It includes case studies on the distribution characteristics and temporal changes of microplastics in China's offshore waters, dynamic monitoring of seagrass bed distribution in China, comprehensive health assessments of typical bay ecosystems in China, and dynamic remote sensing monitoring and analysis of nearshore raft aquaculture. Based on the research presented in this chapter, we make the following recommendations.

(1) Utilize the 2023 mid-term evaluation of the SDGs to reassess and adjust relevant indicator systems. Currently, five specific targets, SDG 14.1, SDG 14.2, SDG 14.3, SDG 14.a, and SDG 14.c, are in the state of "having methods but lacking data", which hinders both the United Nations and national governments from fully grasping the status of these targets and effectively implementing related management policies. It is necessary to adjust the existing indicator system appropriately, based on current concerns and the mid-term evaluation status, in order to collect more monitoring data for SDG assessments. For example, SDG 14.1 could include additional indicators for microplastic pollution and marine radioactive contamination, SDG 14.2 could add evaluation indicators for the restoration of typical marine ecosystems (such as mangroves, seagrass beds, and coral reefs), and SDG 14.7 could be revised to focus on the growth of sustainable fisheries in terms of scale or output value in the country.

(2) Focus on achievements and challenges in marine sustainable development in China. Given China's achievements and challenges in marine sustainable development, particular emphasis should be placed on marine system management in the context of the "Dual Carbon" goals and global climate change. More in-depth and comprehensive scenario analyses are needed, providing additional research cases to explore the pathways for coastal health and sustainable development. This will offer scientific support and decision-making assistance for further efforts in this area.

In the future, we will continue to enhance the capacity for sharing and applying Big Earth Data in marine sustainable development. This includes building data-sharing platforms, online computing platforms, and data service platforms to promote the timely sharing and dissemination of data and knowledge. We aim to elevate the development of the blue economy and marine technological innovation, contributing to the implementation of the 2030 Agenda.

References

Bureau of Fisheries, Ministry of Agriculture and Rural Affairs of the People's Republic of China, National Fisheries Technology Extension Center, China Society of Fisheries (2021) China Fishery statistical yearbook 2020. China Agriculture Press, Beijing (In Chinese)

Ding JF, Sun CJ, Li JX et al (2022) Microplastics in global bivalve mollusks: a call for protocol standardization. J Hazard Mater 438:129490

FAO (2016) The state of world fisheries and aquaculture 2016: contributing to food security and nutrition for all. Food and Agriculture Organization of the United Nations, Rome

FAO (2018) The state of world fisheries and aquaculture. https://openknowledge.fao.org/items/432390cf-4480-48aa-b8ee-9ecd7ee8c58e [2024-08-01]

Gao FL, Li JX, Hu J et al (2023) A review of microplastics in China marine waters. J Ocean Univ China 22:1326–1340

General Office of the State Council of the People's Republic of China (2022) Action plan for new pollutant control. https://www.gov.cn/zhengce/content/2022-05/24/content_5692059.htm [2022-05-04] (In Chinese)

Gu X, Liu SS, Li Y et al (2022) A review of sources, status, and risks of microplastics in the largest semi-enclosed sea of China, the Bohai Sea. Chemosphere 306:135564

Li Q, Jin RJ, Ye ZJ et al (2022) Mapping seagrass meadows in coastal China using GEE. Geocarto Int 37(26):12602–12617

Liu H, Huang XP, Wang YL et al (2016) Newly discovered seagrass bed and its ecological characteristics in the coastal area of Caofeidian, Bohai Sea. Chin J Ecol 35(7):1677–1683 (In Chinese)

Liu S, Pan YF, Li HX et al (2022) Microplastic pollution in the surface seawater in Zhongsha Atoll, South China Sea. Sci Total Environ 822:153604

Logan M, Hu ZY, Brinkman R et al (2020) Ecosystem health report cards: an overview of frameworks and analytical methodologies. Ecol Ind 113:105834

Ministry of Ecology and Environment of the People's Republic of China (2019) China marine ecological environment status bulletin 2018. https://www.mee.gov.cn/hjzl/sthjzk/jagb/201905/P020190529532197736567.pdf [2024-05-30] (In Chinese)

Ministry of Ecology and Environment of the People's Republic of China (2020) China marine ecological environment status bulletin 2019. https://www.mee.gov.cn/hjzl/sthjzk/jagb/202006/P020200603371117871012.pdf [2021-05-24] (In Chinese)

Ministry of Ecology and Environment of the People's Republic of China (2021) China marine ecological environment status bulletin 2020. https://www.mee.gov.cn/hjzl/sthjzk/jagb/202105/P020210526318015796036.pdf [2021-05-24] (In Chinese)

Ministry of Ecology and Environment of the People's Republic of China (2022) China marine ecological environment status bulletin 2021. https://www.mee.gov.cn/hjzl/sthjzk/jagb/202205/P020220527579939593049.pdf [2022-05-22] (In Chinese)

SDG Working Group of CASEarth (2019) Ecosystem health assessment in Jiaozhou Bay, China (2019). http://sdgs.casearth.cn/en/cs/lbw/202009/t20200910_580029.html [2024-07-24]

Sun CJ, Ding JF, Gao FL et al (2022) Challenges and perspectives of marine microplastic research. Adv Marine Sci 40(4):717–724 (In Chinese)

Sun S, Sun XX et al (2015) Theory and practice of gulf ecosystem: a case study of Jiaozhou Bay. Science Press, Beijing (In Chinese)

Sun XX, Liang JH, Zhu ML et al (2018) Microplastics in seawater and zooplankton from the Yellow Sea. Environ Pollut 242:585–595

Wu YJ, Zhang HK (2018) Changes and protection strategies of seaweed beds in Hepu, Guaugxi. China Sci Technol Inf 22:68–69 (In Chinese)

Yu X, Huang W, Wang YJ et al (2022) Microplastic pollution in the environment and organisms of Xiangshan Bay, East China sea: an area of intensive mariculture. Water Res 212:118117

Zhang WW, Zhang SF, Zhao Q et al (2020a) Spatio-temporal distribution of plastic and microplastic debris in the surface water of the Bohai Sea, China. Marine Pollut Bull 158:111343

Zhang Y, Wang CY, Chen JB et al (2022) Shape-constrained method of remote sensing monitoring of marine raft aquaculture areas on multitemporal synthetic Sentinel-1 imagery. Remote Sens 14(5):1249

Zhang Y, Wang CY, Ji Y et al (2020b) Combining segmentation network and nonsubsampled contourlet transform for automatic marine raft aquaculture area extraction from Sentinel-1 images. Remote Sens 12(24):4182

Zhou Y, Jiang ZJ, Qiu GL et al (2023) Distribution status, degradation reasons and protection countermeasures of seagrass resources in China. Oceanologia Et Limnol Sin 54(5):1248–1257 (In Chinese)

Zhu L, Bai HY, Chen BJ et al (2018) Microplastic pollution in North Yellow Sea, China: observations on occurrence, distribution and identification. Sci Total Environ 636:20–29

8 SDG 15 Life on Land

8.1 Background

China attaches great importance to ecological and environmental issues, upholds the concept "lucid waters and lush mountains are invaluable assets", and coordinates the systematic management of mountains, water, forests, fields, lakes, grasses, and sand (Huang 2022), and comprehensively promotes the construction of the nature reserve system, which has steadily improved the quality of ecosystems, effectively mitigated the aggravation of biodiversity loss, significantly enhanced the ability to cope with climate change, and made promising progress in the implementation of SDG 15 targets.

In order to promote the better implementation of SDG 15 targets in China, it is urgent to improve the scientific and technological content of ecosystem conservation and restoration strategies, and to promote ecological assessment, conservation, and restoration with technological innovation as the driving force to maximize the benefits. Conducting large-scale and spatially explicit SDG 15 indicator state assessment and clarifying the corresponding contributions of climate change and human activities are important to guide the formulation of scientific conservation and restoration policies. However, ecosystem change is influenced by multifactor interactions at multiple scales, and traditional data or analytical methods are clearly beyond their reach. Big Earth Data, with its characteristics of exploring correlations among multidisciplinary data and further discovering new models, rules, and knowledge, already has great potential for research work related to SDG 15. In 2023, we focused on three SDG 15 specific objectives, using Big Earth Data as a means to conduct three case studies to better explore the value of Big Earth Data in the service of SDG 15.

8.2 Main Contributions

In response to the data and methodological gaps faced by the SDG 15 progress assessment, we analyzed the habitat dynamics and conservation gaps of the black-necked crane, a keystone species on the Qinghai-Xizang Plateau, monitored the spatial distribution patterns and dynamic changes of oases in the arid region of Northwest China, and evaluated the effectiveness of ecological restoration in the Saihanba region. It can provide important support for the assessment of SDG 15 at the regional and global scales (Table 8.1).

H. Guo, *Big Earth Data in Support of the Sustainable Development Goals (2023)—China*, Sustainable Development Goals Series, https://doi.org/10.1007/978-981-95-2636-9_8

Table 8.1 Cases and their main contributions

Targets	Case studies	Contributions
SDG 15.1 By 2020, ensure the conservation, restoration, and sustainable use of terrestrial and inland freshwater ecosystems and their services, in particular forests, wetlands, mountains, and drylands, in line with obligations under international agreements	Habitat dynamics and conservation gaps of the black-necked crane, a keystone species on the Qinghai-Xizang Plateau	Decision support: the priority conservation areas (PCAs) plan for black-necked cranes was developed, and its spatial extent was compared with that of existing nature reserves at national, provincial, municipal, and county levels across the Qinghai-Xizang Plateau. This analysis aimed to identify conservation gaps and provide a scientific basis for the future effective conservation of black-necked cranes
SDG 15.3 By 2030, combat desertification, restore degraded land and soil, including land affected by desertification, drought, and floods, and strive to achieve a land degradation-neutral world	Spatial distribution pattern and dynamic change of oases in the arid region of Northwest China	Data product: oasis distributions in the arid region of Northwest China from 2000 to 2020
SDG 15 Protect, restore, and promote sustainable use of terrestrial ecosystems, sustainably manage forests, combat desertification, and halt and reverse land degradation and biodiversity loss	Evaluation of ecological restoration in the Saihanba Region	Decision support: explore the geological suitability for vegetation restoration in the Saihanba region and propose a series of countermeasures and recommendations

8.3 Case Study

8.3.1 Habitat Dynamics and Conservation Gaps of the Black-Necked Crane, a Keystone Species on the Qinghai-Xizang Plateau

Target: SDG 15.1 By 2020, ensure the conservation, restoration, and sustainable use of terrestrial and inland freshwater ecosystems and their services, in particular forests, wetlands, mountains, and drylands, in line with obligations under international agreements.

- **Background**

Biodiversity is the basis for human survival and sustainable development, and biodiversity conservation is of great significance to ecosystem stability and ecological security. The main reason for biodiversity decline is habitat loss and fragmentation, so conducting wildlife habitat research is crucial for biodiversity conservation (Pereira et al. 2012). In recent years, the assessment of wildlife habitat suitability has received increasing attention from scholars (Peterson 2006).

The black-necked crane (*Grus nigricollis*) belongs to the genus *Grus* in the biological family *Gruidae* of the order *Gruiformes*, and is a national Class I protected animal in China (Zheng and Wang 1998), classified as near threatened (NT) by the International Union for Conservation of Nature (IUCN). In the early 1990s, the number of black-necked cranes was 5,000–6,000 (Archibald and Meine 1996). In the late 1990s, poaching activities, agricultural development, grassland degradation, and swamp drought led to a decline in the local wildlife population in the Zoige Wetland. With subsequent monitoring, more black-necked cranes have been observed, and their numbers have been increasing further. The total number of black-necked cranes globally has reached 10,000–10,200 (Li 2014), including 6,600–6,800 adult individuals. There are more than 2,600 black-necked cranes in the Zoige Wetland in Sichuan, China, which is the largest breeding site of black-necked cranes (Liu et al. 2014). However, with the rapid economic growth in the early 21st century, tourism became popular, and infrastructure was continuously built. At the same time, global warming exacerbated the rapid melting of glaciers, leading to an increase in wetland areas. Human interference and climate change still remain to be great threats to the distribution of black-necked cranes (Li 2019).

The black-necked crane is the only one of the world's 15 species of cranes that grows, breeds, and overwinters on the plateau, mainly on the Qinghai-Xizang Plateau and the Yunnan-Guizhou Plateau in China. Most black-necked cranes breed in the Zoige Wetland at the northeastern end of the Qinghai-Xizang Plateau and overwinter at lower altitudes on the Qinghai-Xizang Plateau, the Yunnan-Guizhou Plateau, southern Xizang, and Bhutan (Li 2014). The Zoige Wetland, situated on the eastern margin of the Qinghai-Xizang Plateau, represents the largest and most ecologically significant swamp wetland within this region (Bai et al. 2022).

Biodiversity serves as the foundation for all life on Earth and has increasingly garnered global attention. Conservation efforts primarily focus on in-situ and ex-situ strategies, with the establishment of protected areas being a critical and effective approach to safeguarding natural ecosystems, wildlife, and their habitats through in-situ conservation. To achieve harmony between humans and nature, it is necessary to strengthen biodiversity conservation while promoting economic development. Therefore, in order to protect the black-necked cranes and their habitat, it is necessary to conduct research on the habitat suitability of the black-necked cranes and their protection status. The habitat suitability evaluation model, also known as the species distribution model, is used to evaluate species' habitat suitability, identify regional changes in the spatio-temporal distribution of species, and reveal possible environmental variables influencing species changes, providing an important scientific basis for biodiversity conservation (Bai et al. 2022). The study investigated the spatio-temporal changes in the extent of suitable habitats for black-necked cranes on the Qinghai-Xizang Plateau and explored the key factors driving these habitat changes.

In this case, the Qinghai-Xizang Plateau in China was selected as the study area. By utilizing geographical information technology, the optimized maximum entropy (MaxEnt) was used to evaluate and analyze the black-necked cranes' habitat and to identify the main environmental variables that affect their distribution. Based on the results predicted by the MaxEnt model, the importance of environmental variables was analyzed to identify the main environmental variables and establish the evaluation index system. The habitat suitability index (HSI) was then applied to assessing the habitat suitability of black-necked cranes and analyzing their dynamic changes over different periods from 1980 to 2020. Using the ZONATION model, the priority conservation areas (PCAs) plan for black-necked cranes was developed, and its spatial extent was compared with that of existing nature reserves at national, provincial, municipal, and county levels across the Qinghai-Xizang Plateau. This analysis aimed to identify conservation gaps and provide a scientific basis for the future effective conservation of black-necked cranes.

- **Data**

- Field distribution data of black-necked cranes: The distribution data of black-necked cranes were obtained by reviewing published literature and previous field surveys (Zhang et al. 2012; Mi et al. 2017), the Global Biodiversity Information Facility (http://www.gbif.org), and China Bird Report (http://www.birdreport.cn), only the data from 2015 to 2020 were retained. Duplicate and redundant points were removed, and the refined distribution data were imported into ArcGIS for further analysis.
- Terrain variables, including aspect, slope, and altitude, the data of which come from the Resource and Environmental Science Data Platform (https://www.resdc.cn/).
- Foraging conditions: Soil moisture data for 1980, 1990, 2000, 2010, and 2020 were obtained from the ERA5 based on the European Centre for Medium-Range Weather Forecasts (https://www.ecmwf.int/). The distance to water sources comes from land use data (https://www.resdc.cn/), which are obtained by calculating Euclidean distances after extracting raster data for paddy fields, rivers, lakes, reservoirs, and marshes. The 1982 and 1990 data of NDVI were from GIMMS NDVI3g v1 (http://data.tpdc.ac.cn/zh-hans/data/9775f2b4-7370-4e5e-a537-3482c9a83d88), and the 2000–2020 data were from the Satellite Application Center for Ecology and Environment, Ministry of Ecology and Environment (http://www.secmep.cn/).
- Human activities: Land use data of 1980, 1990, 2000, 2010, and 2020 were from the Resource and Environmental Science Data Platform of the Chinese Academy of Sciences (http://www.resdc.cn).
- Climate data, including factors such as average temperature and average precipitation in 1980, 1990, 2000, 2010, and 2020, the data of which were obtained from the China 1 km resolution month-by-month precipitation and temperature dataset (1901–2020) from the National Tibetan Plateau Data Center (http://data.tpdc.ac.cn/) (Peng 2020).

- **Method**

This case evaluated the habitat suitability of the black-necked crane in the Qinghai-Xizang Plateau of China from 1980 to 2020, and studied its dynamic changes based on a multimodel combination approach, coupling its suitable habitat areas and existing nature reserves, and carrying out a conservation gap analysis by using spatial analysis methods.

Firstly, for data collection and processing, the distribution data of black-necked cranes from 2015 to 2020 and the factors related to the breeding sites, foraging sites, and water sources of black-necked cranes were collected. These factors specifically included terrain, foraging conditions, human activities, and climatic conditions within the study area, and the data were preprocessed. The ArcGIS plug-in SDM Toolbox (v2.5) was used to spatially filter the distribution data of the black-necked crane, while the ENMTools package (v1.0.6) was employed to analyze correlations among environmental variables.

Secondly, in terms of model optimization and construction, the MaxEnt model was optimized using the Enmeval package (v2.0.3) to determine the parameters for feature combinations and regularization multipliers (RM). Based on the results of the MaxEnt model, a long-term series of black-necked crane habitat evaluation index systems was established.

Finally, in the analysis of the distribution range of suitable habitats and protection gaps in different periods, the HSI model was applied by integrating evaluation indices and weights of various environmental factors to assessing the spatio-temporal dynamics of black-necked crane habitat suitability from 1980 to 2020. The ZONATION spatial planning model was used to plan the priority areas for black-necked crane conservation and analyze the conservation gaps.

- **Results and Analysis**

The highly suitable habitats in China from 1980 to 2020 are located at the northeastern end of the Qinghai-Xizang Plateau, specifically the Zoige Wetland in northern Sichuan Province, near Qinghai Lake in southern Qinghai Province, and central and southern Xizang. The highly suitable habitat area for the black-necked cranes on the Qinghai-Xizang Plateau remained basically unchanged from 1980 to 1990, but decreased from 1990 to 2000. The highly suitable habitat area for black-necked cranes on the Qinghai-Xizang Plateau continued to increase from 2000 to 2020, increasing by 26.34% in 2020 compared with 1980. The increased area of highly suitable habitat was 79,378 km^2, accounting for 2.53% of the study area, mainly in the Nagqu City in central Xizang, the Zhaqu River and Jinsha River Basin, the Zoige Wetland, and the surroundings of Qinghai Lake (Figs. 8.1 and 8.2). From 1980 to 2020, the center of mass coordinates of the black-necked crane's habitat was located in the southern part of Qinghai Province, with a general trend of migration to the lower latitudes in the south and to the hinterland of the Qinghai-Xizang Plateau (Fig. 8.3).

The conservation priority of the black-necked crane on the Qinghai-Xizang Plateau is generally higher in the east than in the central and western parts of the plateau, with the highest PCAs concentrated in the eastern part of the plateau, central Xizang, the Yarlung Zangbo River Basin, and the Hengduan Mountains. In particular, the first-level PCAs were concentrated in the Zoige Wetland, near Qinghai Lake and Selin Co Lake in central Xizang, especially in the Qinghai Lake National Nature Reserve, the First Meander of Yellow River National Nature Reserve, the Zoige Wetland National Nature Reserve, the Gansu Gahai Zecha National Nature Reserve, the Selin Co Black-necked Crane National Nature Reserve in Xizang, and the Black-necked Crane National Nature Reserve in the Middle Yarlung Zangbo River Valley. Therefore, the conservation gaps were mainly located in the northern part of the Zoige Wetland and the counties around the Qinghai Lake. There were small portions of conservation gaps in the Yarlung Zangbo River Basin and Nima County of Xizang, Maduo County and Dulan County of Qinghai Province, Shiqu County of Sichuan Province, and Shangri-La City of Yunnan Province. The second-level PCAs and the third-level PCAs are scattered across the central Qinghai-Xizang Plateau. According to the statistical results in Table 8.2, these areas exhibit a low conservation coverage and contain unprotected regions, indicating the need for further optimization of conservation planning.

Highlights

- Using the results of the MaxEnt model, the weights and grading standards of various environmental factors were established. These factors were integrated through the HSI model to assess the habitat suitability of the black-necked crane on the Qinghai-Xizang Plateau across five periods from 1980 to 2020, resulting in distribution maps for each time period.
- The spatio-temporal changes in habitat suitability of black-necked cranes and the patterns of change in suitable habitat areas were analyzed in five periods from 1980 to 2020.
- Based on the analysis of conservation priority areas and gaps for black-necked cranes, relevant recommendations were proposed.

- **Discussion and Outlook**

Following the results of priority areas and protection gaps for black-necked crane conservation, the main recommendations are as follows. (1) Optimize and adjust the range of Qinghai Lake National Nature Reserve by considering the actual and potential distribution of several key protected waterbirds. It is recommended to build ecological corridors for wetland waterbirds between different protected areas in the Zoige Wetland, and to implement joint conservation measures. (2) In the central and western parts of the Qinghai-Xizang Plateau, it is recommended to carry out a comprehensive evaluation to establish nature reserves or small nature protected areas. In the Xinjiang Altun Mountain National

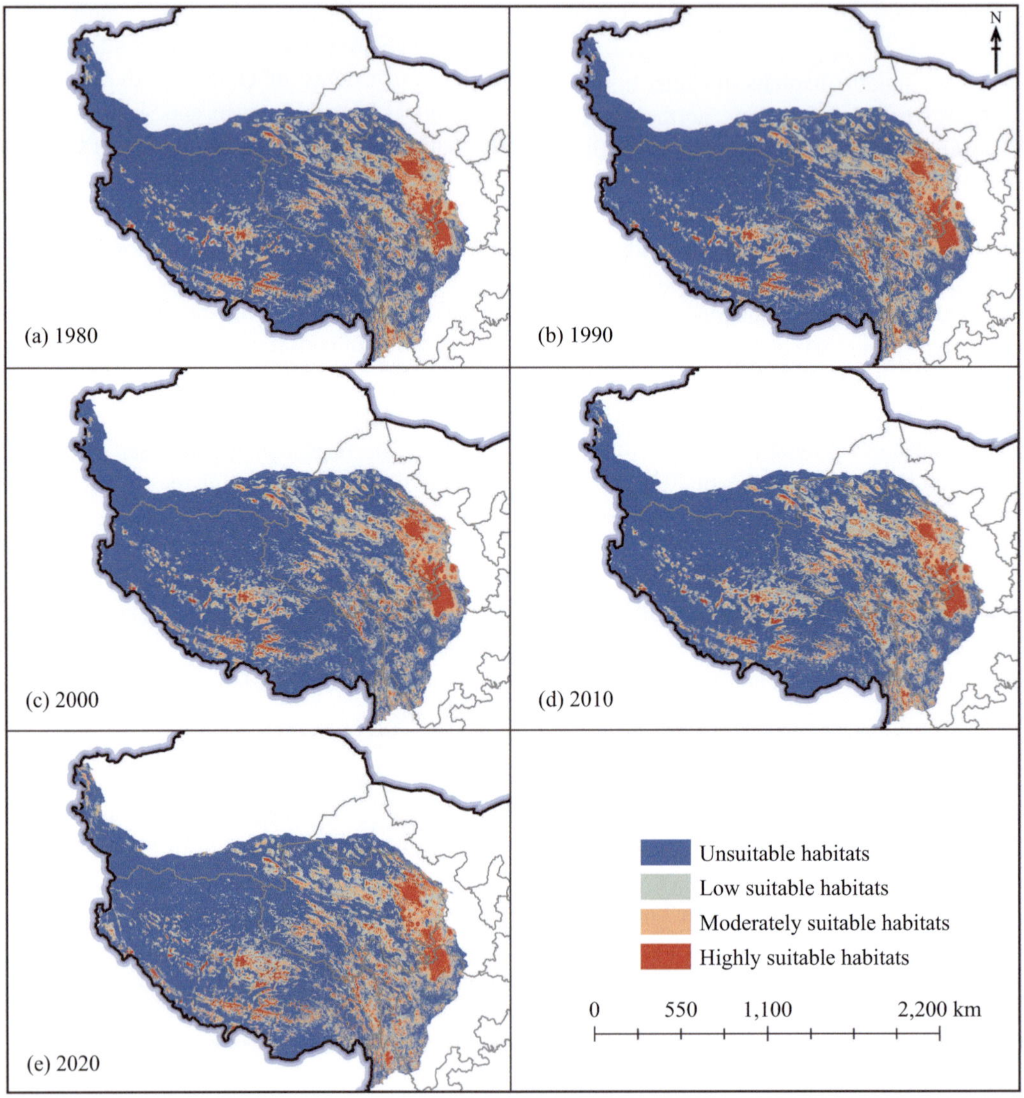

Fig. 8.1 Habitat suitability evaluation results of black-necked cranes

Nature Reserve and the Qiangtang National Nature Reserve, it is suggested to focus on the elements related to the protection of wetland birds and their habitats, and to timely supplement and learn relevant management experience. (3) Strengthen the protection and restoration of black-necked crane habitats, enhance monitoring and patrolling efforts in conservation gap areas, and establish a scientific monitoring system to promptly acquire data on ecosystems, key protected wildlife, and other related aspects within nature reserves.

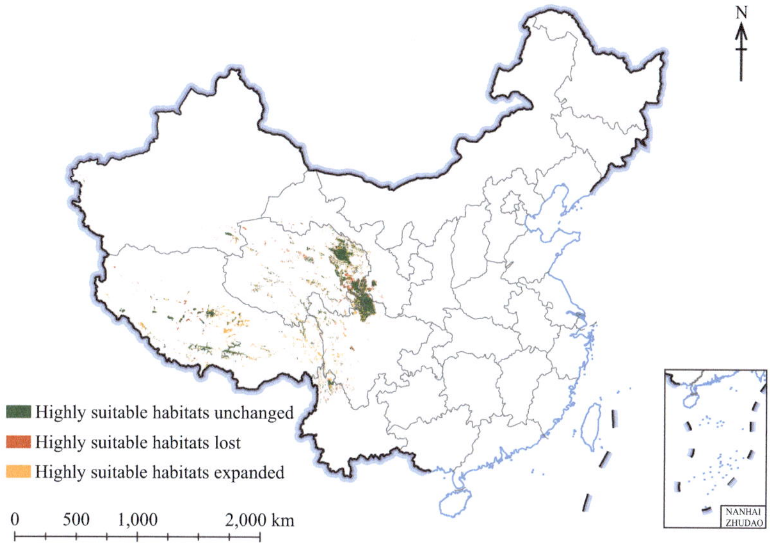

Fig. 8.2 Changes in highly suitable habitat range between 1980 and 2020

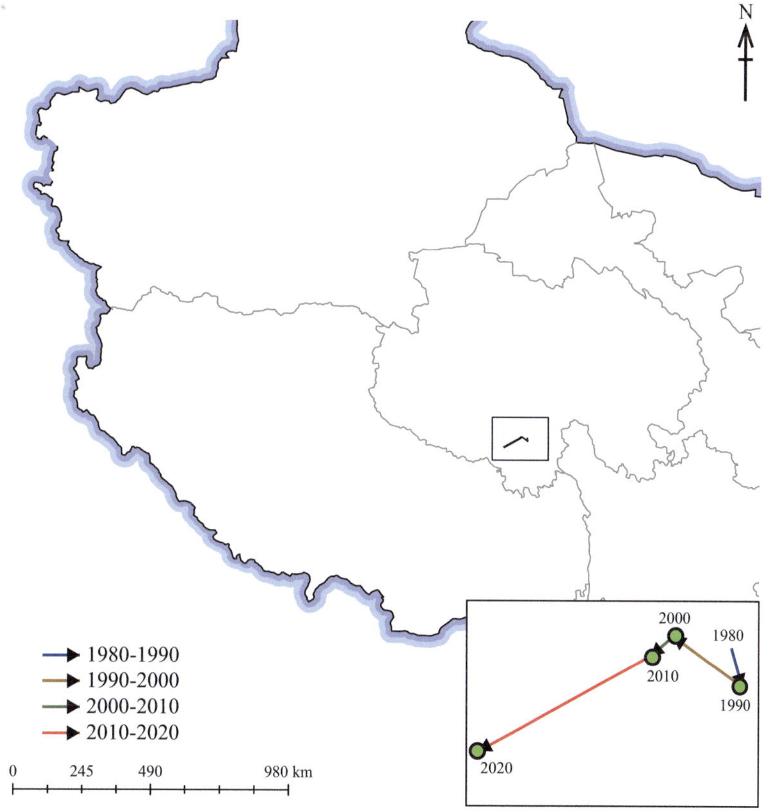

Fig. 8.3 Schematic diagram of the superposition of PCAs and nature reserves

Table 8.2 The protection ratio of different levels of PCAs in the nature reserves

PCAs	Area/km^2	Proportions of different levels of PCAs in nature reserves directly or indirectly protecting black-necked cranes/%	Proportions of different levels of PCAs in all nature reserves in the Qinghai-Xizang Plateau/%
1st level	63,428	29.27	30.86
2nd level	205,670	8.83	13.02
3rd level	365,809	4.10	11.61

8.3.2 Spatial Distribution Pattern and Dynamic Change of Oases in the Arid Region of Northwest China

Target: SDG 15.3 By 2030, combat desertification, restore degraded land and soil, including land affected by desertification, drought, and floods, and strive to achieve a land degradation-neutral world.

- **Background**

The arid region of Northwest China is located in central Asia, with a geographic location at 73°–107°E and 35°–50°N. It is an inland arid region bordered by the Helan Mountains to the east, and the Kunlun Mountains, Altun Mountains, and Qilian Mountains to the south. Its northern and western sides are bounded by international borders, covering an area of approximately 2.09×10^6 km^2 (Guo et al. 2020). The region comprises the Nei Mongol Plateau, Altai Mountains, Tianshan Mountains, Kunlun Mountains, Altun Mountains, Qilian Mountains, Junggar Basin, Turpan Basin, and Tarim Basin. The overall topography displays a landscape pattern with alternating mountains and basins, while the mountain-oasis-desert complex ecological system is its fundamental characteristic. The mountainous areas have large modern glaciers, and glacial meltwater accounts for more than 25% of the total surface runoff in the arid region of Northwest China, playing a significant role in river flow composition and stability (Qin and Ding 2022). The climate in the arid region of Northwest China is dry with scarce precipitation, and its water resources amount to only 9.9557×10^{10} m^3, which is 3.46% of the total water resources in China (Chen et al. 2023). It is the region with the most severe water scarcity among China's resource-based regions (Chen et al. 2023). The oasis is the main area for human habitation and development in the arid region of Northwest China. Although its area accounts for only 3%-5% of the total arid area, it supports over 90% of the population and produces more than 95% of the country's GDP. It is the main carrier that sustains human activities and economic and social development in the arid region of Northwest China (Shen et al. 2001).

The desert and Gobi areas in the arid region of Northwest China account for approximately 80% of the national total (Chen et al. 2023), making it a particularly fragile ecological environment area with prominent desertification issues. However, it is also the main distribution area of oases in China. Desertification and oasis formation are opposing processes in arid regions, and efforts to combat desertification are often accompanied by changes in oasis formation. Analyzing the spatial distribution and changes of oases in Northwest China, and conducting multi-scale dynamic assessments of their social, economic, and environmental impacts, are important for understanding their societal and ecological values, and objectively evaluating the effectiveness of desertification prevention and control. Existing research on oases in Northwest

China mainly focuses on specific oases, and there is relatively little research on their overall spatial pattern and dynamic changes. Moreover, there are issues with the spatial resolution and timeliness of existing oasis research. In response to these, combined with the goal of SDG 15.3 "formulate voluntary targets to achieve land degradation neutrality (LDN)", we carried out an assessment of the spatial distribution pattern and dynamic changes of oases in the arid region of Northwest China from 2000 to 2020. This might provide a decision-making reference for relevant national departments to scientifically grasp the development and changes of oases and take countermeasures.

- **Data**

- GDEMV2 DEM data for thc arid region of Northwest China from the Geospatial Data Cloud (http://www.gscloud.cn/) with 30 m spatial resolution (2020).
- Temperature and precipitation data for the arid region of Northwest China from the National Tibetan Plateau Data Center (https://data.tpdc.ac.cn/) with 1 km spatial resolution (1991–2020).
- Snow depth water equivalent data for the arid region of Northwest China from the Climate Data Store of Copernicus, Programme of the European Union (https://www.copernicus.eu/) with 0.1°spatial resolution (1991–2020).
- Landsat TM/ETM+/OLI data for the northeast of the arid region of Northwest China from the U.S. Geological Survey (https://earthexplorer.usgs.gov/) with 30 m spatial resolution (2000–2020).
- Land cover data for the arid region of Northwest China from the Resource and Environmental Science Data Center of the Institute of Geographic Sciences and Natural Resources Research, CAS (https://www.resdc.cn/) with 30 m spatial resolution (2020).
- Population and economic data for the arid region of Northwest China, published by Statistic Bureau of Xinjiang Uygur Autonomous Region, Gansu Province Bureau of Statistics, and Statistic Bureau of Nei Mongol Autonomous Region (2000–2020).

- **Methodology**

This case mainly used the decision tree classification method for oasis remote sensing classification, using soil adjusted vegetation index (SAVI), normalized difference built-up index (NDBI), index-based built-up index (IBI), MNDWI, and DEM data to classify cropland, forest land, grassland, building land, water bodies, and other land features in the oasis area.

$$\mathrm{SAVI} = (1 + L)(\rho_{\mathrm{NIR}} - \rho_{\mathrm{R}})/(\rho_{\mathrm{NIR}} + \rho_{\mathrm{R}} + L) \tag{8.1}$$

$$\mathrm{NDBI} = (\rho_{\mathrm{SWIR}} - \rho_{\mathrm{NIR}})/(\rho_{\mathrm{SWIR}} + \rho_{\mathrm{NIR}}) \tag{8.2}$$

$$\mathrm{MNDWI} = (\rho_{\mathrm{G}} - \rho_{\mathrm{SWIR}})/(\rho_{\mathrm{G}} + \rho_{\mathrm{SWIR}}) \tag{8.3}$$

$$\mathrm{IBI} = [\mathrm{NDBI} - (\mathrm{SAVI} + \mathrm{MNDWI})/2]/[\mathrm{NDBI} + (\mathrm{SAVI} + \mathrm{MNDWI})/2] \tag{8.4}$$

where ρ_{NIR}, ρ_{R}, ρ_{SWIR}, and ρ_{G} are the reflectance of the near-infrared, red, shortwave infrared, and green bands, respectively. L is the soil adjustment coefficient. Based on the decision tree classification, the oasis classification results for the years 2000, 2005, 2010, 2015, and 2020 were visually verified using Google Earth. The results showed that the overall classification accuracy of the oasis was 90.48% (114/126). When verified with the 2020 Chinese land use data provided by the Resource and Environmental Science Data Center of the Chinese Academy of Sciences, the overall classification accuracy of the oasis was 86.51% (109/126). The verification accuracy of other years using remote sensing images and visual interpretation was also over 83.33%. Therefore, the classification results of the oases in the arid region of Northwest China from 2000 to 2020 obtained using the same classification method have a high accuracy.

The land use change rate was used to analyze and evaluate the oasis changes in the arid region of Northwest China from 2000 to 2020.

$$R_t = \frac{U_\mathrm{b} - U_\mathrm{a}}{U_\mathrm{a}} \times \frac{1}{t} \times 100\% \tag{8.5}$$

where U_a and U_b are the areas of a certain land use type in the early and late stages, respectively. t represents the length of the study period, and R_t is the land use change rate corresponding to time period t.

In the stability analysis of oasis landscape pattern changes, indicators such as number of patches (NP), mean patch size (MPS), patch density (PD), and landscape shape index (LSI) were selected to calculate and evaluate the changes in oasis patches and landscape evaluation from 2000 to 2020. Besides, the coefficient of variation (CV) in statistics was used to calculate the dispersion degree of the data on the time series. The larger the CV is, the higher the dispersion degree of the data is, indicating that the data are approximately unstable on the time scale.

$$\sigma = \sqrt{\frac{1}{n}\sum_{i=1}^{n}(a_i - \overline{a})^2} \ (i = 1, 2, 3 \ldots n) \tag{8.6}$$

$$\mathrm{CV} = \sigma/\mu \tag{8.7}$$

where σ is the standard deviation, μ is the average of the array, and a is the array.

The trend of changes in temperature (TEMP), precipitation (PRCP), and snow depth/snow water equivalent (SWE) in the arid region of Northwest China from 1991 to 2020 was analyzed using a simple linear regression trend line, where slope > 0 and slope < 0 indicate an increasing and decreasing trend in TEMP, PRCP or SWE, respectively.

$$\mathrm{slope} = \frac{\sum_{i=1}^{n}(x_i - \overline{x})(y_i - \overline{y})}{\sum_{i=1}^{n}(x_i - \overline{x})^2} (i = 1, 2, 3 \ldots n) \tag{8.8}$$

where "slope" is the trend slope, x is the time, and y is the TEMP, PRCP, or SWE variables.

- **Results and Analysis**

According to the monitoring results of oasis distribution in the arid region of Northwest China from 2000 to 2020 (Fig. 8.4), the results showed that oases were mainly distributed in Xinjiang Uygur Autonomous Region (Northern and Southern Xinjiang), Gansu Province (Hexi Corridor), and Nei Mongol Autonomous Region (Alxa). The oasis area in Southern Xinjiang had the highest proportion (48.67%), followed by Northern Xinjiang (33.38%). The oasis areas in the Hexi Corridor and Alxa were relatively small, accounting for 14.61% and 3.34% of the total oasis area, respectively. The oasisification achievement in the arid region of Northwest China from 2000 to 2020 was significant, with a stable increasing trend in the oasis area. The average annual growth rate was 5,573 km^2/a, but the growth rates varied among different time periods and regions (Table 8.3). The oasis areas in Southern Xinjiang and Hexi Corridor maintained a high growth rate in 2000–2005 and 2005–2010, respectively, while the oasis area in Alxa had the highest growth rate in 2010–2020, with an average annual increase rate higher than 1.05%. The oasis area growth rate was relatively stable in Southern Xinjiang and the Hexi Corridor, while the oasis area growth rate in Northern Xinjiang had more variations in 2000–2020. From 2000 to 2020, the annual change rate of oasis area in the arid region of Northwest China was the highest in Alxa oasis (0.97%), followed by Southern Xinjiang oasis (0.70%) and Hexi Corridor oasis (0.65%), while Northern Xinjiang oasis was relatively lowest (0.45%).

There are significant differences in the NP, PD, MPS, and LSI indices of the oasis landscape in the arid region of Northwest China (Fig. 8.5). From 2000 to 2020, the NP of oasis in the arid region of Northwest China showed a decreasing trend, and the NP of oases in the Hexi Corridor was relatively stable, and its CV was 0.041. The PD of the oasis also showed a decreasing trend, with the PD of the oasis in Southern Xinjiang being relatively stable (CV = 0.051), while the

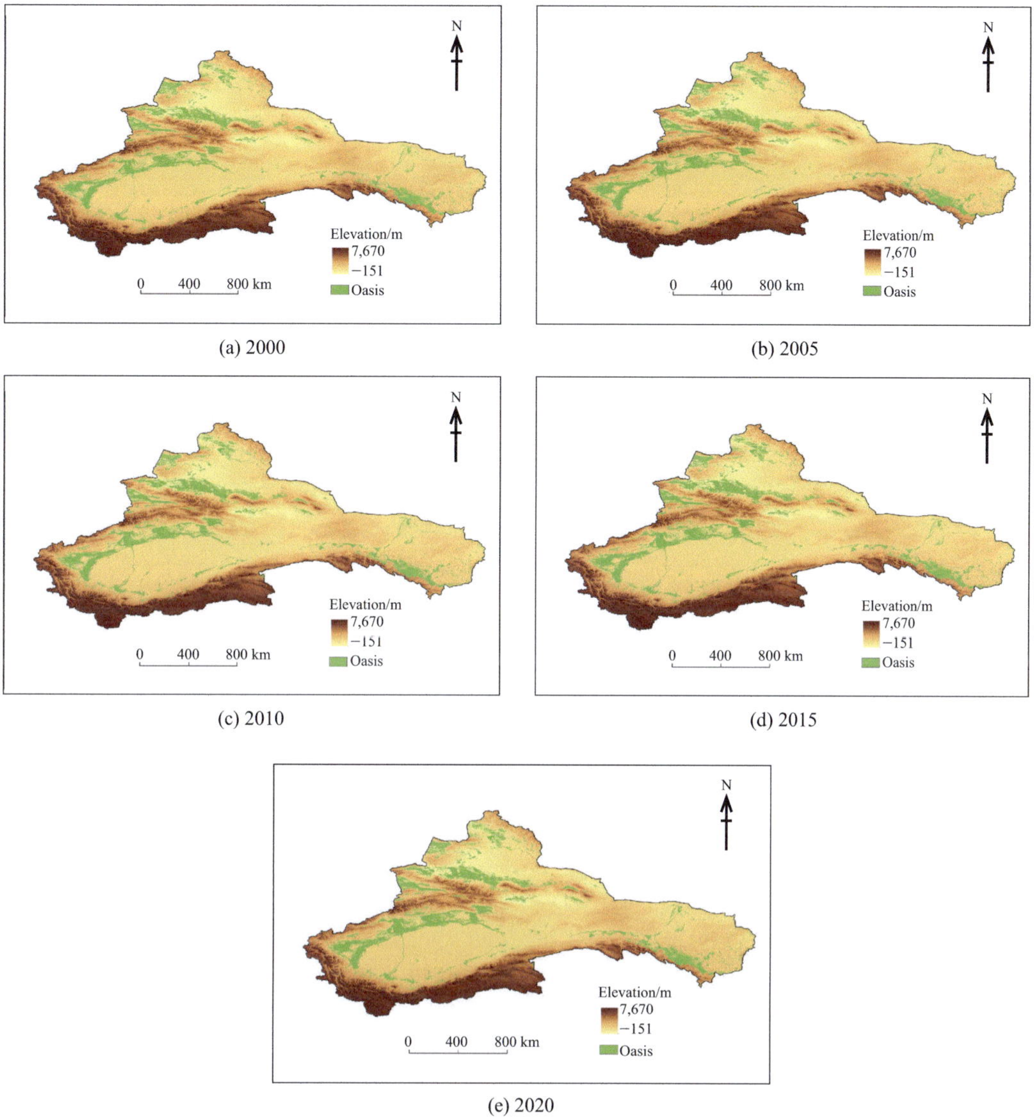

Fig. 8.4 Oasis distributions in the arid region of Northwest China from 2000 to 2020

Table 8.3 Land use change rates of oasis from 2000 to 2020

Period	Southern Xinjiang	Northern Xinjiang	Hexi Corridor	Alxa
2000–2005	0.78	0.34	0.39	0.63
2005–2010	0.53	0.75	0.78	0.64
2010–2015	0.74	0.72	0.67	1.05
2015–2020	0.61	−0.04	0.66	1.32

Unit: %

PD of the oasis in Alxa was significantly higher than those of other oases, indicating a higher degree of fragmentation. From 2000 to 2020, the MPS of the oasis showed an increasing trend, with the MPS of the oasis in the Hexi Corridor being relatively highest, the MPS of the oasis in Southern Xinjiang being relatively stable (CV = 0.048), and the MPS of the oasis in Alxa being

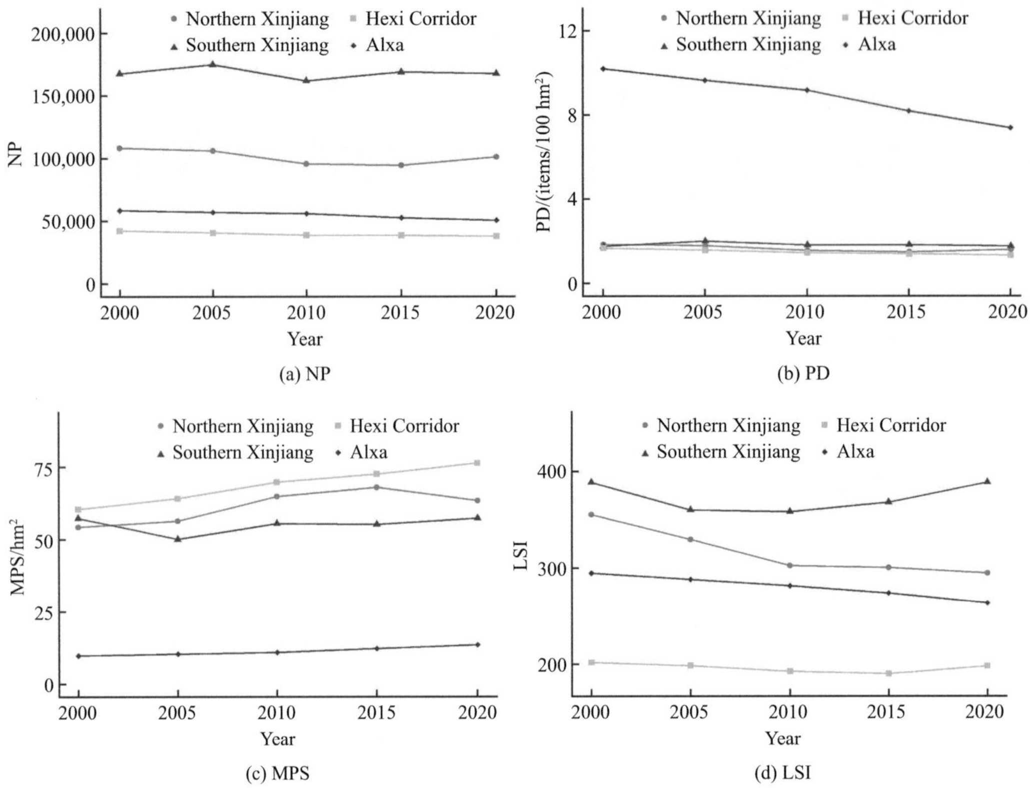

Fig. 8.5 Variations of oasis landscape indices from 2000 to 2020

relatively small and showing significant changes (CV = 0.119). From 2000 to 2020, the LSI of the oasis was the highest in Southern Xinjiang, with the most complex oasis shape, followed by the oasis in Northern Xinjiang, and the LSI of the oasis in the Hexi Corridor was relatively lowest, with a relatively simple oasis shape. The LSI of the oasis in the Hexi Corridor showed the most stable changes (CV = 0.022), while the LSI of the oasis in Northern Xinjiang showed significant changes (CV = 0.072).

The expansion and contraction of oasis areas in the arid region of Northwest China are largely affected by atmospheric precipitation and surrounding melting snow and ice. From 1991 to 2020, the mean annual temperature in the region was 6.13 °C and increased at a rate of 0.0306 °C/a, with 99.76% of the region experiencing a warming trend (Fig. 8.6a). The mean annual precipitation was 115.47 mm and slowly increased at a rate of 0.1881 mm/a. While the precipitation decreased in the northern, central, and southwestern parts of the region (39.47% of the area), it increased in other areas (60.53% of the area, Fig. 8.6b). The areas with mean annual SWE > 200 mm in the arid region of Northwest China from 1991 to 2020 were mainly distributed in the northern, western, and southwestern mountainous areas of Xinjiang. SWE in the other regions was relatively low. The area with a mean annual SWE < 20 mm accounts for 60.84% of the total area. The proportion of areas with decreasing SWE is 83.12%, while the SWE in the northwestern, western, and southwestern Xinjiang is decreasing (Fig. 8.6c). The increasing temperature directly leads to higher surface evaporation, which, to some extent, makes oasis maintenance and development more dependent on water resources. The development of oases, especially in areas with declining precipitation, will increasingly rely on surrounding high mountain snow and ice

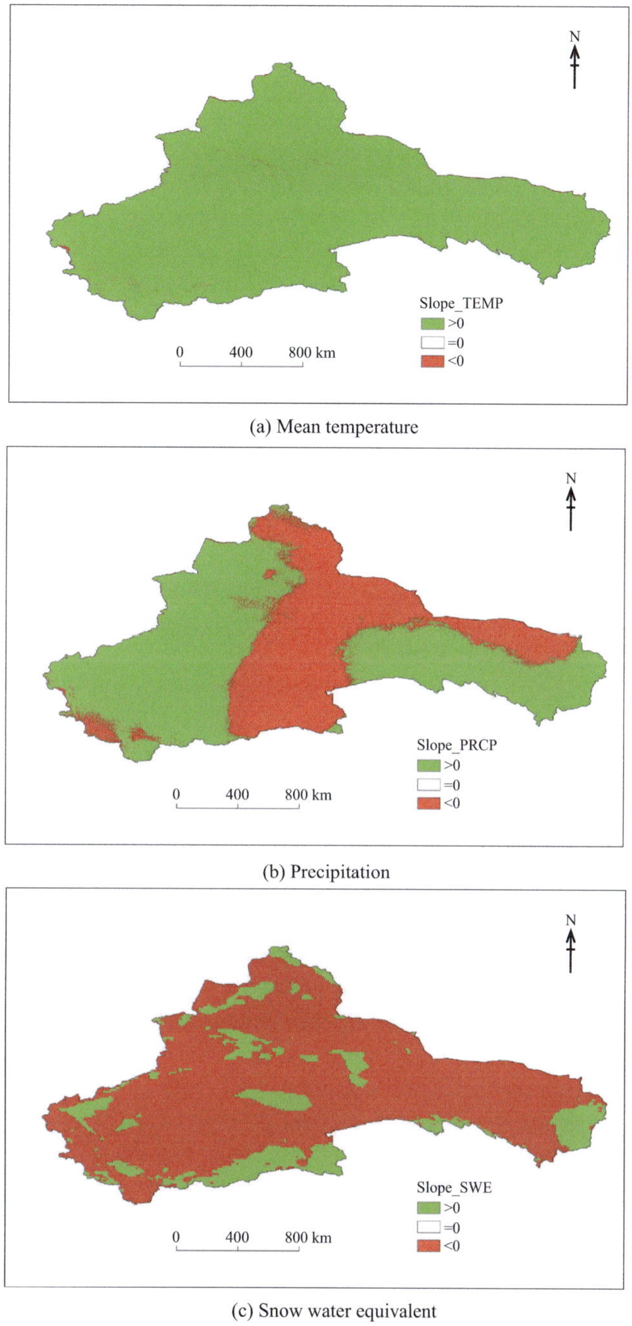

Fig. 8.6 Spatial distributions of mean temperature, precipitation, and snow water equivalent trends from 1991 to 2020

meltwater, upstream river flows, or groundwater supply. In cases of insufficient water resources in upstream rivers, reduction of ice and snow melting water from high mountains, and over-exploitation of groundwater, maintaining and developing oases in these regions will face severe challenges.

In terms of population changes, the population in the arid region of Northwest China was 28.3462 million in 2000 and 35.8918 million in 2020, with a population growth rate (26.62%) much higher than that of the oasis area (12.38%). The rapid increase in population has directly exerted greater pressure on local water and land resources. Therefore, based on the ecological carrying capacity of oases in the arid region of Northwest China, it is important to control the development scale of oases, scientifically regulate the local population size and the scale of industrial and agricultural development to maintain the stability and achieve sustainable development of oases in the region.

Highlights

- Based on the framework of the United Nations Convention to Combat Desertification (UNCCD) and Big Earth Data, we conducted remote sensing monitoring and evaluation of the spatial distribution pattern and dynamic changes of oases in the arid region of Northwest China from 2000 to 2020.
- Using Landsat images and DEM data, the decision tree classification method was used to extract an oasis in the arid region of Northwest China. A 30 m spatial resolution oases dataset of the arid region of Northwest China from 2000 to 2020 was established.
- From 2000 to 2020, oases in the arid region of Northwest China were mainly distributed in the southern and northern Xinjiang, the Hexi Corridor, and Alxa. Since 2000, significant achievements have been made in oasis development in the arid region, and the oasis area has shown a stable increasing trend, with an average annual growth rate of 5,573 km^2/a.
- From 2000 to 2020, the annual change rate of oasis area in the arid region of Northwest China was highest in Alxa oasis (0.97%), followed by Southern Xinjiang oasis (0.70%) and Hexi Corridor oasis (0.65%), while Northern Xinjiang oasis was relatively lowest (0.45%).

- **Discussion and Outlook**

This case refers to the Good Practice Guidance on SDG Indicator 15.3.1 of UNCCD. Using internationally and domestically shared scientific datasets, a study was conducted on the spatial distribution pattern of oases in the arid areas of Northwest China from 2000 to 2020. The dynamic changes of oases were evaluated at multiple scales, including oasis society, economy, and environment.

The coexistence of oases and deserts is a unique geographic landscape in arid areas and the core area for agricultural development in these regions. Oasis formation and desertification are basic geographic processes that are fundamentally opposite to each other, and oasis formation, as a reversal of desertification under specific conditions, occurs locally. In the process of oasis formation, especially in the arid region of Northwest China, water resources play a crucial role and exhibit considerable instability, and local changes in water resources directly affect the development and transformation of oases. Moreover, recent advancements in technology and the increased ability of humans to intervene on the Earth's surface have achieved significant results in the construction of artificial oases and the development of potential oases, enhancing the controllability of oasis development. In the future, the prevention and control of desertification in arid areas should be based on the framework of the Earth's surface system, with scientific control of oasis size and reasonable promotion of desertification prevention and control work based on the carrying capacity of water and land resources in the region. This will be an important direction for achieving sustainable development of oases in the arid region of Northwest China.

8.3.3 Evaluation of Ecological Restoration in the Saihanba Region

Target: SDG 15 Protect, restore, and promote sustainable use of terrestrial ecosystems, sustainably anage forests, combat desertification, and halt and reverse land degradation and biodiversity loss.

- **Background**

With the excessive exploitation and destruction of natural resources by humans, the problem of global land degradation and desertification is becoming increasingly severe, which directly affecting economic and social development, and food security at global and regional scales (Haregeweyn et al. 2023; Bai et al. 2008). The Saihanba region, located in the northern part of Hebei Province, is the largest artificial forest area in northern China, and serves as a pivotal region for China's ecological civilization construction with continuous vegetation restoration and ecological restoration. To achieve SDG 15.3, the government restored degraded land and prevented desertification, and long-term vegetation restoration efforts, which had been implemented in the Saihanba region and demonstrated remarkable success (Xing et al. 2023; Tan et al. 2013).

In 2022, the Saihanba forest area expanded to 1.151 million mu, and the forest coverage rate reached 82% (Ma 2023). Additionally, the increase in precipitation in the Saihanba region is significantly greater than that in surrounding areas, effectively regulating runoff during both wet and dry seasons. The interannual variability of regional water resources has shifted from negative to positive and continues to rise, while groundwater levels have also increased, accompanied by enhanced water conservation capabilities. The ecological environment in the Saihanba area has been effectively restored and protected, making positive contributions to the restoration and sustainable development of the ecosystem. It has become one of the typical cases of SDGs of the United Nations and holds significant demonstration and reference value for achieving the objective of zero growth in land degradation. However, vegetation restoration requires consideration of multiple factors, including geology, hydrology, and human activities. To better guide vegetation restoration efforts, in-depth research is needed on the geological suitability of the Saihanba region.

This case study aims to facilitate the scientific greening of the Bashang Plateau through a comprehensive ground substrate survey. By calculating the water balance of small watersheds, reconstructing ancient environmental and vegetation communities, conducting spatial layout surveys suitable for afforestation, and selecting optimal areas for afforestation and suitable vegetation types, we analyze the coupling relationship between surface substrate types and plant communities. Additionally, we further explore the geological suitability for vegetation restoration in the Saihanba region. This study proposes a series of countermeasures and recommendations to address critical scientific questions, such as "what trees to plant, where to plant, and how many to plant", to ensure the long-term benefits and sustainable development of ecological restoration projects.

- **Data**

The temperature, precipitation, cross-section runoff, evapotranspiration from remote sensing retrieval, vegetation, and the Third National Land Survey data in the Saihanba region from 2001 to 2021. The temperature and precipitation data are sourced from actual observations at the Saihanba Mechanical Forest Farm Meteorological Station. The runoff data of the river section comes from the observation of the national control section of the Xiaoluan River. Remote sensing images are obtained by downloading from NASA's official website and the Fengyun Satellite Data Service network, and then the actual evapotranspiration is inverted.

- **Method**

According to the technical framework for water balance research, data on temperature, precipitation, cross-section runoff, and evapotranspiration obtained through remote sensing, as well as vegetation and the Third National Land Survey data in the Saihanba area from 2001 to 2021, were systematically collected. By using the

InVEST model, changes in precipitation, actual evapotranspiration, runoff, and water storage from 2001 to 2021 were analyzed. Based on calculating ecological water consumption of the forest and grassland ecosystems in the Saihanba area from 2001 to 2021, the analysis of the relationship between the age and species of trees and actual evapotranspiration, along with the ecological water consumption characteristics of different tree species at various growth stages. Furthermore, an analysis of actual evapotranspiration and the spatial distribution characteristics of different land cover types was performed to investigate the impact of large-scale afforestation on temperature, precipitation, actual evapotranspiration, and water storage in the region. The effects of artificial forest ecosystems on regional climate improvement and water conservation were also evaluated.

Focusing on the future scientific greening needs of Saihanba and its surrounding areas, such as "where to plant, what trees to plant, and how many to plant", we will deepen the survey of the coupling between ground substrates and suitable forest and grassland ecosystems. We investigate types of native tree species present in the Saihanba region since the Holocene and propose optimal afforestation areas on the Bashang Plateau based on geological conditions. At the same time, by using survey data from four vegetation quadrats and regional groundwater level measurements on the Bashang Plateau, the mutual constraint relationship between NDVI and groundwater level was analyzed from the perspective of biodiversity and "how many trees to plant", and the geological site suitability of rebuilding three "Saihanba" in Hebei Province was evaluated.

Based on the analysis of the formation of Moon Lake in the Saihanba region and the changes in its surface area since 2013, combined with the characteristics of the water-bearing rock formation at the bottom of the lake, and considering the seepage communication problem between the shallow lake basalt and clay distribution zone and the Ruyi River, as well as the phenomenon of reduced effective surface runoff caused by hardened road surfaces entering the lake, this case proposes an ecological restoration plan for Moon Lake from two aspects, namely surface water extraction and groundwater regulation, to support the management of natural lake shrinkage in the Bashang Plateau.

- **Results and Analysis**

The hydrological regime of the Saihanba region is primarily recharged by meteoric precipitation, while water consumption takes place through various pathways such as soil evaporation, vegetative transpiration, surface runoff, and the artificial exploitation of water resources. Meteorological monitoring from 1991 to 2020 indicates an increasing trend in annual precipitation at the rate of 1.69 mm/a, compared to the period from 1961 to 1990, yielding a net accumulation of 50 mm over three decades. The increase in water availability has significantly contributed to the enhancement of local ecological conditions.

Throughout the Holocene epoch (approximately 10,000 years before present), the Saihanba region has experienced distinct climatic oscillations, characterized by cold and dry, warm and wet, and fluctuations in temperature. During the Holocene Optimum, which occurred 6,000–2,000 years before present, regional temperatures were comparable to modern climate conditions under the influence of global warming. The regional native vegetation exhibited a diverse assemblage during this period. The woody vegetation comprised arboreal taxa such as *Quercus*, *Betula*, *Ulmus*, *Pinus*, *Juglans*, *Larix*, and *Carpinus*. The understory vegetation was characterized by *Prunus sibirica*, *Hippophae*, and *Ericaceae* shrubs, complemented by a herbaceous layer dominated by species from the *Artemisia*, *Chenopodiaceae*, and *Poaceae* (Fig. 8.7).

Deciduous broad-leaved forests thrive optimally at groundwater depths of 2–6 m, while herbaceous plants are best suited to groundwater depths of 1–3 m (Fig. 8.8). Vegetation degradation is evident in areas where the groundwater depth exceeds 10 m, such as Daqinggou Village to Dasuji Village in Shangyi County, the

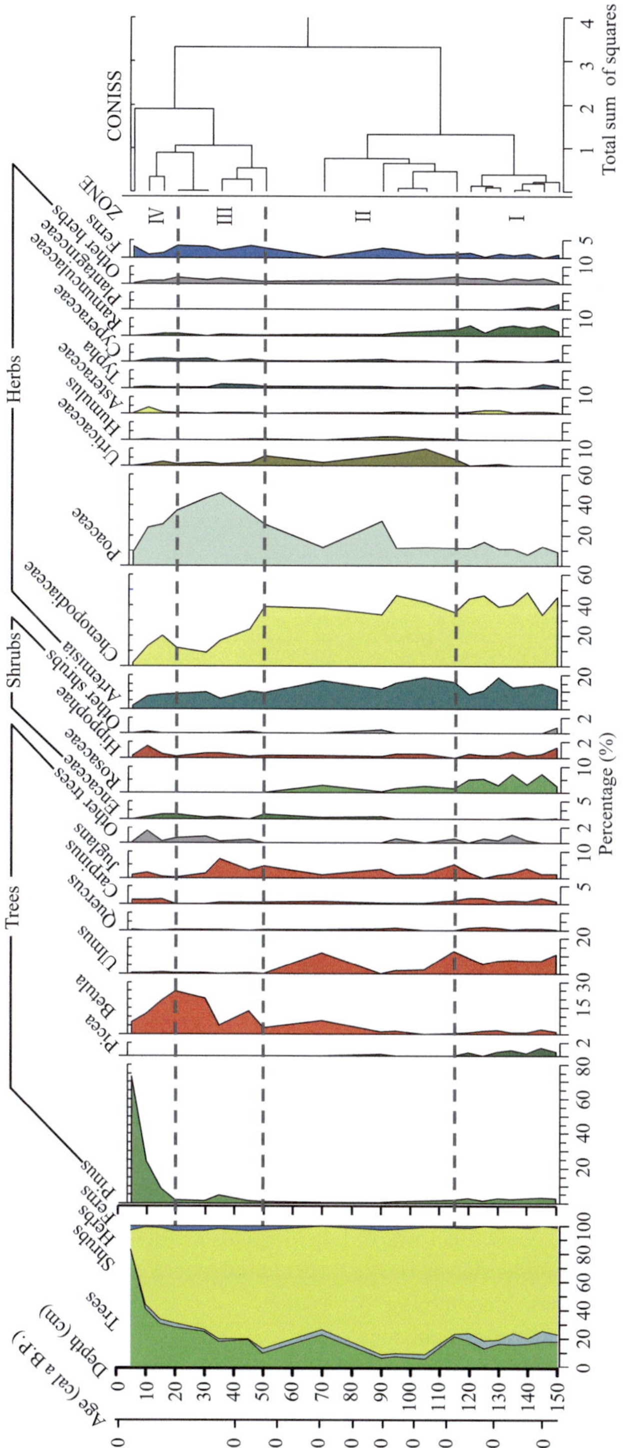

Fig. 8.7 The Holocene vegetational evaluation history of the Saihanba region

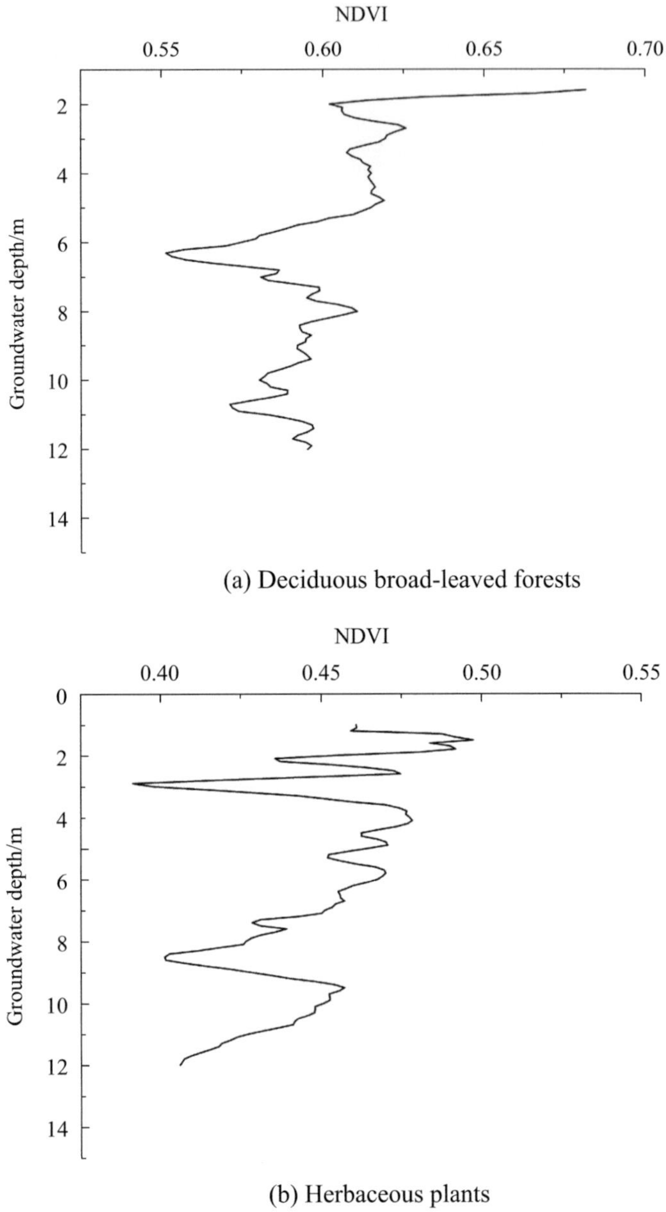

Fig. 8.8 Groundwater depth of deciduous broad-leaved forests and herbaceous plants

Chabei Management Zone and Xiaoertai Town in Zhangbei County, and the Shandian River Basin in Guyuan County. Therefore, maintaining a groundwater depth of less than 6 m can enhance ecological resilience and promote sustainable vegetation development in the Bashang Plateau region.

Highlights

- The effective improvement of the local climate and ecological environment by the artificial forest plantation in the Saihanba region has been determined.

- The geological suitability of scientific greening site selection was analyzed.

- **Discussion and Outlook**

Artificial forests have a significant impact on regional climate improvement and water conservation with global warming. Firstly, artificial forests can absorb substantial amounts of carbon dioxide, alleviate the problem of global warming, and enhance the regional climate by releasing oxygen and regulating temperature and humidity. Secondly, they can increase soil retention and water storage capacity, reduce soil erosion, and protect surface and groundwater resources. Research indicates that the artificial forest system in the Saihanba area plays a crucial role in conserving and protecting regional water resources, contributing to "peak shaving and drought replenishment", and improving the stability and sustainability of water resources. Based on existing research results, this case proposes the following suggestions for ecological restoration in the Saihanba area.

(1) The restoration and protection of ecological systems should be implemented. The Saihanba area has greatly improved the local ecological environment through the vigorous development of artificial forests, but the restoration and protection of the ecosystem still need to be further strengthened. It is recommended that the area of afforestation continue to be expanded, especially in areas that are vulnerable to drought and other extreme weather events. In addition, mixed planting of multiple vegetation types can be used to increase the stability of the ecosystem.
(2) Water resource management in the region should be strengthened. It is suggested that more stringent water resource management policies be formulated, including restricting water resource exploitation, enhancing water utilization efficiency, encouraging water conservation measures, etc., to ensure the sustainable utilization of water resources and protect the local ecosystem.
(3) Scientific monitoring and evaluation should be executed. To gain a better understanding of the changing trends of the ecological environment and the effectiveness of ecological restoration work, it is imperative to intensify scientific monitoring and evaluation. It is recommended to establish a complete ecological environment monitoring system, adopt modern monitoring approaches, promptly obtain various data on the ecological environment, and conduct scientific analysis and assessment.

In the future, the patterns of plant communities and the sustainable development of artificial forests in the Saihanba region should be based on enhancing the adaptability and stability of the ecosystem. This approach must consider water conservation functions and the types of surface substrates while adhering to principles such as tree species diversity and regional suitability. The aim is to achieve the dual goals of ecological environment protection and sustainable development. We can draw upon the ecological restoration experience in the Saihanba region to explore more appropriate vegetation restoration solutions for different regions, and further facilitate the realization of global SDGs.

8.4 Summary

This chapter makes full use of the characteristics of Big Earth Data and, at the Chinese scale, focuses on the three themes of species habitat protection, oasis distribution monitoring, and ecosystem restoration. Through case studies, it identifies problems, explores the potential, innovates technologies, and demonstrates applications. It achieves an analysis of habitat changes and protection gaps for typical species (black-necked cranes) on the Qinghai-Xizang Plateau, monitors spatial distribution patterns and dynamic changes of oases in the arid regions of Northwest China, and evaluates the effectiveness of ecological restoration in the Saihanba region of Chengde. This provides critical support for ecosystem restoration, biodiversity conservation, and land degradation management in China.

China's terrestrial biological sustainability goals continue to develop in a positive direction. However, as China is a developing country, the pressures we face remain significant. At the same time, the implementation of national strategies such as the GDI and the Belt and Road Initiative requires us to export successful experience and resources to the international community. Under the dual pressures of domestic and international demands, there is an urgent need to better leverage the value of Big Earth Data to innovatively support the implementation of terrestrial biological sustainability goals from a scientific and technological perspective.

References

Archibald GW, Meine CD (1996) The cranes: status survey and conservation action plan. IUCN, Gland

Bai JJ, Hou P, Zhao YH et al (2022) Research progress of species habitat suitability models and their verification. Chin J Ecol 41(7):1423–1432 (In Chinese)

Bai ZG, Dent DL, Olsson L et al (2008) Global assessment of land degradation and improvement: 1. Identification by remote sensing. ISRIC World Soil Information, Wageningen

Chen YN, Li ZQ, Xu JH et al (2023) Changes and protection suggestions in water resources and ecological environment in arid region of Northwest China. Bull Chin Acad Sci 38(3):385–393 (In Chinese)

Guo ZC, Wei W, Shi PJ et al (2020) Spatio-temporal changes of land desertification sensitivity in the arid region of Northwest China. Acta Geogr Sin 75(9):1948–1965 (In Chinese)

Haregeweyn N, Tsunekawa A, Tsubo M et al (2023) Progress and challenges in sustainable land management initiatives: a global review. Sci Total Environ 858:160027

Huang RQ (2022) Huang Runqiu: in-depth study and implementation of the spirit of the 20th National Congress of the Communist Party of China and strive to build a beautiful coexistence between man and nature. https://www.gov.cn/xinwen/2022-12/26/content_5733590.htm [2024-08-09] (In Chinese)

Li FS (2014) IUCN Black-necked Crane (*Grus nigricollis*) conservation plan. Zool Res 35(S1):3–9

Li FS (2019) IUCN SSC crane specialist group: crane conservation strategy. International Crane Foundation. Baraboo

Liu Q, Wang K, Li FS et al (2014) Numbers and distribution of black-necked cranes (*Grus nigricollis*) at Zoige Wetland on the eastern Qinghai-Xizang Plateau. Zool Res 35(S1):134–138 (In Chinese)

Ma AP (2023) From the vast wilderness to the vast forest sea, artificial restoration creates the ecological miracle of Saihanba. http://www.xinhuanet.com/2023-12/07/c_1130012829.htm [2024-09-01] (In Chinese)

Mi CR, Huettmann F, Guo YM et al (2017) Why choose Random Forest to predict rare species distribution with few samples in large undersampled areas? Three Asian crane species models provide supporting evidence. PeerJ 5:e2849

Peng SZ (2020) 1-km monthly precipitation dataset for China (1901–2020). https://data.tpdc.ac.cn/zh-hans/data/faae7605-a0f2-4d18-b28f-5cee413766a2 [2025-03-21]

Pereira HM, Navarro LM, Santos MI (2012) Global biodiversity change: the bad, the good, and the unknown. Annu Rev Environ Resour 37:25–50

Peterson AT (2006) Uses and requirements of ecological niche models and related distributional models. Biodivers Inform 3(1):59–72

Qin DH, Ding YJ (2022) Assessment report on climate and ecological environment evolution in China—evolution of climate and ecological environment in China: 2021 (Comprehensive Volume). EDP Sciences, Beijing (In Chinese)

Shen YC, Run JW, Wu GH et al (2001) Oasis of China. Henan University Press, Kaifeng (In Chinese)

Tan WB, Zhou LP, Liu KX (2013) Soil aggregate fraction-based ^{14}C analysis and its application in the study of soil organic carbon turnover under forests of different ages. Chin Sci Bull 58:1936–1947

Xing B, Wang YN, Yin ZQ et al (2023) Influence of forest ecosystem on regional temperature and precipitation variation in the Saihanba region. Geol Bull China 42(12):2174–2184 (In Chinese)

Zhang T, Ma M, Zhang X et al (2012) Distribution and population size of black-necked cranes in autumn in East Kunlun and Altun Mountains, Xinjiang. Chin J Zool 47(6):31–35 (In Chinese)

Zheng GM, Wang QS (1998) China red data book of endangered animals: aves. Science Press, Beijing (In Chinese)

9 Integrated Evaluations and Interactions Among SDGs

9.1 Background

The SDGs encompass a wide array of intricate and multifaceted issues, with interwoven relationships among different goals, primarily manifested as synergistic and trade-off effects. Specifically, synergistic effects denote the phenomenon where achieving a particular goal simultaneously enhances the improvement of other goals, highlighting the mutually beneficial relationships among them. Conversely, trade-off effects signify that pursuing one goal might necessitate sacrificing other goals, leading to a delicate balance (Pradhan et al. 2017; Zhang et al. 2022). As the centers of regional population aggregation and economic activity, cities are crucial for achieving sustainable development. Conducting an SDG cross-sectoral and integrated assessment for cities at the prefectural level across China helps clarify the interconnections among SDGs in different cities. By dynamically simulating the achievement of SDGs under various future development scenarios, this approach supports the formulation of targeted strategies for sustainable urban development. It also promotes the optimal allocation of urban resources, maximizes the overall benefits of SDGs in cities, and enables the rational adjustment of sustainable development pathways. This is of great significance for effectively achieving the 2030 Agenda.

Big Earth Data consist of large datasets associated with spatial attributes and Earth sciences, including data on land, oceans, atmosphere, and human activities. It is characterized by massive volume, multi-source, and multi-temporal dimensions (Guo et al. 2021; Dong et al. 2021), offering significant advantages in interaction and integrated studies of SDGs at the urban scale. On the one hand, Earth observation and geographic information data provide continuous spatial and temporal coverage, addressing inconsistencies in the standards and quality of statistical data at the urban scale. These data can promptly capture the changes in urban surface elements, and their high spatial resolution offers more detailed and precise urban surface information, providing critical insights for SDG monitoring at the urban scale. On the other hand, geographic information modeling and simulation methods, based on spatial analysis and other techniques, can effectively uncover the interactions among different urban SDGs, revealing their potential mechanisms and key factors (Xiao et al. 2022). Furthermore, constructing predictive models for future sustainable development that account for inter-city SDG interactions and dynamically simulating the achievement of SDGs under different future development scenarios can provide scientific support for future urban planning and decision-making.

H. Guo, *Big Earth Data in Support of the Sustainable Development Goals (2023)—China*, Sustainable Development Goals Series, https://doi.org/10.1007/978-981-95-2636-9_9

The *Big Earth Data in Support of the Sustainable Development Goals (2022)—China* provides a comprehensive assessment of the sustainable development progress in typical regions of China. It explores the synergies and trade-offs among SDGs at the provincial level in China, analyzes the shifts in these trade-offs across provinces, and identifies exemplary positive cycles in different provinces. The report also selects key indicators that promote the collaborative development of SDGs, offering decision-making guidance to help establish prioritized development goals for each province and address SDG trade-off issues in the development process. Moreover, the report considers the social, economic, and natural environmental characteristics of different regions, establishes an SDG evaluation system tailored to the development of each region in China, and assesses the sustainable development progress across various regions of the country. This year's report will focus on evaluating the sustainable development progress at the prefecture-level city scale in China, examining the synergies and trade-offs among SDGs in different cities, and forecasting the future trajectory and trends of SDG achievement in these cities. The aim is to provide decision-making support and exemplary references for the sustainable development of prefecture-level cities across China. Additionally, the report discusses and analyzes the ecological status, biodiversity distribution, and the direct synergies and trade-offs with human health at the spatial grid scale in China.

9.2 Main Contributions

To clarify the complex synergies and trade-offs among SDGs, this report primarily focuses on the integrated evaluations and interactions among SDGs in Chinese cities at the prefectural level and above, and multi-objective trade-offs and synergies for "One Health" in China. It showcases the technical advantages of Big Earth Data in supporting SDG integration and interactions on a global scale. The report contributes mainly in two areas: method and model, and decision support (Table 9.1).

9.3 Case Study

9.3.1 Integrated Assessment of Progress on SDGs for Cities at or Above the Prefecture Level of China

Target: SDG 1, SDG 2, SDG 3, SDG 4, SDG 5, SDG 6, SDG 7, SDG 8, SDG 9, SDG 10, SDG 11, SDG 12, SDG 13, SDG 15, SDG 16, SDG 17.

- **Background**

Cities are considered key factors in achieving over 60% of the SDGs (Lafortune et al. 2019). Assessing city-level progress toward the SDGs not only helps identify the priorities for development in various cities but also promotes effective resource allocation and rational distribution of industries. A comprehensive multi-year assessment of SDG progress in cities at or above the prefecture level in China can effectively facilitate the overall and balanced realization of the SDGs at national and sub-national (provincial, urban agglomeration, and city) scales.

Currently, the SDG comprehensive index is widely used for comprehensive assessments and comparisons of SDG progress across different countries and cities. However, this index represents overall progress toward the SDGs by calculating the arithmetic mean of the scores of the 17 goals, neglecting the consideration of balanced development among different SDGs. If there is a significant disparity in the scores of the 17 SDGs for a city, relying solely on the SDG comprehensive index may result in an overly optimistic view of the city's sustainable development. Given that all 17 SDGs should be treated equally and that improvements in a single SDG may facilitate or hinder the achievement of others, it is particularly important to quantify the differences in the development of the 17 SDGs. The concept of evenness originates from ecology, which characterizes biodiversity

Table 9.1 Cases and their main contributions

Themes	Case studies	Contributions
Integrated and interaction analysis of SDGs in cities at or above the prefecture level of China	Integrated Assessment of Progress on SDGs for Cities at or above the prefecture level of China	Method and model: construct a city-level indicator system of 16 SDGs (excluding SDG 14) with traditional statistical data and Big Earth Data, including 61 assessment indicators; assess the city-level progress toward achieving SDGs along two dimensions, namely SDG comprehensive index performance and balanced development; solve the problem of insufficient information caused by the use of the SDGs comprehensive index only in other assessments Decision support: identify development priorities and development pathways to guide Chinese cities in formulating development strategies; provide a reference for cities in other countries with similar scales and development statuses
	Assessing spatial spillover effects of SDGs for cities at or above the prefecture level in China	Method and model: for the first time, traffic travel time data matrices are used as the inter-city weight matrix in spatial econometric models. Compared to traditional spatial weight matrices that only consider the straight-line distance between two cities, the traffic travel time matrix better reflects the degree of connection between cities and has more practical significance Decision support: interactions among Chinese cities generate positive spatial spillover effects, thereby promoting the achievement of SDGs across cities. Specifically, a 10% increase in the SDG comprehensive index of neighboring regions leads to a 2.80% increase in the SDG comprehensive index of the local region. Moreover, the spatial spillover effect of urban clusters is significantly higher than that of non-urban clusters
	Analysis of synergy trade-offs and projection of future development scenarios for SDGs for cities at or above the prefecture level in China	Method and model: using a combination of simultaneous equation model and Granger causality analysis, the study explores the synergies and trade-offs among 16 SDGs across 285 cities at or above the prefecture level in China. A future development iterative prediction model, which accounts for the interactions between urban SDGs, was constructed Decision support: regions with closely interconnected SDGs interactions are mostly located in urban agglomeration areas, where the development of these urban clusters can enhance inter-city interactions. As learning costs rise, cities achieve sufficient development, transitioning from continued development to full development, leading to an increasing number of cities reaching SDGs. Most cities need to prioritize development in areas such as food security, quality education, and climate change to accelerate progress toward achieving the SDGs
Multi-objective trade-offs and synergies for "One Health" in China	SDGs trade-offs and synergies for "One Health" in China	Decision support: from the perspective of "One Health", the trade-offs and synergies among relevant SDGs, including China's ecological status, biodiversity, and zoonotic diseases, are revealed. It is demonstrated that changes in ecological conditions can have a positive impact on human health

by quantifying the distribution characteristics of different species within the same community. Introducing the balanced score to quantify the balanced development of SDGs at the city scale not only helps visualize the progress disparities among the 17 SDGs within the same city but also provides a more comprehensive comparative result of sustainable development processes across different cities. This can serve as a scientific reference for optimizing the development pathways of case cities and identifying development priorities for other cities.

- **Data**

– Statistical data of SDGs for 285 Chinese cities released by the National Bureau of Statistics from 2010 to 2020 (http://www.stats.gov.cn/).

– Land use classification data from 2010 to 2020 with a spatial resolution of 30 m (Yang and Huang 2021).

- **Methods**

Based on the SDG framework proposed by the United Nations and informed by China's sustainable development practices, this case study establishes a comprehensive set of 61 indicators under 16 SDGs (excluding SDG 14) for 285 cities at or above the prefecture level of China. This indicator framework is designed according to the principles of adaptability, measurability, comparability, wide coverage, and multiple timeframes. Through this framework, the assessments of comprehensive SDG progress are conducted. Referencing the SDG progress chart methodology of the United Nations (United Nations 2020), the status of each SDG score for urban clusters and non-urban clusters in China for the year 2020 is calculated, along with the distance to achieving SDGs and the development trends of each SDG from 2010 to 2020. The status of the 2020 SDG score is divided into five levels based on the 20% threshold: target met or almost met, close to target, moderate distance to target, far from target, and very far from target. The development trends of each SDG from 2010 to 2020 are classified into four levels based on the SDG score growth rate: substantial progress/on track, fair progress but acceleration needed, limited or no progress, and deterioration. Additionally, the SDG comprehensive index score for each city is calculated, and an improved radar chart method is used to calculate the balance score for different cities (Liu et al. 2020). SDG balance is measured based on the score differences among SDGs, reflecting the level of balance in SDG development.

- **Results and Analysis**

From 2010 to 2020, China's 285 cities at or above the prefecture level made significant progress in sustainable development. Urban clusters outperformed non-urban clusters in achieving SDGs, with a more pronounced growth trend. Overall, these 285 cities have made good progress toward SDGs, whether located in urban clusters or not. SDG 1 (No Poverty) has been essentially achieved, and SDG 12 (Responsible Consumption and Production) is close. Challenges exist in achieving SDG 6 (Clean Water and Sanitation), SDG 10 (Reduced Inequalities), SDG 11 (Sustainable Cities and Communities), and SDG 13 (Climate Action). Notably, SDG 13 shows limited progress due to its primary indicator being carbon emissions. While SDG 2 (Zero Hunger), SDG 4 (Quality Education), SDG 5 (Gender Equality), and SDG 8 (Decent Work and Economic Growth) face significant challenges, they exhibit a rapid development trend. Both SDG 3 (Good Health and Well-being) and SDG 7 (Affordable and Clean Energy) show fast development trends in both urban and non-urban clusters, with urban clusters having a more favorable status compared to non-urban clusters and a higher likelihood of achieving them by 2030 (Fig. 9.1). SDG 9 (Industry, Innovation and Infrastructure) and SDG 17 (Partnerships for the Goals) both pose great challenges in achieving the 2030 goals in both urban agglomerations and non-urban agglomerations, but they are both showing a rapid development trend.

Chinese cities at or above the prefecture level have higher scores for balanced SDG development than for the comprehensive index. However, the growth trend of the comprehensive index scores is more pronounced, especially among cities in urban clusters. Urban clusters not only enhance the overall performance of SDGs but also possess the capability to promote balanced development across SDGs. From 2010 to 2020, urban clusters in China demonstrated higher average scores in both the SDG comprehensive index and balance level, compared to non-urban clusters (Fig. 9.2), and exhibited a more significant growth trend. For most cities, the growth rate of the SDG comprehensive index score exceeded that of the balance score, particularly for cities with lower initial comprehensive index scores. Spatially, cities in the eastern regions not only had higher SDG comprehensive index scores, but also paid more attention to balanced development among SDGs, as evidenced by the most notable growth trend in balance scores. Cities in the northeast saw moderate improvements in both the

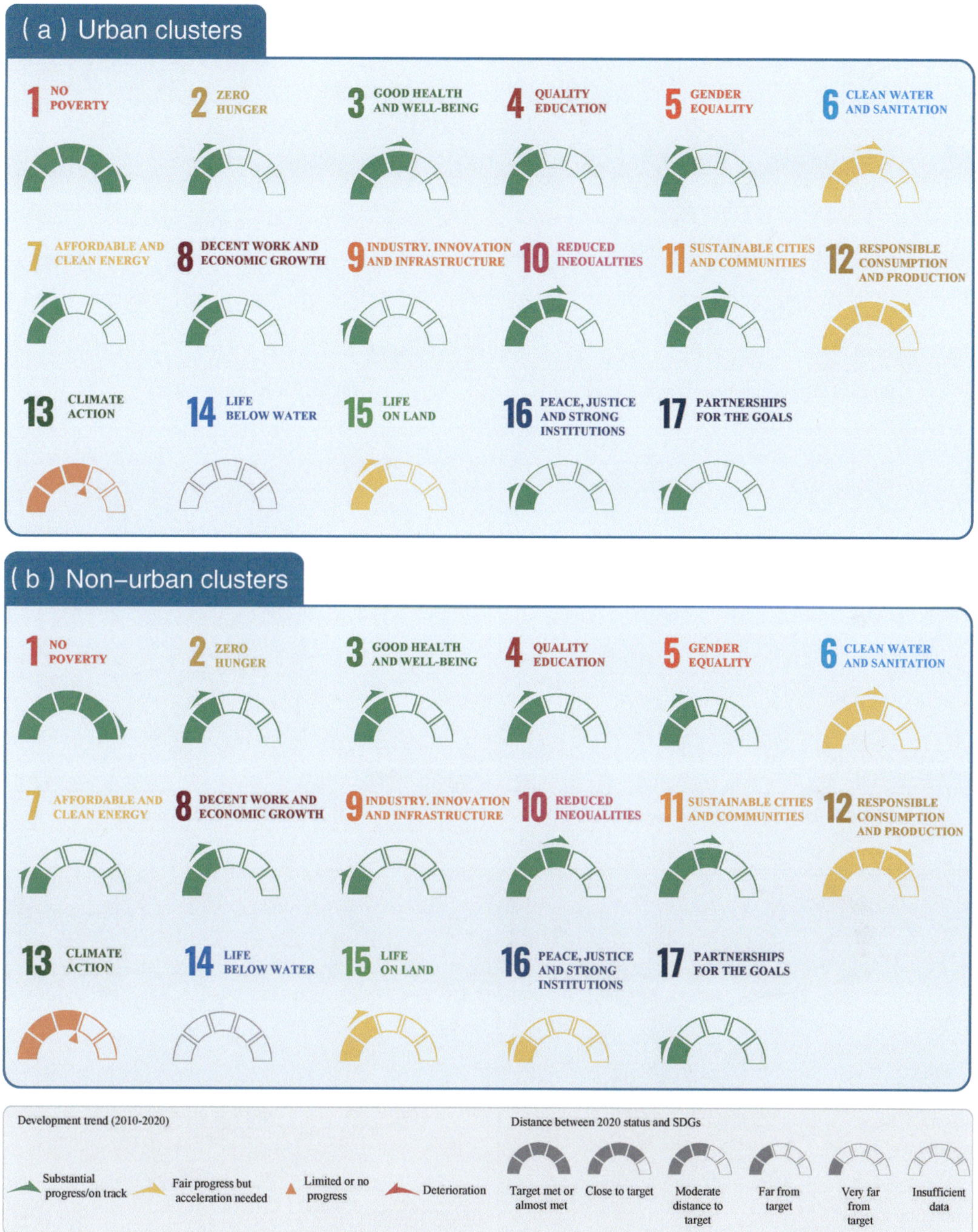

Fig. 9.1 SDG progress dashboard of China's 285 cities at or above the prefecture level in 2010–2020

SDG comprehensive index and balance scores, while cities in the central and western regions achieved more remarkable progress in enhancing the overall performance of SDGs. Generally, urban cluster cities with close socioeconomic ties, well-established infrastructure, balanced industrial distribution, emphasis on technological innovation, and effective environmental management are more likely to achieve SDGs. Additionally, different cities should formulate

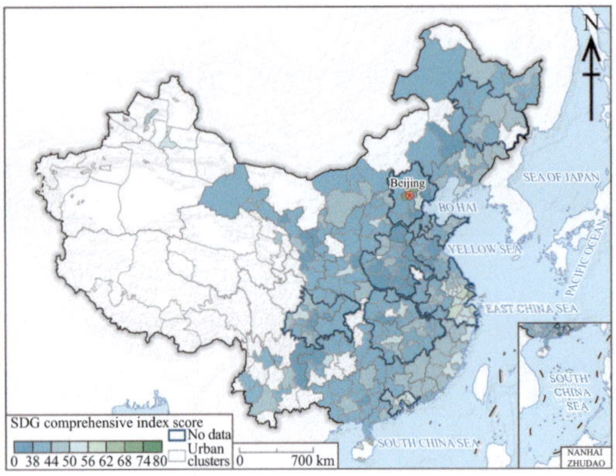

(a) 2010 SDG comprehensive index score

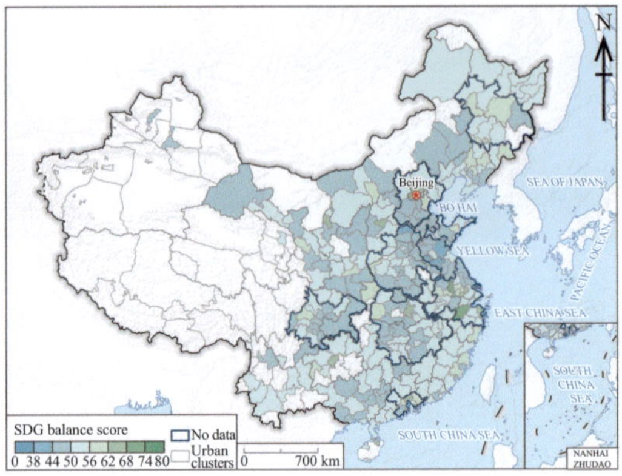

(b) 2010 SDG balance score

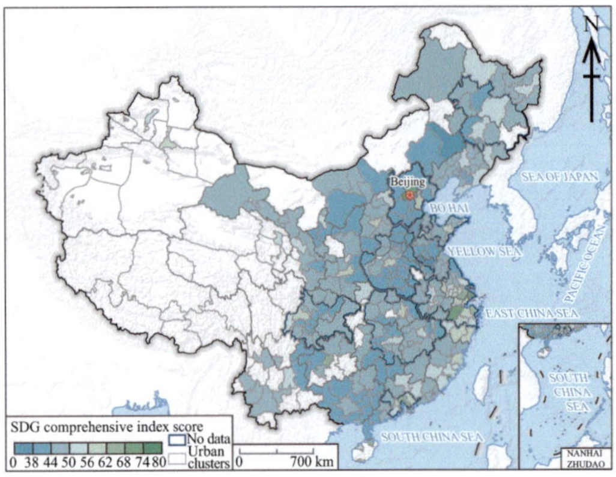

(c) 2015 SDG comprehensive index score

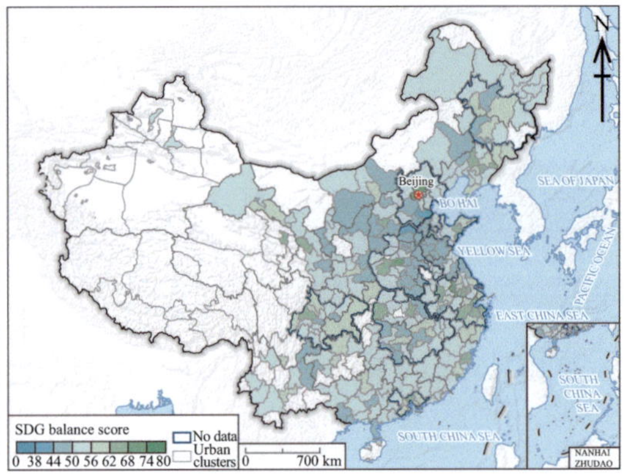

(d) 2015 SDG balance score

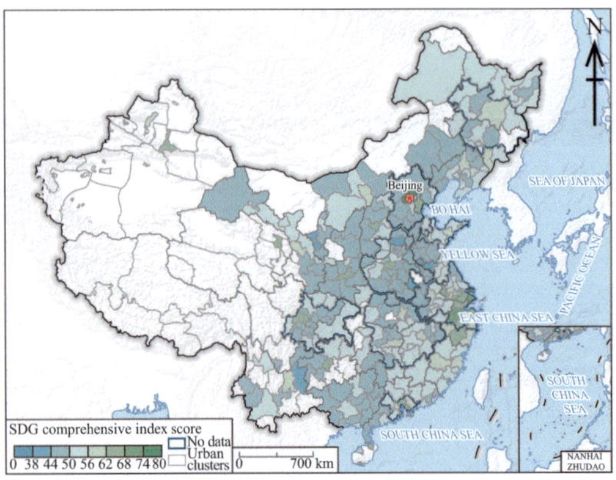

(e) 2020 SDG comprehensive index score

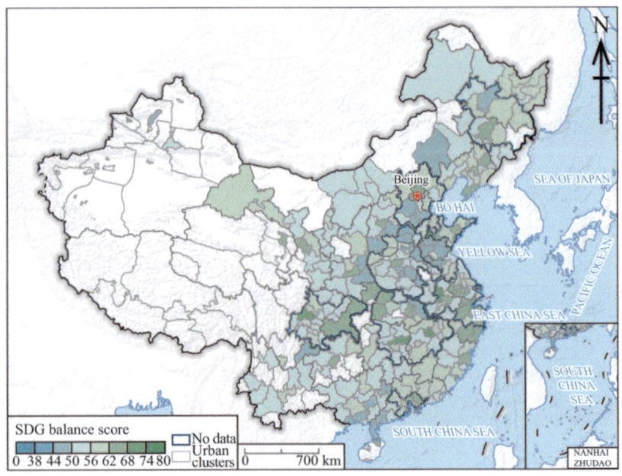

(f) 2020 SDG balance score

Fig. 9.2 Integrated assessment of SDG progress in 285 cities at or above the prefecture level in 2010–2020 development strategies and paths that align with their geographical location, advantageous industries, foundational environment, developmental priorities, and the socioeconomic and environmental conditions of surrounding cities.

Highlights

- Construct a city-level indicator system of 16 SDGs (excluding SDG 14) with traditional statistical data and Big Earth Data, including 61 assessment indicators.
- Assess the city-level progress toward achieving SDGs along two dimensions, namely SDG comprehensive index performance and balanced development; solve the problem of insufficient information caused by the use of the SDGs comprehensive index only in other assessments.

- **Discussion and Outlook**

The SDGs emphasize holistic and balanced development and achievement. This thematic study uses both Big Earth Data and traditional statistical data to assess comprehensive SDG progress in 285 cities at or above the prefecture level in China, encompassing both SDG comprehensive index performance and balanced development. The methods used in this study can be applied to assessing progress toward SDGs for other cities and scales (e.g., global scale) to identify development priorities and pathways. The findings of this study not only provide guidance for the formulation of development strategies in Chinese cities but also serve as a reference for cities in other countries with similar scales and development statuses.

Achieving the SDGs at the city scale plays a critical role in bridging local and national efforts, providing data and case support for provincial and national sustainable development strategies, while also offering adaptive suggestions for localizing the SDGs at the county level. However, it is important to recognize that data gaps still exist, and there is a need to establish a more comprehensive and broadly applicable city-level evaluation indicator system to provide stronger support for the holistic realization of the SDGs.

9.3.2 Assessing Spatial Spillover Effects of SDGs for Cities at or Above the Prefecture Level in China

Target: SDG 1, SDG 2, SDG 3, SDG 4, SDG 5, SDG 6, SDG 7, SDG 8, SDG 9, SDG 10, SDG 11, SDG 12, SDG 13, SDG 14, SDG 15, SDG 16, SDG 17.

- **Background**

In 2015, the United Nations proposed 17 SDGs as an extension of the MDGs, aiming for global sustainability progress by 2030 (Fukuda-Parr 2016). These SDGs focus on addressing a wide array of global challenges, such as poverty, hunger, education, health, gender equality, clean water and sanitation, clean energy, economic growth, industrial innovation, reducing inequalities, sustainable cities, climate action, and the protection of marine and terrestrial ecosystems, etc. (Georgeson and Maslin 2018).

As the largest developing country and the second-largest economy in the world, China plays a pivotal role on the global stage, with its influence on economic, social, political, and cultural areas steadily growing. Therefore, China bears a significant responsibility in advancing the SDGs (Feng et al. 2013). While the 17 SDGs emphasize the role of national governments in planning and implementation, local governments are often at the forefront of delivering the necessary progress (Wiedmann and Allen 2021). With nearly 300 prefecture-level or above cities, these Chinese cities are expected to act as primary drivers in China's efforts to achieve the SDGs.

The rapid development of globalization has led to increasingly tight connections between regions and their neighbors, rather than allowing them to exist as independent entities with no interaction (Xue et al. 2024). The progress of SDGs in a particular area can be affected by various factors from its neighboring regions, such as trade, population movement, and water pollution (Schmidt-Traub et al. 2017; Xiao et al. 2023). Positive spatial spillover effects (synergistic effects) refer to the ways in which other regions can promote the achievement of SDGs in the focal region, while negative spatial spillover effects (trade-off effects) indicate that other regions may hinder the achievement of SDGs in the focal region. To better and more quickly facilitate the progress of SDGs in Chinese cities, it is particularly necessary to explore spatial spillover effects on the progress of SDGs between cities.

- **Data**

– SDG index covers 17 SDGs, 285 Chinese cities (including 136 megalopolis cities and 149 non-megalopolis cities), and 14 years (2007–2020), with each indicator sourced from Xiao et al. (2022).
– Control variables are obtained from the China Urban Construction Statistical Yearbook, China Statistical Yearbook for Regional Economy, various city statistical yearbooks, and provincial statistical yearbooks.
– The method for constructing the governance efficiency data is derived from Xiao et al. (2021).

- **Methods**

1. **Spatial econometrics model**

In spatial econometrics, models like the SDM, SLM, and SEM were used to capture different types of spatial spillover effects. To select the most appropriate model for regression, a series of related tests were conducted. The process unfolded as follows.

Firstly, this study performed LR test to examine whether spatial-fixed effects, time-fixed effects, or spatial- and time-fixed effects were included in the model. According to Table 9.2, for three datasets (Chinese cities, megalopolis cities, and non-megalopolis cities), both the null hypotheses that "spatial-fixed effects were jointly insignificant" and "time-fixed effects were jointly insignificant" were rejected at the 1% significance level. This indicated that both spatial- and time-fixed effects should be adopted in the model for these datasets. Secondly, under the SDM, the Hausman test was applied to deciding which of the fixed and random effects

Table 9.2 Test results of model selection

Tests	Chinese cities	Megalopolis cities	Non-megalopolis cities
Joint significance test for spatial-fixed effect	719.0256 [0.0000]	354.9542 [0.0000]	354.9500 [0.000]
Joint significance test for time-fixed effect	940.8178 [0.0000]	441.9069 [0.0000]	441.9100 [0.000]
LR test spatial lag	20.3272 [0.0024]	13.2921 [0.0386]	31.8489 [0.0000]
LR test spatial error	13.8417 [0.0315]	17.4847 [0.0077]	25.2483 [0.0003]
Wald test spatial lag	19.5355 [0.0033]	15.1888 [0.0188]	31.2643 [0.0000]
Wald test spatial error	12.1397 [0.0589]	16.7260 [0.0103]	20.5196 [0.0022]
Hausman test	413.7260 [0.0000]	218.5737 [0.0000]	225.1435 [0.0000]

Note Values in the "[]" are the probability values, denoting the rejection of the null hypothesis at the significance level should be included in the model. As shown in Table 9.2, for the three datasets, the null hypothesis of the existence of random effects was rejected at the 1% level of significance. This suggested that the fixed effect model should be selected in the model. Lastly, to test whether SDM can be simplified into SLM or SEM, the LR test and the Wald test were employed. From Table 9.2, for all three datasets, the null hypotheses stating that "SDM can be simplified to SLM" and "SDM can be simplified to SEM" were both rejected at a significant level. This implied that SDM was the suitable model. Based on the results of all tests, the SDM with both spatial- and time-fixed effects was the suitable model for Chinese cities, megalopolis cities, and non-megalopolis cities.

Based on the results of the model selection mentioned above, this study employed the SDM to assess the spatial spillover effects of the SDG progress between focal cities and their neighboring cities in China. The specific expression formula of the SDM was as follows:

$$\ln\left(\frac{y_{i,t}}{y_{i,t-1}}\right) = \alpha \ln\left(y_{i,t-1}\right) + \rho \sum_{l=1}^{N} w_{il}$$

$$\ln\left(\frac{y_{l,t}}{y_{l,t-1}}\right) + X_{it}\beta + \sum_{l=1}^{N} w_{il}X_{lt}\gamma + \mu_i + \varepsilon_{it} \tag{9.1}$$

where $y_{i,t}$ is the SDG index of city i at the year t; $\propto$ is the coefficient to examine whether there is a convergence trend in the SDG index across cities. If $\propto$ is significantly negative, it indicates there is β-convergence. β-convergence means a negative correlation between the growth rate and the initial level of the SDG index. In other words, the SDGs index growth rate of cities with lagging SDGs performance is higher than that of cities with good SDGs performance; ρ is spatial lag regression coefficient. If ρ is positive, the SDG progress of neighboring cities can promote the SDG progress of the focal city; X_{it} representes control variables related to the environment, society, economy, etc.; $\sum_{l=1}^{N} w_{il}X_{lt}$ suggestes spatial spillover effects of the control variables; γ is the vector of spatial autocorrelation coefficients of the control variables; W_{il} indicates the spatial weight between city i and city l. The distance between cities was not solely determined by the straight-line geographical distance between two locations. It was also influenced by transportation conditions and road networks. Compared to the geographic distance, between cities calculated based on centroids, the driving time matrix took into account not only geographical distance, but also traffic conditions, congestion levels, and road quality. Therefore, the driving time matrix provided a more accurate reflection of the true "distance" between cities.

2. **Variable description**

This study aimed to examine the spatial spillover effects between focal cities and their neighboring cities on the SDG progress. Therefore, the SDG progress was set as the dependent variable, represented by the logarithmic growth rate of the SDG index. As for the independent variables, some studies suggested that growth rate was negatively correlated with initial development levels, while others argued that regions with initial advantages can accumulate more benefits, whereas those in relatively disadvantaged positions can fall further behind (Cheng et al. 2020; Han et al. 2018; Sala-i-Martin 1994). Accordingly, in this study, the initial SDG level was set as an explanatory variable, represented by the logarithm of the initial SDG index. Moreover, since SDGs encompass economic, social, and environmental aspects, economic development, social performance, and environmental protection were also treated as explanatory variables. These were represented by per capita GDP with a logarithm in the current year minus that in the previous year, unemployment rate with a logarithm in the current year minus that in the previous year, and the ratio of labor engaged in water and environmental management with a logarithm in the current year minus that in the previous year. Additionally, given the importance of effective public governance for the achievement of the SDGs by 2030, governance efficiency progress was included as an explanatory variable. This was represented by the governance efficiency with logarithm in the current year minus that in the previous year.

- **Results and Analysis**

1. **Spatial autocorrelation test**

In this study, the local Moran's *I* for each city was calculated to reflect the spatial correlation between focal cities and their neighboring cities. Figure 9.3b, c show scatter plots of the local Moran's *I* for 2010 and 2020, respectively, highlighting two key findings: focal cities with high SDG progress are surrounded by cities with similarly high SDG progress, while focal cities with low SDG progress are surrounded by cities with similarly low SDG progress. Furthermore, Fig. 9.3a provides a more intuitive representation of the spatial correlation distribution for cities in 2020. In general, this study should consider incorporating spatial factors into the spatial regression model in the following analysis.

2. **Spatial spillover effects on SDG progress across Chinese cities**

Figure 9.3d showed that for Chinese cities, the spatial autoregressive coefficient was 0.2800, and the null hypothesis that the spatial autoregressive coefficient was equal to 0 was rejected at the 1% significance level. This implied that when the progress of the SDG index of neighboring cities increases by 10%, the progress of the SDG index of the focal cities would increase by 2.80%. Therefore, the interactions between cities created a positive spatial spillover effect (synergetic effects), promoting mutual progress toward the SDGs. This suggested that the SDG progress was not only driven by internal factors within central cities but also influenced by spatial spillover effects from neighboring cities. If the spatial autoregressive coefficient were negative, it would indicate that the interactions between cities might generate a negative spatial spillover effect (trade-off effects), hindering the SDG progress of neighboring cities. Some studies argue that focal cities can absorb various production resources from neighboring cities, and as a result, the stronger the focal city, the larger the regional disparity, leading to a siphon effect (Lu et al. 2022; Zhou et al. 2019). However, this study indicated that the SDG progress in Chinese cities showed a significant transboundary synergetic effect, and that the SDG progress was facilitated between focal cities and their neighboring cities.

3. **Spatial spillover effects on SDG progress by megalopolis and non-megalopolis cities**

From Fig. 9.3d, both spatial spillover effects in megalopolis cities and non-megalopolis cities were positive (synergistic effects). Specifically, in megalopolis cities, when the progress of the SDG index in neighboring cities increased by 10%, the progress of the SDG index of the focal cities can rise by 2.940%. In contrast, in non-megalopolis cities, a 10% increase in the progress of the SDG index in neighboring cities can contribute to a 1.600% improvement in the progress of the SDG index in a focal city. Consequently, the magnitude of the spatial spillover effect in megalopolis cities was higher than that in non-megalopolis cities, with the magnitude in megalopolis cities being 83.75% stronger than that in non-megalopolis cities.

Highlights

- The study reveals transboundary synergistic effects, their magnitude, and the spatial heterogeneity across 285 Chinese cities from 2007 to 2020, offering newly feasible approaches for accelerating the collaborative achievement of China's 17 SDGs.
- For the first time, the driving time matrix is used as the weight matrix between cities in spatial econometric models. Unlike traditional spatial weight matrices that only consider the straight-line distance between cities, the driving time matrix accounts for not only

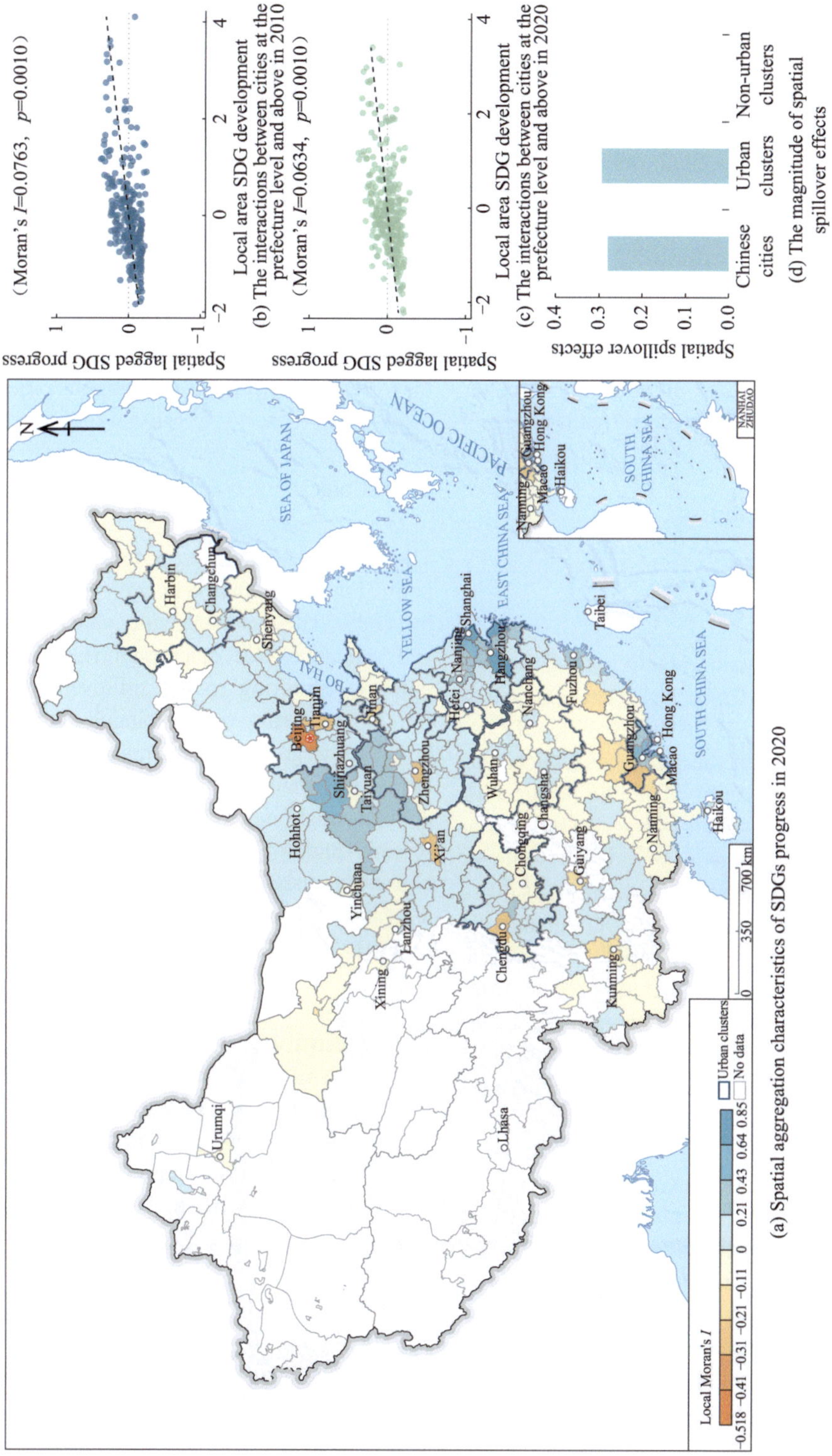

Fig. 9.3 Spatial spillover effects on SDG progress across Chinese cities

geographical distance but also road conditions, traffic congestion, and road quality. This driving time matrix more accurately reflects the true "distance" between cities, providing a valuable reference for relevant studies.

- The interactions between Chinese cities generate positive spatial spillover effects, promoting the achievement of SDGs in neighboring regions. Specifically, a 10% improvement in the progress of the SDG index of neighboring cities can boost a 2.80% increase in the progress of the SDG index of focal cities. Moreover, the spatial spillover effects between megalopolis cities are significantly stronger than those between non-megalopolis cities.

- **Discussion and Outlook**

1. **Discussion**

In light of the regression results, transboundary interactions between Chinese cities generate positive spatial spillover effects (synergistic effects). When the progress of the SDG index of neighboring cities increases by 10%, the progress of the SDG index of focal cities can rise by 2.80%. Moreover, the magnitude of spatial spillover effects (synergistic effects) of megalopolis cities is 83.75% stronger than that of non-megalopolis cities. This implies that megalopolis cities play a more important role in accelerating the achievement of China's 2030 SDG targets. To better and more rapidly advance the SDG progress of Chinese cities, the following recommendations are proposed.

First, since the progress of the SDG index of neighboring cities has a synergistic effect on the progress of the SDG index of the focal city, it is recommended to strengthen cooperation and collaboration between cities. This can be achieved through cross-city project partnerships, resource sharing, and information exchange to collectively promote the realization of SDGs.

Second, considering the spatial heterogeneity of Chinese cities (the spatial spillover effects of megalopolis and non-megalopolis cities), it is suggested to accelerate the SDG implementation in megalopolis cities, where more substantial positive spatial spillover effects (synergistic effects) occur, and expand the scope of megalopolis cities to drive the SDG progress of neighboring cities. Priority can be given to piloting policies and projects within megalopolis cities to more effectively enhance the SDG progress. On this basis, spatial structure optimization and integration of urban functions can be carried out to maximize the synergistic effects of megalopolis cities.

2. **Outlook**

The recommendations proposed in this study will help China collaboratively achieve the 17 SDGs by 2030. They also provide valuable insights and references for other countries and global cities in exploring spatial spillover effects. As the global focus on sustainable development continues to grow, it is anticipated that the framework presented in this study will be widely applied and promoted in practice. Through further research and implementation, this framework can be continuously improved and optimized, making a greater contribution to global sustainable development.

9.3.3 Analysis of Synergy Trade-Offs and Projection of Future Development Scenarios for SDGs for Cities at or Above the Prefecture Level in China

Target: SDG 1, SDG 2, SDG 3, SDG 4, SDG 5, SDG 6, SDG 7, SDG 8, SDG 9, SDG 10, SDG 11, SDG 12, SDG 13, SDG 15, SDG 16, SDG 17.

- **Background**

The SDGs of the United Nations cover a broad and complex range of themes, and there is an intricate interrelationship between the goals, mainly characterized by synergies and trade-offs. Specifically, synergies refer to the phenomenon whereby the realization of a particular objective simultaneously contributes to the improvement of other objectives, a relationship that refers specifically to the mutual reinforcement of objectives. The trade-off effect, on the other hand, implies that in the pursuit of one objective, other objectives may have to be sacrificed, creating a situation of mutual checks and balances (Pradhan et al. 2017; Zhang et al. 2022). This synergy and trade-off effect is not limited to one region, but is widespread among different geographic units due to the role of people-to-people exchanges, economic trade, cultural communication, material transfers, etc., and is known as the spatial spillover effect. An urban cluster city is a whole with compact spatial organization and close economic ties. The internal interaction of urban cluster cities is frequent, and there is a significant spatial spillover effect. However, the current research emphasizes the spillover effect of the whole region and lacks the exploration of inter-cities. At the same time, due to the differences in natural resource endowment and level of social development, each region faces unique SDG interactions in the development process. Therefore, it is of great significance to carry out a multi-indicator intersectional and comprehensive assessment of the SDGs, sort out the intersection between the SDGs indicators in different regions, and dynamically simulate the realization of the SDGs under different development scenarios in the future, in order to formulate sustainable development paths tailored to the local conditions, and to effectively realize the 2030 Agenda.

Big Earth Data in Support of the Sustainable Development Goals (2022)—China has already provided a comprehensive assessment of the sustainable development process in typical regions of China, exploring the synergies and trade-offs between SDGs indicators at the provincial level in China. The report further extends it to the level of China's prefectural cities, evaluates the process of sustainable development and the status of balanced development, and digs deeper into the interactions of SDG indicators among different cities. Also, this case predicts the future development process and trend of China's cities at or above the prefecture level, aiming to provide decision-making support and demonstration reference for the sustainable development of cities, and contribute to the realization of the SDGs on a national scale.

- **Data**

- Time series data of SDGs from 285 Chinese cities at or above the prefecture level (2000–2020) from the National Bureau of Statistics (http://www.stats.gov.cn/).
- Land use classification data (2000–2020) with a spatial resolution of 30 m (http://irsip.whu.edu.cn/resources/CLCD.php).

- **Methods**

Combining the methods of simultaneous equation and Granger causality, we explored the synergies and trade-offs among the 16 SDGs of 285 Chinese cities, and constructed an iterative prediction model of future development that takes into account the interactions of the SDGs between cities.

Through the establishment of a simultaneous equation model, structural vector autoregressive model, and Granger causality, this research explores the synergy and trade-off effect of SDGs within and between cities. Based on the idea of learning from similar neighboring cities with better SDGs, we integrated the synergy and trade-off interactions of the SDGs in cities, and constructed a similar growth prediction model that took into account the interactions of SDGs. Combined with the annual average compound growth rates of SDGs, five distinct development paths are set, including the continuation development path, the moderate development path, the appropriate development path, the active development path, and the full development path. We predict the process of realizing the SDGs and the development trend of Chinese cities in 2030 under different paths, and explore the scenario differences and spatial differences in the process of realizing urban SDGs. Based on the principle of minimum cost of urban development, optimize the sustainable development path of urban SDGs, and provide decision-making support for cities to achieve SDGs.

- **Results and Analysis**

Typical interactions among SDGs between Chinese cities from 2000 to 2020 are shown in Fig. 9.4. Overall, the SDG comprehensive index has synergy and trade-off relationships between cities, with the number of synergy relationships between cities accounting for 60.04% and the number of trade-off relationships accounting for 39.96%. Synergy relationships outnumbered trade-off relationships by approximately 1.5 times. In the process of SDG development, the relationships among Chinese cities are interconnected, emphasizing that urban development is not an isolated effort of individual cities but rather influenced by the spatial spillover effects of neighboring cities. Spatially, regions with close interactions among SDGs between cities are predominantly located in urban cluster areas, especially within urban cluster regions like the Yangtze River Delta, Chengdu-Chongqing, and the Pearl River Delta. Synergies among SDG interactions between cities are significantly higher in quantity and intensity compared to trade-offs, with urban cluster development facilitating positive synergistic interactions among cities. The comparison results indicated that both the proportion and intensity of synergy relationships in the SDG comprehensive index among urban cluster cities are higher than those in non-urban cluster cities, while the proportion and intensity of trade-off relationships are lower. In summary, the development strategy of urban clusters in China effectively promotes positive synergistic development among cities, while also restraining the counteracting effects of trade-offs among cities. This approach contributes to the holistic and sustainable development of cities.

Estimated results of SDG progress in Chinese cities at or above the prefecture level by 2030 are shown in Fig. 9.5. Substantial disparities in SDGs progress under the appropriate development path exist among different cities by 2030. The urban clusters such as the Yangtze River Delta and Chengdu-Chongqing exhibit more achievements in SDGs, while some cities in the northeastern and southwestern regions show lower SDGs completion, necessitating tailored strategies based on regional conditions. When categorizing SDGs into basic needs, economic development, environmental protection, and social progress, the completion of SDGs of basic needs remains favorable across different development paths. As development costs increase, there is an increase in the number of cities achieving SDGs across the five development paths, most notably in the completion of social progress goals. This suggests that appropriately raising the development cost can accelerate urban social sustainability. In terms of individual SDGs, it is anticipated that by 2030, most cities will achieve goals such as SDG 1 (No Poverty), SDG 2 (Zero Hunger) and SDG 17 (Partnerships for the Goals), while SDG 4 (Quality Education), SDG 11 (Sustainable Cities and Communities), SDG 12 (Responsible Consumption and Production), and SDG 15 (Life on Land) still require focused efforts. Thus, rational policies are needed to promote quality education, actively build sustainable cities and communities, encourage the shift toward green production and consumption, and foster the positive development of life on land, all of which will contribute to the overall realization of SDGs.

The optimal implementation paths for China's cities at or above the prefecture level are depicted in Fig. 9.6. The majority of cities in China can achieve SDG 1 (No Poverty), SDG 5 (Gender Equality), and SDG 7 (Affordable and Clean Energy) simply by continuing their current development paths or slightly increasing their development efforts. However, for SDG 2 (Zero Hunger),

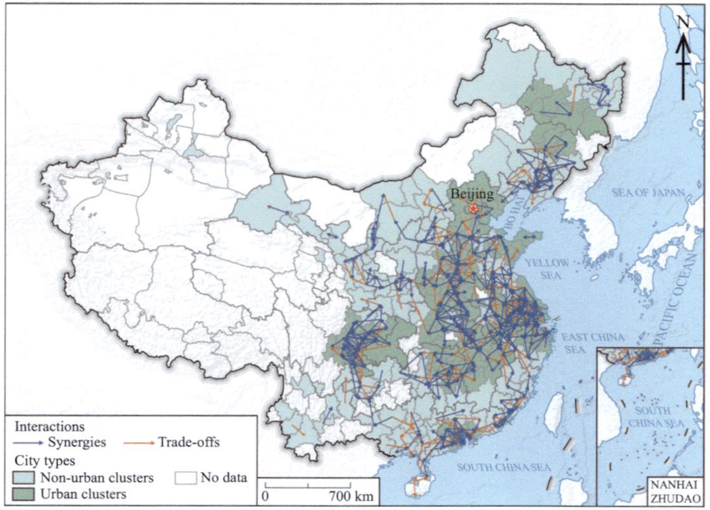

(a) Spatial distribution of top 1% typical interactions

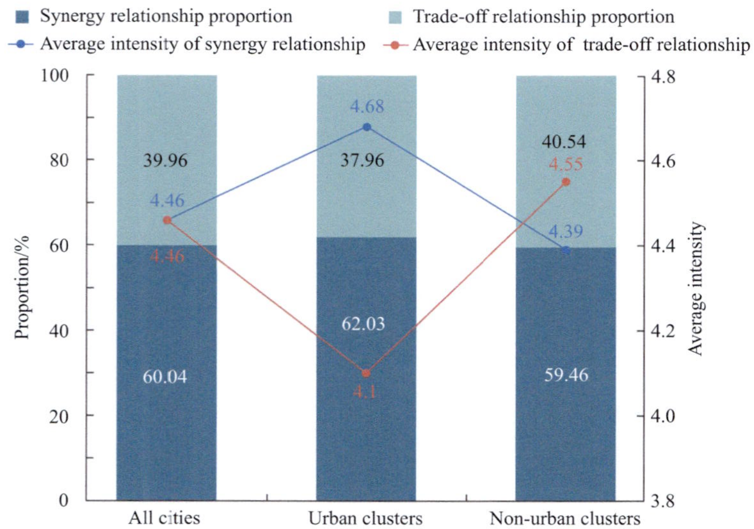

(b) Proportion of total interactions

Fig. 9.4 Interactions among SDGs in Chinese cities at or above the prefecture level in 2000–2020 (Top 1%)

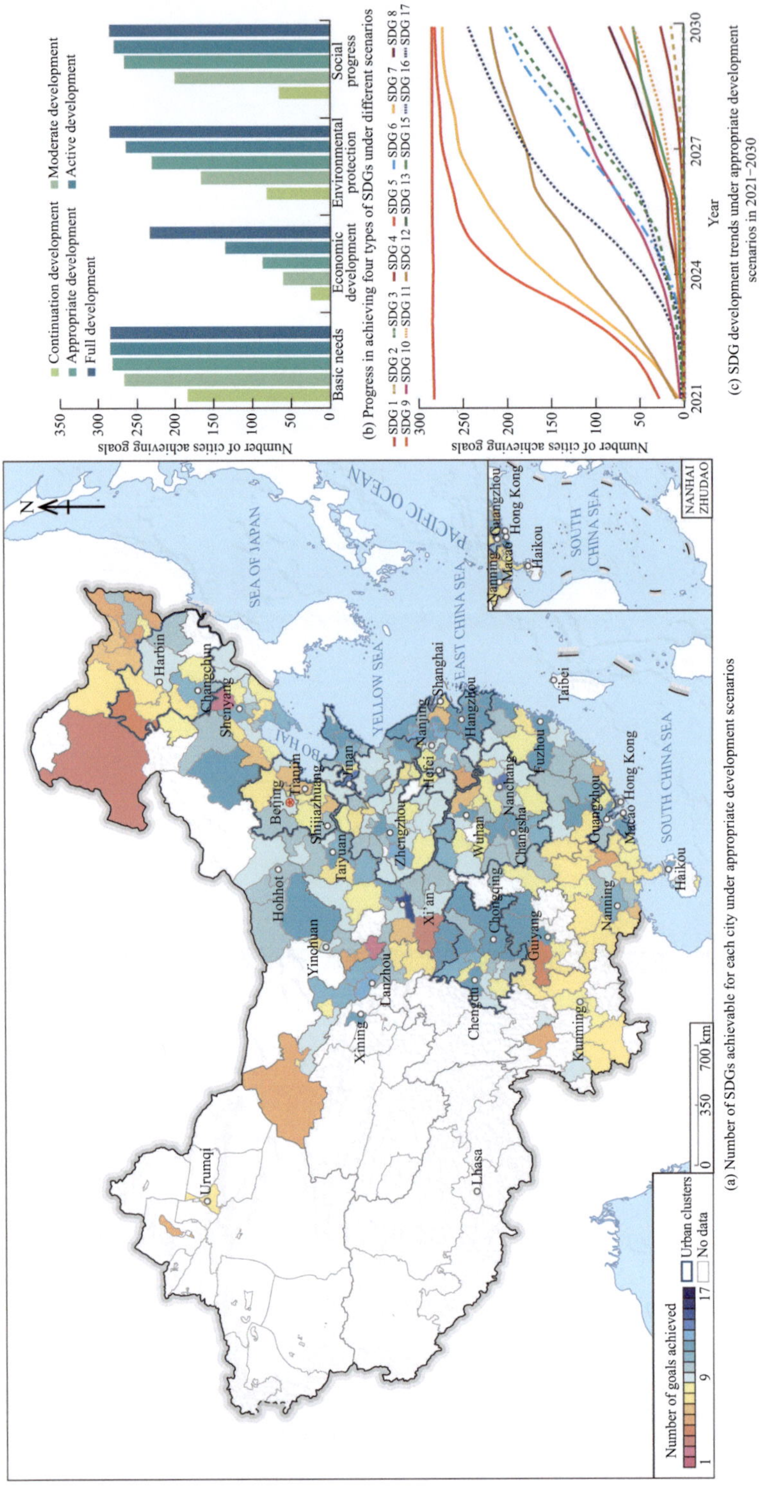

Fig. 9.5 Estimated results of SDG progress in Chinese cities at or above the prefecture level by 2030

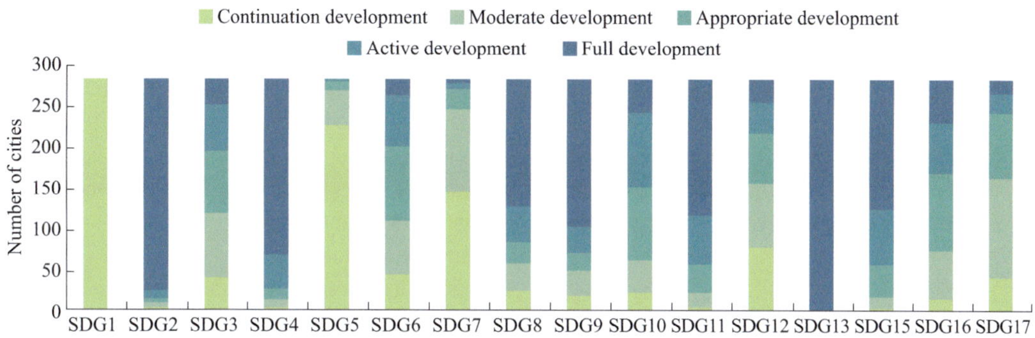

Fig. 9.6 Number of cities at or above the prefecture level achieving each SDG under different development scenarios

SDG 4 (Quality Education), and SDG 13 (Climate Action), most cities need to increase investment in urban development and formulate rational and feasible policies. Meanwhile, the optimal implementation paths for China's cities at or above the prefecture level exhibit significant spatial variations, with urban cluster development strategies effectively promoting the attainment of SDGs. In urban clusters like the Yangtze River Delta and the Pearl River Delta, achieving SDG 8 (Decent Work and Economic Growth) and SDG 9 (Industry, Innovation and Infrastructure) requires merely continuing the current development paths or adopting a moderate development path with limited additional investments. Conversely, cities in other regions may need higher investments (taking the active or full development paths) to realize SDG 8 and SDG 9.

Highlights

- The SDGs have synergy and trade-off relationships between cities; the areas with intensive interaction of SDGs targets are mostly distributed in urban cluster cities; the development of urban cluster cities can promote the interaction between cities.
- With the increase in learning costs, cities have been fully developed, and the number of cities that have achieved SDGs has gradually increased from continuous development to full development; most cities need to prioritize development in food security, quality education, and climate change to facilitate the city's achievement of the SDGs.

- **Discussion and Outlook**

In this case, SDG synergy and trade-off relationship analysis and SDG future scenario simulation are carried out at the scale of Chinese prefectural cities. It is found that the SDG development of urban cluster cities is significantly better than that of non-urban cluster regions, and there are more synergies than trade-offs in SDGs among cities, and that the SDG target sustainability of Chinese cities shows an upward trend in 2030. The findings can provide decision-making references for accelerating the construction of urban cluster cities in China, promoting synergistic development of inter-city SDGs, suppressing the impact of inter-city trade-offs, and optimizing the path of sustainable urban development. However, due to the limitations of the SDGs indicator data used and the complexity of multi-indicator interactions of SDGs between cities, the analysis results may have certain limitations.

Based on the findings of this chapter, we propose the following recommendations.

(1) Strengthen the capacity for the acquisition of Big Earth Data on SDG indicators; develop an SDG indicator system suitable for different administrative levels in China, such as city and county, aligning with the SDG indicators; establish standardized data collection processes to support multi-indicator interactions, integrated SDG assessment, and spatial spillover effects among cities.
(2) Conduct spatial decomposition research for non-spatial statistical SDG indicators;

spatialize non-spatial statistical SDG indicators based on factors such as population and land use, to enable an integrated assessment of SDGs with high spatial resolution, providing technical support for adopting the right SDG implementation paths in China's cities at or above the prefecture level.

The exploration of multi-indicator interactions of SDGs among Chinese cities at or above the prefecture level, comprehensive assessment, and future scenario simulation are important tools for exploring the optimization of the sustainable development implementation paths at the prefectural and municipal levels in China, as well as for comprehensively assessing the process of sustainable development of urban cluster cities. China's regions need to focus on the balanced development of SDG goals in the process of implementing sustainable development, inhibiting the trade-off role of inter-city SDGs, and promoting the synergy of inter-city SDGs, especially those of urban cluster cities.

9.3.4 SDGs Trade-Offs and Synergies for "One Health" in China

Target: SDG 3, SDG 6, SDG 11, SDG 13, SDG 15.

- **Background**

Under the framework of "One Health", the interconnected health of animals, humans, and the environment has garnered increasing attention, becoming a focal point in global sustainable development research (Saha and Davis 2022). The COVID-19 pandemic has caused delays—and in some cases, regressions—in global progress toward achieving the United Nations' SDGs. Specifically, the pandemic has expanded the scope of SDG 3 (Good Health and Well-being) from a focus on diseases like HIV/AIDS and tuberculosis to heightened awareness of zoonotic diseases, which account for 75% of emerging infectious diseases (Plowright et al. 2017). The occurrence of zoonotic diseases is closely linked to the health of humans, livestock, wildlife, plants, and ecosystems, underscoring the critical role of the "One Health" framework in advancing the SDGs.

Environmental changes, such as shifts in climate and land use, can impact ecosystems, altering wildlife population sizes and habitats, reshaping biodiversity and endangered species distributions, and even leading to wildlife encroaching into human settlements. These changes may facilitate animal pathogens crossing species barriers to infect humans (Carlson et al. 2022; Gibb et al. 2020). In 2022, FAO, United Nations Environment Programme (UNEP), WHO, and the World Organisation for Animal Health jointly released the One Health Joint Plan of Action, aiming to improve the health of humans, animals, plants, and the environment while promoting sustainable health development globally (WHO 2022). In terms of environmental health, remote sensing-based ecological index (RSEI) data and similar metrics provide excellent spatial and temporal continuity, effectively revealing long-term ecological changes (He et al. 2021). Regarding animal health, biodiversity levels significantly influence the risk of viral spillover between hosts, ultimately affecting human health. As for human health, the majority of emerging infectious diseases in humans are zoonotic, with regions prone to outbreaks concentrated in tropical areas and densely populated regions outside the tropics, such as Europe, the United States, Asia, and Latin America (Allen et al. 2017; Olival et al. 2017). China is one of the global hotspots for infectious diseases, with common examples including plague and hemorrhagic fever. However, the trade-offs and synergies between multiple SDGs under the "One Health" framework remain unclear in the context of China.

The overarching goal of this case study is to evaluate the relationships between environmental health, animal health, and human health. For environmental health, metrics such as ecological quality and the RSEI are utilized.

For animal health, data from the IUCN Red List is employed. For human health, nationwide incidence rates of Class A and Class B infectious diseases are analyzed. The primary scientific questions addressed are as follows: How do changes in ecological conditions affect the distribution of biodiversity? How does biodiversity change, in turn, influence the occurrence and transmission of zoonotic diseases hosted by animals?

- **Data**

- MODIS 09A1 v6: MODIS/Terra Land Surface Temperature/Emissivity 8-Day L3 Global 1 km SIN Grid (https://lpdaac.usgs.gov/products/mod11a2v006/).
- MOD11A2 v6: ODIS/Terra Land Surface Temperature/ Emissivity 8-Day L3 Global 1 km SIN Grid (https://lpdaac.usgs.gov/products/mod11a2v006/).
- 2000–2020 RSEI dataset (He et al. 2021).
- IUCN Red List of Threatened Species: Biodiversity data (https://www.iucnredlist.org/).
- 2005–2020 national Class A and Class B infectious disease data in China from the China Health Statistics Yearbook.

- **Methods**

First, the RSEI product, which represents ecological quality and biodiversity data for birds, amphibians, and reptiles from the IUCN Red List, is used to explore how changes in ecological conditions affect biodiversity distribution. On this basis, nationwide Class A and Class B infectious disease incidence data are integrated to further investigate how changes in biodiversity influence the occurrence and spread of zoonotic diseases hosted by animals. The primary research method employed in this case is regression analysis. Linear regression analysis, in particular, can help identify the multi-year variation trends in Class A and Class B infectious diseases across provinces, facilitating trade-off and synergy analyses with ecological conditions and biodiversity.

- **Results and Analysis**

From 2000 to 2020, China's overall ecological security has shown a gradual improvement trend. Ecological restoration projects have played a critical role in advancing environmental improvement. As shown in Fig. 9.7, changes in the RSEI exhibit a clear east–west spatial differentiation. Specifically, significant ecological improvements have been observed in the central and northeastern regions, while limited improvements have occurred in the southwestern and northwestern regions due to natural environmental constraints. Over the past two decades, regions with notable ecological improvements include the Loess Plateau, the southern and northern parts of the North China Plain, the Northeast China Plain, and the northern areas of the Middle and Lower Reaches of the Yangtze River Plain. These regions have achieved significant results through ecological restoration projects and the promotion of ecological civilization, leading to marked enhancements in local environmental quality. In contrast, regions with deteriorating ecological conditions are primarily located in the Tianshan Mountains, the central Qinghai-Xizang Plateau, the central North China Plain, and the Yangtze River Delta. These areas face harsh natural conditions, such as mountainous terrain, desertification, and severe soil erosion. Coupled with unsustainable human activities, these factors have led to noticeable ecological degradation.

Studying the macro-patterns of biodiversity provides crucial insights for ecological conservation and management. According to the results in Fig. 9.8, the distribution of endangered mammals in China exhibits a spatial pattern of "high in the south, low in the north". This distribution may be attributed to factors such as complex terrain, warm and humid climate, and diverse vegetation types in southern regions. The southwestern areas, including Sichuan, Yunnan, and eastern Xizang, are hotspots for

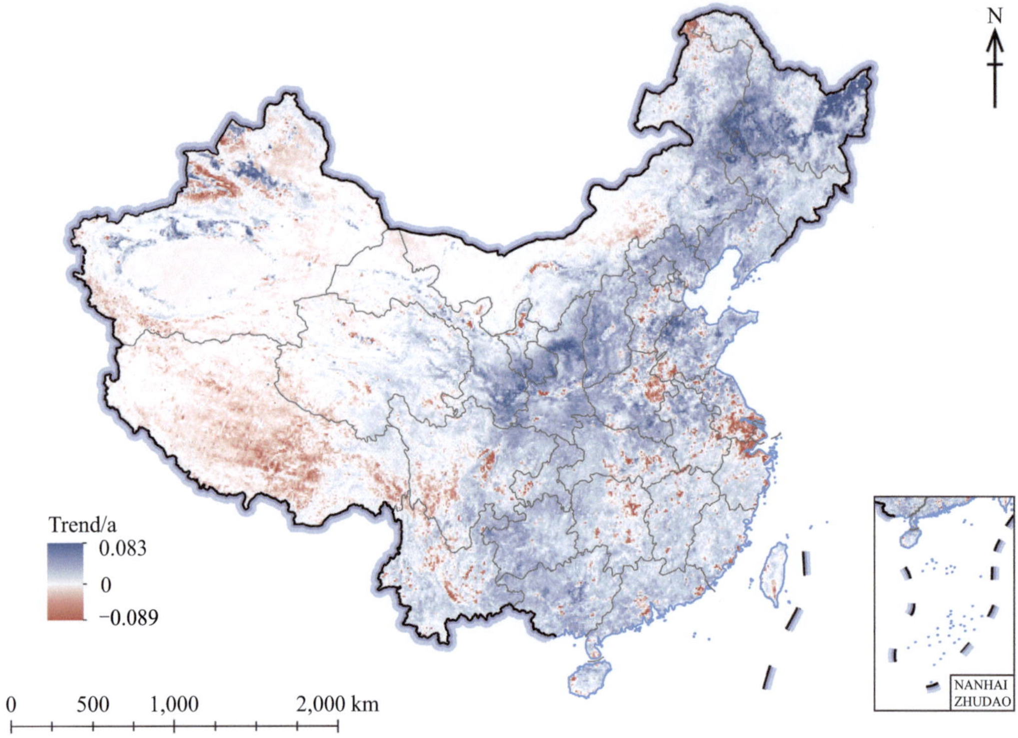

Fig. 9.7 The spatial pattern of RSEI in China since 2000

mammal abundance. These regions feature unique geographical and climatic conditions, diverse ecosystems, and vegetation types that create favorable habitats for various mammals. In contrast, the northern regions show lower mammal abundance, mainly in Northeast China, North China, and Northwest China. These areas are characterized by cold and arid climates, relatively uniform terrain, and limited vegetation diversity, offering less suitable ecological conditions for animal survival and reproduction. Bird distribution, on the other hand, demonstrates a spatial pattern of "high in the east, low in the west". Southeastern coastal areas, the Yellow Sea and the Bohai, southwestern regions, and northern Xinjiang are the primary areas of high bird abundance. These regions feature warm and humid climates, diverse vegetation types, and complex ecosystems, providing ideal habitats and breeding grounds for birds. Conversely, the northwestern region shows low bird abundance due to arid and cold conditions, limited vegetation types, and relatively simple ecosystems, which are less favorable for avian survival and reproduction. Similarly, amphibian distribution follows a spatial pattern of "high in the south, low in the north". Sichuan, Yunnan, and Guizhou are key hotspots for amphibian abundance. These areas boast complex terrain, warm and humid climates, diverse vegetation types, and abundant water resources, creating optimal living conditions for amphibians.

Human health is a critical component of the ecological environment, with infectious diseases being one of the most significant threats to human well-being. Among the 40 legally classified infectious diseases in China, the majority of zoonotic diseases fall into Class A and Class B categories. From 2005 to 2020, the overall trend of the incidence rates of infectious diseases across provinces has shown a decline nationwide. However, certain regions have exhibited a noticeable increase in incidence rates, including Xizang, Hunan, Anhui, and Shandong (Fig. 9.9). This highlights the

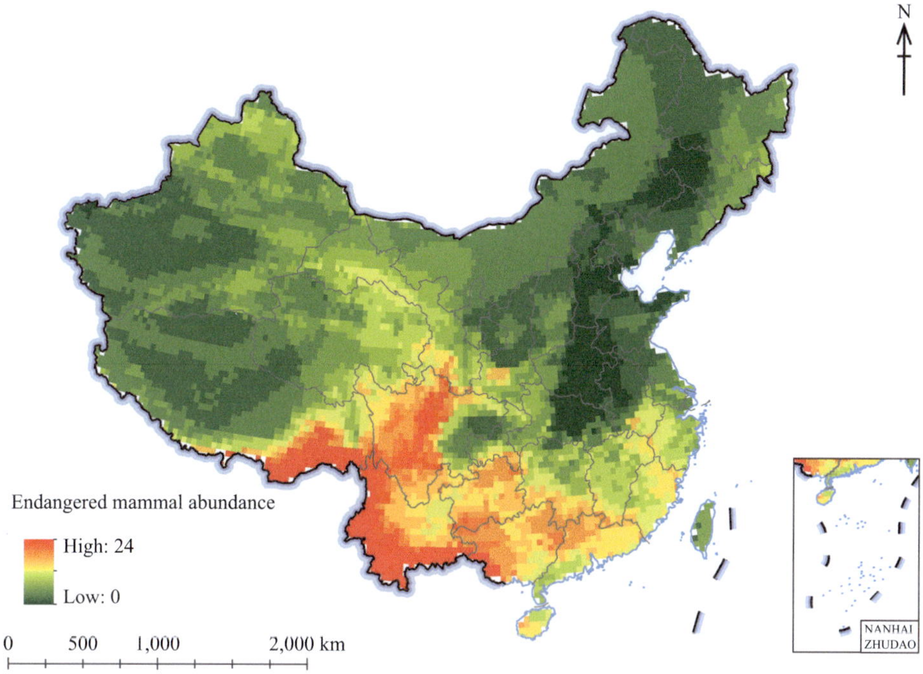

(a) Endangered mammal abundance

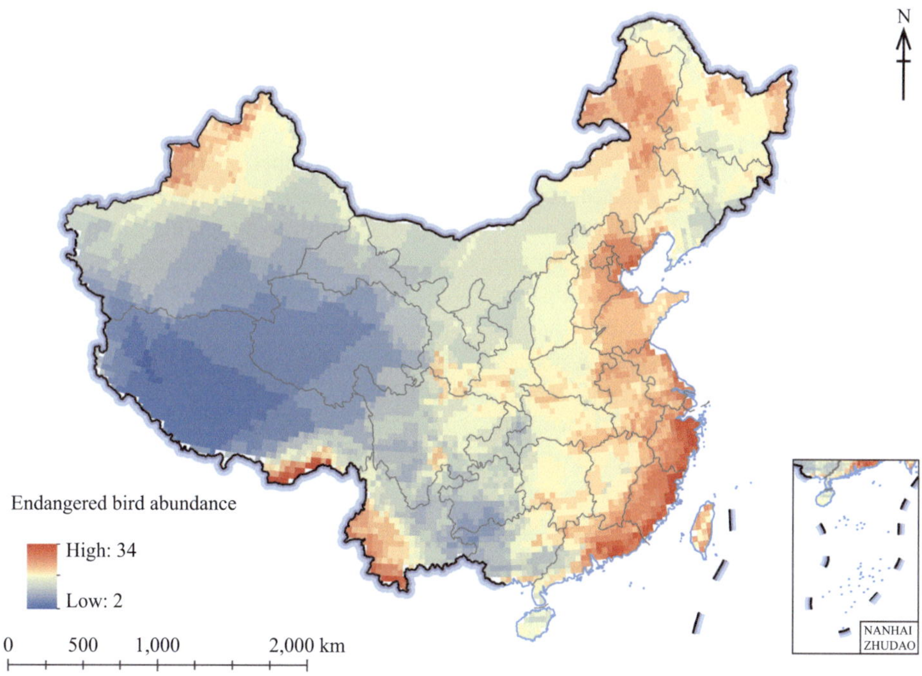

(b) Endangered bird abundance

Fig. 9.8 The distribution of the abundance of endangered mammals, endangered birds, and endangered amphibians in China

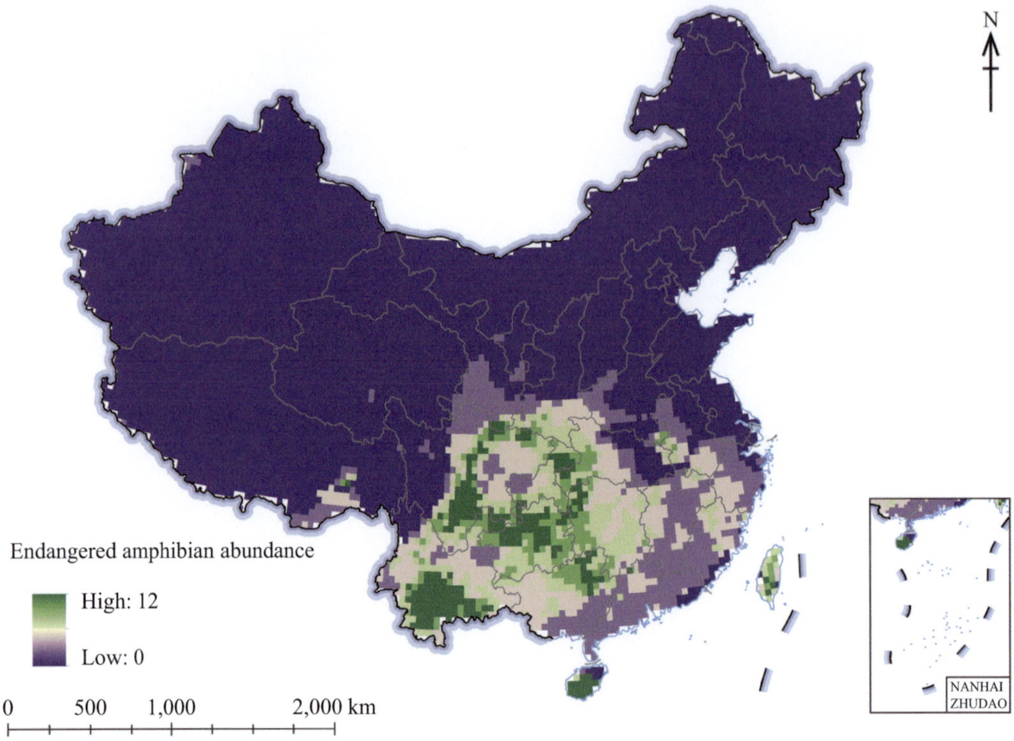

(c) Endangered amphibian abundance

Fig. 9.8 (continued)

need for region-specific strategies in addressing infectious diseases, especially those of zoonotic origin.

Focusing on zoonotic diseases, hemorrhagic fever with renal syndrome serves as a representative example. The study reveals that the incidence rate has significantly decreased in northern regions, while showing an increasing trend in southern and western regions (Fig. 9.10). This indicates that the spread and occurrence of infectious diseases vary under different geographical and ecological conditions. In areas undergoing ecological restoration projects, such as the Loess Plateau, the Yunnan-Guizhou Plateau, and Nei Mongol, a marked decline in the incidence of hemorrhagic fever has been observed. This suggests that improvements in ecological conditions can have a positive impact on human health by potentially reducing the prevalence of such diseases.

Through further overlay analysis of the data, this case study provides deeper insights into the relationship between the ecological environment and biodiversity. Firstly, the biodiversity of endangered mammals and amphibians is higher in southern regions, including Sichuan, Yunnan, Guangxi, and Guangdong. This can be partially attributed to the relatively good ecological conditions in these areas. These regions are known for their beautiful ecological environment, abundant natural resources, and unique biological communities. However, they also face challenges in balancing development and conservation. To protect these species, it is crucial to control human activities that may damage the natural environment while strengthening ecological restoration and conservation efforts to preserve the natural ecosystems in these areas. Secondly, the biodiversity of endangered mammals is lower in the northeastern, northern, and northwestern regions,

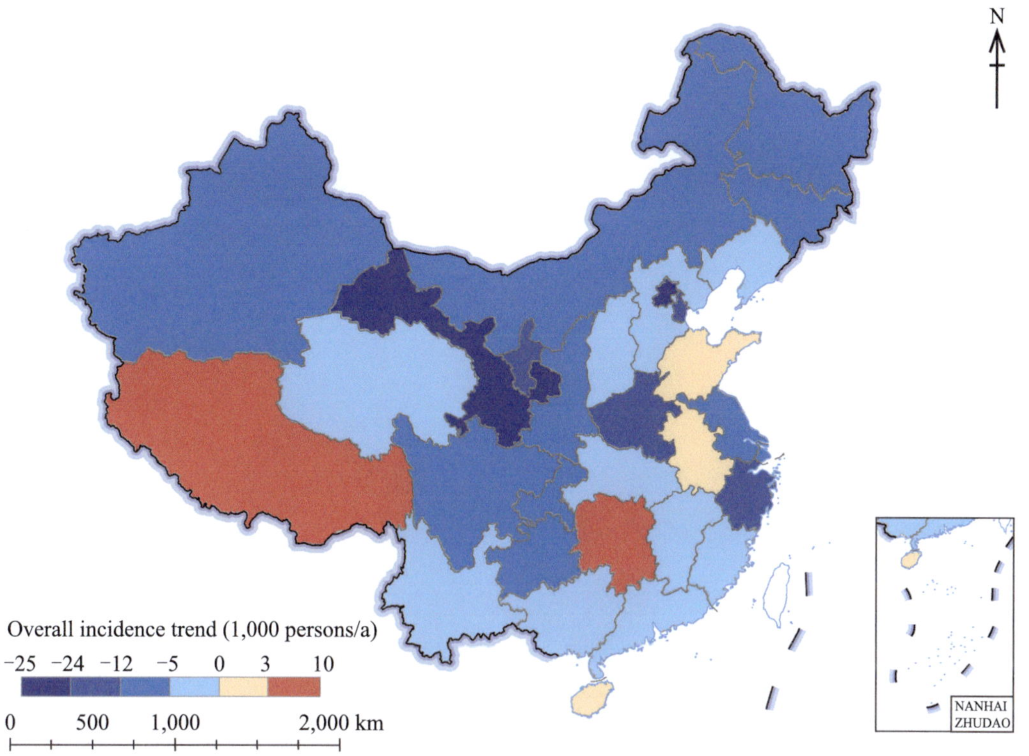

Fig. 9.9 Trend distribution of the total incidence of Class A and Class B infectious diseases in China in 2005–2020. *Note* No data for Taiwan Province

which is primarily linked to poor ecological security in these areas. To protect these species, it is necessary to strengthen ecological governance and conservation efforts in these regions, control human impact on the natural environment, and promote ecological restoration and improvement. Additionally, the study found that the biodiversity of endangered bird species is higher in the southeastern coastal areas, the Yellow Sea and the Bohai, the southwest, and northern Xinjiang. Over the past two decades, these regions have shown significant improvements in their ecological security. On the other hand, regions with lower biodiversity of endangered bird species are mainly concentrated in the northwest, where ecological security has been declining over the past 20 years. Finally, the biodiversity of endangered amphibians is concentrated in the southern regions, such as Sichuan, Yunnan, and Guizhou, which also have relatively good ecological security that has been improving over the past 20 years. In contrast, the biodiversity of endangered amphibians is lower in the northeastern, northern, and northwestern regions. This highlights the importance of focusing on the protection of these species, taking measures to safeguard the natural environments in these areas, increasing conservation efforts, and preventing species extinction.

Highlights

- Reveal the spatio-temporal patterns of biodiversity abundance in endangered mammals, amphibians, and birds.
- Unveil the trade-offs and synergies between SDG-related goals, such as ecological status, biodiversity, and Class A and Class B infectious diseases in China, from the "One Health" perspective.

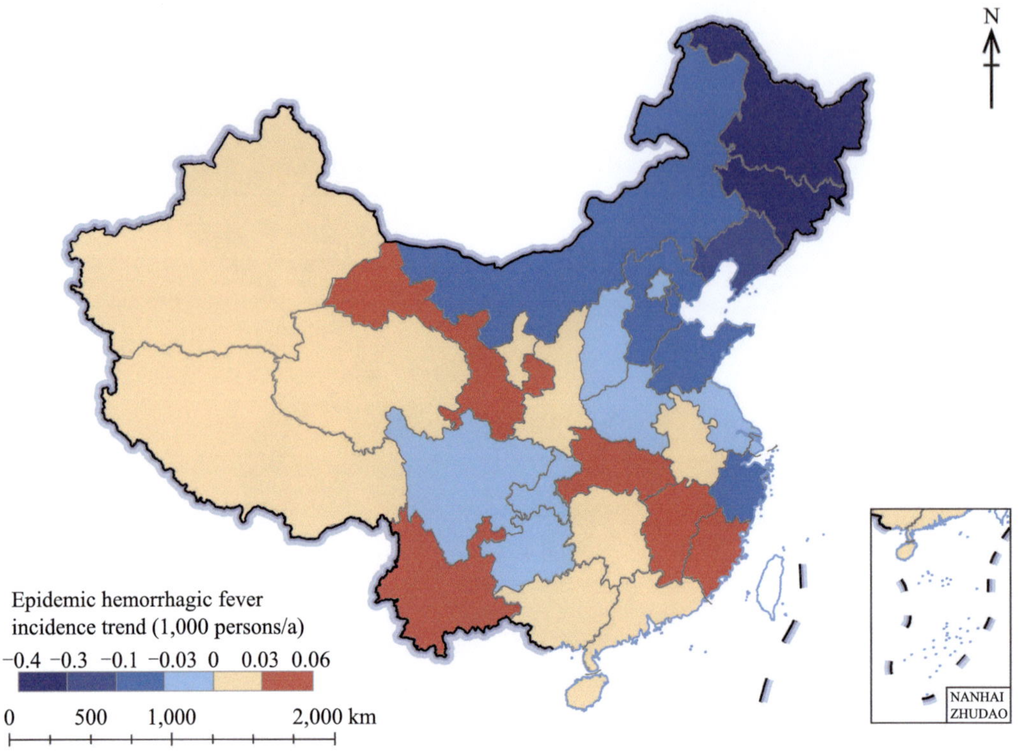

Fig. 9.10 Prevalence trend map of epidemic hemorrhagic fever in China from 2005 to 2020. *Note* No data for Taiwan Province

- **Discussion and Outlook**

The SDGs emphasize overall, balanced, and synergistic development. By utilizing remote sensing data, biodiversity, and statistical data, this case study reveals the trade-offs and synergies between ecological status, biodiversity, zoonotic diseases, and other SDG-related goals in China from the "One Health" perspective. The research findings can provide a reference for achieving the SDGs, as well as for predicting zoonotic diseases in China.

However, this case study has the following limitations: (1) Data acquisition is challenging, as the infectious disease incidence data come from provincial level statistics and do not cover all types of zoonotic diseases; (2) biodiversity data lack long-term monitoring sources, and currently mainly rely on IUCN Red List data. Future research should further improve the data sources, using observation data such as from GBIF to enhance biodiversity data analysis. Based on data availability, the research should focus on specific infectious diseases, host species, and their corresponding ecological indicators, and strengthen the analysis of the internal linkages between different SDGs.

9.4 Summary

This chapter first conducts a comprehensive SDG progress assessment for cities at or above the prefecture level of China, analyzes the synergies and trade-offs of SDGs among cities, and predicts future development scenarios. It also explores the SDGs' multi-objective trade-offs and synergies from the "One Health" perspective. The findings reveal significant spatial differences and associations in the overall SDG scores and balanced development across Chinese prefecture-level cities. Urban agglomerations show better SDG development than non-agglomeration regions. The synergies between SDGs are more prevalent than trade-offs between cities, and by 2030, the sustainable development of SDGs across Chinese cities is projected to show an upward trend.

Next, the chapter reveals the trade-offs and synergies between ecological status, biodiversity, zoonotic diseases, and other SDG-related goals from the "One Health" perspective, demonstrating that changes in ecological status can have a positive impact on human health. The conclusions of this chapter provide decision-making references for different cities in China to prioritize development goals, accelerate SDG construction in urban agglomerations, promote SDG collaborative development between cities, mitigate trade-off impacts between cities, and optimize sustainable development pathways for cities. However, due to the limitations of the SDG indicator data used and the complexity of the multi-indicator cross-relationships, the analysis results may have certain limitations.

Based on the research in this chapter, we make the following recommendations.

(1) Strengthen the capacity building for acquiring SDG-related Big Earth Data, construct an SDG indicator system for prefecture-level cities and counties in China that is compatible with the United Nations' SDG framework, and improve standardized processes for data collection. This would provide data support for the multi-indicator cross-analysis and comprehensive assessment of SDGs across cities, as well as the spatial spillover effects of SDGs between cities.
(2) Deepen the research into the analysis methods for the multi-indicator interaction and spillover effects of SDGs between cities, using exploratory spatial data analysis, spatial econometric models, and deep learning methods. This would explore the multi-indicator interaction and spatial spillover effects of SDGs between prefecture-level cities in China and provide methodological references for promoting SDG regional collaboration in cities.
(3) Conduct research on the spatial decomposition of non-spatial statistical SDG indicators. Based on spatial data such as population and land use, spatialize other non-spatial statistical SDG indicators to achieve a high spatial resolution comprehensive SDG assessment. This would provide technical support for refined decision-making on SDG implementation paths for prefecture-level cities in China.

The study of multi-indicator interactions, comprehensive assessments, and future scenario simulations of SDGs across prefecture-level cities in China is an important means of exploring and optimizing sustainable development implementation paths for cities and assessing the progress of sustainable development in urban agglomerations. In the process of implementing sustainable development, prefecture-level cities in China need to focus on the balanced development of SDGs, mitigate the trade-off interactions of SDGs between cities, and promote collaborative development of SDGs, especially in urban agglomerations.

References

Allen T, Murray KA, Zambrana-Torrelio C et al (2017) Global hotspots and correlates of emerging zoonotic diseases. Nat Commun 8(1):1124

Carlson CJ, Albery GF, Merow C et al (2022) Climate change increases cross-species viral transmission risk. Nature 607(7919):555–562

Cheng ZH, Liu J, Li LS et al (2020) Research on meta-frontier total-factor energy efficiency and its spatial convergence in Chinese provinces. Energy Econ 86:104702

Dong JW, Chen Y, Zhou Y et al (2021) Big Earth Data supports synergies and trade-offs of sustainable development goals: progress and prospect. Bull Chin Acad Sci 36(8):950–962 (In Chinese)

Feng KS, Davis SJ, Sun LX et al (2013) Outsourcing CO_2 within China. Proc Natl Acad Sci USA 110(28):11654–11659

Fukuda-Parr S (2016) From the millennium development goals to the sustainable development goals: shifts in purpose, concept, and politics of global goal setting for development. Gend Dev 24(1):43–52

Georgeson L, Maslin M (2018) Putting the United Nations sustainable development goals into practice: a review of implementation, monitoring, and finance. Geogr Environ 5(1):e00049

Gibb R, Redding DW, Chin KQ et al (2020) Zoonotic host diversity increases in human-dominated ecosystems. Nature 584(7821):398–402
Guo HD, Liang D, Chen F et al (2021) Big Earth Data facilitates sustainable development goals. Bull Chin Acad Sci 36(8):874–884 (In Chinese)
Han L, Han B, Shi X et al (2018) Energy efficiency convergence across countries in the context of China's Belt and road initiative. Appl Energy 213:112–122
He YL, You NS, Cui YP et al (2021) Spatio-temporal changes in remote sensing-based ecological index in China since 2000. J Nat Resour 36(5):1176–1185 (In Chinese)
Lafortune G, Zoeteman K, Fuller G et al (2019) SDG index and dashboards report for European cities. https://sdgtransformationcenter.org/reports/sdg-index-for-eu-cities [2024-09-01]
Liu YL, Du JQ, Wang YF et al (2020) Evenness is important in assessing progress towards sustainable development goals. Natl Sci Rev 8(8):nwaa238
Lu H, Lu X, Jiao L et al (2022) Evaluating urban agglomeration resilience to disaster in the Yangtze Delta city group in China. Sustain Cities Soc 76:103464
Olival KJ, Hosseini PR, Zambrana-Torrelio C et al (2017) Host and viral traits predict zoonotic spillover from mammals. Nature 546(7660):646–650
Plowright RK, Parrish CR, McCallum H et al (2017) Pathways to zoonotic spillover. Nat Rev Microbiol 15(8):502–510
Pradhan P, Costa L, Rybski D et al (2017) A systematic study of sustainable development goal (SDG) interactions. Earth's Future 5(11):1169–1179
Saha S, Davis WW (2022) The need for a "One Health" approach for influenza surveillance. Lancet Glob Health 10(8):e1078–e1079
Sala-i-Martin X (1994) Cross-sectional regressions and the empirics of economic growth. Eur Econ Rev 38(3–4):739–747
Schmidt-Traub G, Kroll C, Teksoz K et al (2017) National baselines for the sustainable development goals assessed in the SDG index and dashboards. Nat Geosci 10:547–555
United Nations (2020) Sustainable development goals progress chart 2020. https://sustainabledevelopment.un.org/content/documents/26727SDG_Chart_2020.pdf [2024-07-24]
WHO (2022) One health joint plan of action (2022–2026): working together for the health of humans, animals, plants and the environment. https://www.who.int/publications/i/item/9789240059139 [2022-10-14]
Wiedmann T, Allen C (2021) City footprints and SDGs provide untapped potential for assessing city sustainability. Nat Commun 12(1):3758
Xiao HJ, Bao S, Ren JZ et al (2023) Transboundary impacts on SDG progress across Chinese cities: a spatial econometric analysis. Sustain Cities Soc 92:104496
Xiao HJ, Wang DP, Qi Y et al (2021) The governance-production nexus of eco-efficiency in Chinese resource-based cities: a two-stage network DEA approach. Energy Econ 101:105408
Xiao HJ, Xu ZC, Ren JZ et al (2022) Navigating Chinese cities to achieve sustainable development goals by 2030. The Innovation 3(5):100288
Xue S, Xiao HJ, Ren JZ (2024) Cross-border interactions on the sustainable development between global countries. Resour Conserv Recycl 204:107525
Yang J, Huang X (2021) The 30 m annual land cover dataset and its dynamics in China from 1990 to 2019. http://irsip.whu.edu.cn/resources/CLCD.php [2024-08-07]
Zhang JZ, Wang S, Pradhan P et al (2022) Untangling the interactions among the sustainable development goals in China. Sci Bull 67(9):977–984
Zhou L, Tian L, Gao Y et al (2019) How did industrial land supply respond to transitions in state strategy? An analysis of prefecture-level cities in China from 2007 to 2016. Land Use Policy 87:104009

10 Summary and Prospects

10.1 Summary

The report details case studies on seven SDGs (Zero Hunger, Clean Water and Sanitation, Affordable and Clean Energy, Sustainable Cities and Communities, Climate Action, Life Below Water, and Life on Land), and the interactions and integrated assessment of multiple SDG indicators at different scales, using Big Earth Data, from the perspectives of data products, methods and models, and decision support.

In terms of data products, 20 datasets at the China scale were developed. Some of them fill gaps in SDG indicator monitoring. For example, the spatial distribution dataset of terraces in China, the proportion of people using clean cooking in China, China's comprehensive air pollution index, China's daily $PM_{2.5}$ gridded dataset, China's seagrass bed distribution dataset, and the arid oasis distribution dataset in Northwest China. Other products improve the spatio-temporal resolution of SDG indicator monitoring and evaluation, such as the 1 m resolution open space information dataset of China's five major urban agglomerations and the 10 m resolution raft aquaculture spatial distribution dataset in China.

In terms of methods and models, 18 Big Earth Data-based models were developed. Some of them provide big data-based solutions for SDG monitoring and evaluation, such as a model for assessing the proportion of China's population relying on clean cooking energy and technology. Additionally, in the context of interactions and integrated assessment of multiple SDG indicators, SDG spatio-temporal variations among different Chinese cities were explored using local spatial autocorrelation and bivariate spatial autocorrelation methods. The synergies and trade-offs of SDGs among cities were explored using the Granger causality test and the simultaneous equation model. By building an iterative prediction model, SDG progress and development trends in different scenarios for China's cities at or above the prefecture level were forecasted.

In terms of decision support, through the integration of the aforementioned data and methods, spatio-temporal analyses of SDG indicators were conducted, yielding 25 recommendations on China's sustainable development efforts. For instance, by identifying the significance and spatial heterogeneity of China's terraced fields and their soil conservation benefits, decision-making in soil conservation and land planning was enhanced; a quantitative vitality framework more suited to the urban scale was proposed, utilizing long-term data to quantitatively analyze the impact of urban vitality on carbon emissions; the changes in China's main flood disaster indicators from 2010 to 2021 showed an overall downward trend, and meteorological, hydrological, and climate information can effectively

H. Guo, *Big Earth Data in Support of the Sustainable Development Goals (2023)—China*, Sustainable Development Goals Series, https://doi.org/10.1007/978-981-95-2636-9_10

help promote disaster prevention and mitigation work; in the future, it is necessary to maintain the healthy and sustainable development of the offshore area in the long term, and achieve high-quality development of the marine economy under effective land-sea coordinated offshore prevention and control measures; the ongoing expansion of urban cluster strategies in China's cities at or above the prefecture level, with a particular focus on SDG 4, SDG 11, SDG 12, and SDG 15, fostering the comprehensive realization of SDGs.

Building on these foundations and utilizing the outcomes of the past four years' reports and national and United Nations statistics, an analysis was conducted of China's progress on primarily environmental SDG indicators from 2010 to 2022. The results demonstrate remarkable progress on environmentally sustainable development indicators and overall stable improvement of all assessed indicators. By 2022, out of the 98 indicators evaluated using Big Earth Data, 52.04% (51 indicators) had already been achieved ahead of schedule.

10.2 Prospects

While SDGs have been significantly promoted by the widespread application of Big Earth Data, their implementation still faces a multitude of challenges, such as the need for higher-quality Big Earth Data, more rational and intelligent methods for SDG indicator assessment, data sharing and utilization, and an integrated evaluation system for multiple indicators (Guo 2022; Guo et al. 2021, 2022). To advance the realization of the 2030 Agenda, the following key actions are recommended.

1. **Promote the calibration and adjustment of the SDG indicator framework**

Currently, some indicators within the UN SDG framework do not adequately capture variations in population, development levels, and geographical environments among countries. Additionally, national-level statistical data fall short of capturing disparities in sustainable development across different geographical regions within nations, thereby impeding the comprehensive realization of the "leave no one behind" principle. Harnessing the robust spatial decomposition capability of Big Earth Data, particularly for environmental indicators, has the potential to enable more comprehensive spatial monitoring and assessment of SDG indicators. To address these limitations, it is recommended that the application of relevant indicators based on Big Earth Data be actively promoted within the SDG indicator framework, serving as a geospatial decomposition complement, facilitating a more precise calibration and enhancement of the SDG indicator system.

2. **Promote the sharing and widespread application of Big Earth Data**

Due to variations in statistical standards and methodologies across nations, there are discrepancies in the standardization and quality of indicator data, and significant disparities in geographical coverage and timeliness exist between data from developed and developing countries. Big Earth Data offers unique advantages in terms of the extensive scope, integration of multiple sources, precision, scientific rigor, and timeliness. Establishing a well-organized management framework and efficient sharing mechanism for this vast pool of Big Earth Data is essential for advancing the SDGs. Currently, the sharing of Big Earth Data faces challenges, including insufficient sharing mechanisms, limited innovation in sharing technologies, homogeneous service models, and inadequate protection of sharing rights. To fully harness the potential of Big Earth Data in achieving global SDGs, it is essential to encourage broader data sharing and utilization, improve the management and integration of spatio-temporal data, establish mechanisms for sharing scientific data, and eliminate data barriers, thus bridging the digital divide.

3. **Drive technological innovation for SDG assessment**

Novel methods and models for monitoring SDG indicators based on Big Earth Data require advancements in automation, intelligence, and timeliness before they can be efficiently utilized. The isolated application of existing mining and analytical theories and methods from a single domain is no longer effective for driving scientific discoveries. Comprehensive application and innovative technological approaches are needed to systematically address significant scientific challenges. While current Big Earth Data analysis methods mainly involve statistical analysis, visualization, and machine learning technologies, we recommend further enhancing the applicability and interpretability of these methods by integrating cutting-edge technologies such as artificial intelligence, deep learning, data fusion, data assimilation, and blockchain, thereby raising the capacity of Big Earth Data to assess SDGs.

4. **Enhance the big data-enabled study of relationships between multiple SDGs**

The 2030 Agenda provides a comprehensive and multi-dimensional development perspective. Due to the interrelations among different SDGs, the pursuit of a single goal may be promoted or constrained by other goals. Big Earth Data's spatio-temporal continuity can complement or replace traditional official statistical data, providing multi-topic information encompassing land, atmosphere, oceans, and population. It is necessary to harness an abundance of multi-source information to promote the cross-sectoral monitoring capabilities across global, national, or regional levels, supporting problem-oriented data mining and exploring the construction of a multi-indicator integrated assessment system. It is essential to thoroughly analyze synergies or trade-offs among different SDGs, precisely track indicator trends, optimize regional SDG synergistic development pathways, and provide a scientific basis for the collective advancement of all SDGs.

5. **Promote the utilization of data from the Sustainable Development Science Satellite**

The Sustainable Development Science Satellite-1 (SDGSAT-1), launched by the CBAS, stands as the world's pioneering scientific satellite tailored to the service of the 2030 Agenda. It offers exclusive data support for SDG indicators that elucidate the interplay between humanity and the environment. In September 2022, the SDGSAT-1 Open Science Program was initiated, engaging over 85 countries in harnessing satellite data until November 2023. It is recommended that the international community widely employ SDGSAT-1 data, a novel data source that can make substantial contributions to SDG research worldwide, particularly in developing nations. Concurrently, exploring the application potential of Big Earth Data in terms of data quality, processing methodologies, spatial precision, and data granularity is recommended to fully leverage its role in evaluating the attainment of SDGs.

References

Guo HD (2022) Big earth data in support of the sustainable development goals (2021): the belt and road. Science Press and EDP Sciences, Beijing

Guo HD, Liang D, Chen F et al (2021) Big earth data facilitates sustainable development goals. Bull Chin Acad Sci 36(8):874–884 (In Chinese)

Guo HD, Liang D, Sun ZC et al (2022) Measuring and evaluating SDG indicators with big earth data. Sci Bull 67(17):1792–1801

MIX
Papier aus verantwortungsvollen Quellen
Paper from responsible sources
FSC® C105338

If you have any concerns about our products,
you can contact us on
ProductSafety@springernature.com

In case Publisher is established outside the EU,
the EU authorized representative is:
Springer Nature Customer Service Center GmbH
Europaplatz 3, 69115 Heidelberg, Germany

Printed by Libri Plureos GmbH
in Hamburg, Germany